Development of Event-Based Pathogen Monitoring Strategies for Watersheds

Awwa Research Foundation

About the Awwa Research Foundation

The Awwa Research Foundation (AwwaRF) is a member-supported, international, nonprofit organization that sponsors research to enable water utilities, public health agencies, and other professionals to provide safe and affordable drinking water to consumers.

The Foundation's mission is to advance the science of water to improve the quality of life. To achieve this mission, the Foundation sponsors studies on all aspects of drinking water, including supply and resources, treatment, monitoring and analysis, distribution, management, and health effects. Funding for research is provided primarily by subscription payments from approximately 1,000 utilities, consulting firms, and manufacturers in North America and abroad. Additional funding comes from collaborative partnerships with other national and international organizations, allowing for resources to be leveraged, expertise to be shared, and broad-based knowledge to be developed and disseminated. Government funding serves as a third source of research dollars.

From its headquarters in Denver, Colorado, the Foundation's staff directs and supports the efforts of more than 800 volunteers who serve on the board of trustees and various committees. These volunteers represent many facets of the water industry, and contribute their expertise to select and monitor research studies that benefit the entire drinking water community.

The results of research are disseminated through a number of channels, including reports, the Web site, conferences, and periodicals.

For subscribers, the Foundation serves as a cooperative program in which water suppliers unite to pool their resources. By applying Foundation research findings, these water suppliers can save substantial costs and stay on the leading edge of drinking water science and technology. Since its inception, AwwaRF has supplied the water community with more than $300 million in applied research.

More information about the Foundation and how to become a subscriber is available on the Web at **www.awwarf.org.**

Development of Event-Based Pathogen Monitoring Strategies for Watersheds

Prepared by:
Paula L. Sturdevant Rees, **Sharon C. Long**, **Rebecca Baker**,
Daniel H. Bordeau, and **Routing Pei**
Department of Civil and Environmental Engineering
University of Massachusetts Amherst, Amherst, MA 01003-5205
and
Paul K. Barten
Department of Natural Resources Conservation
University of Massachusetts Amherst, Amherst, MA 01003

Jointly sponsored by:
Awwa Research Foundation
6666 West Quincy Avenue, Denver, CO 80235-3098
and
U.S. Environmental Protection Agency
Washington DC

Published by:

Awwa Research Foundation

American Water Works Association

IWA Publishing

DISCLAIMER

This study was jointly funded by the Awwa Research Foundation (AwwaRF) and the U.S. Environmental Protection Agency (USEPA) under Cooperative Agreement No. CR827268-01. AwwaRF and USEPA assume no responsibility for the content of this report or for the opinions or statements of fact expressed in the report. The mention of trade names for commercial products does not represent or imply the approval or endorsement of AwwaRF or USEPA. This report is presented solely for information purposes.

CONTENTS

LIST OF TABLES

LIST OF FIGURES

FOREWORD

The Awwa Research Foundation is a nonprofit corporation that is dedicated to the implementation of a research effort to help utilities respond to regulatory requirement and traditional high-priority concerns of the industry. The research agenda is developed through a process of consultation with subscribers and drinking water professionals. Under the umbrella of a Strategic Research Plan, the Research Advisory Council prioritizes the suggested projects based upon current and future needs, applicability, and past work; the recommendations are forwarded to the Board of Trustees for final selection. The foundation also sponsors research projects through the unsolicited proposal process; the Collaborative Research, Research Applications, and Tailored Collaboration programs; and various joint research efforts with organizations such as the U.S. Environmental Protection Agency, the U.S. Bureau of Reclamation, and the Association of California Water Agencies.

This publication is a result of one of these sponsored studies, and it is hoped that its findings will be applied in communities throughout the world. The following report serves not only as a means of communicating the results of the water industry's centralized research program but also as a tool to enlist the further support of the nonmember utilities and individuals.

Projects are managed closely from their inception to the final report by the foundation's staff and large cadre of volunteers who willingly contribute their time and expertise. The foundation serves a planning and management function and awards contracts to other institutions such as water utilities, universities, and engineering firms. The funding for this research effort comes primarily from the Subscription Program, through which water utilities subscribe to the research program and make an annual payment proportionate to the volume of water they deliver and consultants and manufacturers subscribe based on their annual billings. The program offers a cost-effective and fair method for funding research in the public interest.

A broad spectrum of water supply issues is addressed by the foundation's research agenda: resources, treatment and operations, distribution and storage, water quality and analysis, toxicology, economics, and management. The ultimate purpose of the coordinated effort is to assist water suppliers to provide the highest possible quality of water economically and reliably. The true benefits are realized when the results are implemented at the utility level. The foundation's trustees are pleased to offer this publication as a contribution toward that end.

Walter J. Bishop
Chair, Board of Trustees
Awwa Research Foundation

Robert C. Renner, P.E.
Executive Director
Awwa Research Foundation

ACKNOWLEDGMENTS

The authors of this report are indebted to the Department of Conservation and Recreation (DCR), Water Supply Division, of the Commonwealth of Massachusetts for participation in this project. The following Commonwealth of Massachusetts DCR personnel provided extensive support throughout the project: Wachusett Branch - Patricia Austin, Dan Clark, David Gettman, Larry Pistrang, Vincent Vignally, Dave Worden and others; Quabbin Branch - Bob Bishop, Pete Deslauriers, Matt Hopkins, Thom Kyker-Snowman, William Pula, Paul Reyes, David Small, and others.

The authors also wish to acknowledge the assistance of Kavitha Ambikadevi, Rebecca Baker, Dan Bourdeau, Darleen Bryant, Dan Buttrick, Frances Cameron, Joshua Cobb, Greg Devine, Autin Falaxa, Elaine Huang, Matt Kennedy, Jun Li, Ruoting Pei, David Senus, Mike Tauche and Tiffany Tauscher, all graduate research assistants who helped with site setup, data collection and analyses. Undergraduates who contributed to the project include Craig Hebert, Tim Kilroy, Randy Cyr, and Kevin Peret. Graduate students who were fully or partially funded by the project include Rebecca Baker, Dan Bourdeau, Frances Cameron, Joshua Cobb, Austin Falaxa, Jun Li, and Ruoting Pei. In particular, Rebecca Baker provided extensive data management, data analysis, and graphical production assistance. Assistance in putting together *Cryptosporidium* and *Giardia* occurrence data was supplied by Patricia Cline, formerly of Golder Associates.

The advice and insightful comments provided by the Project Advisory Committee (PAC) – including Ariamalar Selyakumar, U.S. Environmental Protection Agency National Risk Management Research Laboratory, Water Supply and Water Resources Division; Roger Edwards, retired, Water Bureau of Portland Oregon; Glenn Patterson, Cooperative Water Program Coordinator, United States Geological Survey; Chris Crockett, Philadelphia Water Department, Office of Watersheds; and Charles Haas, Betz Chair Professor of Environmental Engineering, Drexel University – are greatly appreciated. In addition, we would like to acknowledge the help of the initial AwwaRF project officer, Misha Hasan. Finally, the project would not have been successful without the support and understanding of Linda Reekie, the main AwwaRF project officer. Her assistance is deeply appreciated.

EXECUTIVE SUMMARY

It is widely known that microorganisms such as bacteria (for example, *E. coli*), *Giardia* (often associated with beaver), and *Cryptosporidium* (sometimes associated with pets, livestock, and wildlife) can enter streams during rain and snowmelt events. It is difficult, however, to accurately determine the magnitude and timing of transport during such events. In addition, accurately identifying the sources of these contaminants is a challenge. Hence, the organizations responsible for public water supplies, outdoor recreation, and environmental quality face substantial challenges in their efforts to protect public health.

RESEARCH OBJECTIVES

Microbial and water-quality measurements were collected during storm events under a variety of meteorological and land use conditions in order to 1) identify risk of *Cryptosporidium* oocysts, *Giardia* cysts and other constituents, including microbial indicator organisms and standard water quality parameters, entering surface waters from various land uses during periods of surface runoff; and 2) optimize storm sampling procedures for these parameters. The resulting database consists of wet weather transport data for twenty-one distinct storm events and three dry weather spatial basin sampling events. The study was designed to attain several specific research objectives, including:

- Collection of data describing the occurrence and transport of *Giardia* cysts, *Cryptosporidium* oocysts, and other microorganisms of fecal origin from select land use areas during wet weather events.
- Collection of data characterizing the seasonal and year-to-year variability of pathogen occurrence during wet weather.
- Evaluation of both the relative magnitude and timing of specific water-quality parameter export from various land use areas.
- Evaluation of factors which affect the spatial, seasonal, and annual variability of pathogen occurrence for a range of watershed conditions, including meteorological (precipitation depth, duration, intensity, time since previous precipitation), hydrologic (stream order), basin (size, average slope, shape), and/or land use characteristics.
- Evaluation of relationships between protozoa and standard water quality parameters, standard indicator organisms, and alternative indicator organisms.
- Evaluation of data to identify the best sampling techniques and number of samples for adequate characterization of pollutant concentrations and loads, particularly human pathogens, during storm events within reasonable economic and technical constraints.

APPROACH

Microbial and physical water-quality data were collected at six sampling locations over a period of two to three years and under a variety of meteorological conditions. Three distinct land uses (wildlife, dairy farm/agricultural, and residential) were isolated by pairing the sampling locations. An upstream – downstream sampling strategy was used to isolate export from the dairy and residential study sites. For the wildlife study site, a paired watershed approach was used, with sampling transects established at the outlets of two distinct watersheds, one influenced solely by ranging wildlife and one influenced by both ranging wildlife and an established beaver colony. Data collection frequency included monthly, storm-event based, and continuous sampling.

The analyzed constituents included pathogens and alternative source-specific indicator organisms in addition to traditional water-quality and microbial analyses. Traditional water-quality analyses included pH, temperature, turbidity, conductivity, total suspended solids, total phosphorus, total Kjeldahl-nitrogen, and ammonia nitrogen. Total coliforms and fecal coliforms were measured at all sites as industry standard microbial analyses. Monitoring for *Giardia* cysts and *Cryptosporidium* oocysts was included at all sites to provide insight into the transport of these human pathogens. Sorbitol-fermenting *Bifidobacteria* were measured as an indicator of human sewage contamination for samples from sites affected by residential development and septic systems. *Rhodococcus coprophilus* was measured as an indicator of grazing animal manure contamination for sites impacted by animal rearing operations and free-ranging wildlife. *Clostridium perfringens* spores were measured as an indicator of the presence of cyst and spore type pathogens for all sample sites. Somatic (F^+) and CN (F^-) specific coliphages were measured as an indicator of fecal contamination in general and as an indicator of the presence of viruses for all sample sites. The final database consists of occurrence data during twenty-one distinct storm events and three dry weather events, in addition to monthly standard water quality data and continuous physical data.

CONCLUSIONS

The data from this study suggest that when present in small order headwater streams, *Cryptosporidium* and *Giardia* have a high potential for transport through the watershed. Even small rainfall events (less than 0.25 inch accumulations) were capable of washing cysts and oocysts into the stream. Saturated or near-saturated ground conditions appeared to be a particularly important factor controlling transport to the stream. *Giardia* and *Cryptosporidium* appear to have a seasonal pattern, with higher levels of occurrence in the winter and early spring. It is hypothesized that this may be related to young beaver dispersal timing. Cycles of animal birth and infectivity are also likely an important factor. In addition, the potential for "rain on snow" events during this period may be of particular importance. Such events have a high potential to result in flushing of stored fecal material from the land surface. Many of the documented waterborne *Cryptospordium* outbreaks have also occurred during this period.

Based on the limited data available from this study, inter-annual hydrologic variability appears to be an important consideration in terms of transport to headwater streams. A higher number of positive detections occurred during wetter than average (in terms of precipitation) water years. While *Cryptosporidium* and *Giardia* were detected in one event during a drought year, this was a rain-on-snow event occurring in the wettest part of that year. As noted above, transport potential appears elevated when saturated soil conditions exist, and remotely based soil moisture indexes may assist in identifying events at greater risk for pathogen transport. The first event occurring following attainment of near saturated conditions is likely to exhibit high detection levels. However, pathogen sources are likely to be flushed following such events, resulting in low occurrence levels for subsequent events.

Relative risk of contamination to downstream surface waters by land use type was inferred based on study results. The results suggested that the highest risk of contamination by *Cryptosporidium* was associated with agriculture land use (cattle), with moderate risk associated with pristine wildlife areas and low risk associated with residential areas. The highest risk of contamination by *Giardia* was associated with beaver influenced watersheds. Residential areas were classified with a moderate to high risk level, while agricultural and ranging wildlife locations were classified with a moderate to low risk level. Additional focused studies are necessary to better assess such risk.

RECOMMENDATIONS

Data suggest that supplementation of routine monitoring with event based monitoring is essential for accurate characterization of watershed based pathogenic risk. However, both hydrologic variability and spatial/temporal patterns of pathogen occurrence in the watershed and near-stream zone make quantification of the long-term risk associated with wet weather difficult. Due to the inherent logistic, personnel and financial costs associated with event based sampling, the goals and methodology for any such required sampling must be carefully addressed. It appears appropriate to classify wet weather sampling schemes under one of two broadly defined goals: (1) characterization for watershed based controls or (2) characterization for specification of treatment processes, either short- or long-term. Such classification is necessary due to underlying differences in the monitoring strategies most appropriate for quantifying the associated pathogenic risk.

For both types of monitoring schemes, the time scales of hydrologic variability alone suggest that emphasis should be placed on developing watershed specific long-term records of event-based risk, rather than monitoring of wet-weather events over a short period of time. Monitoring of "rain on snow" wet weather events or wet weather events with moist antecedent rainfall conditions is likely to produce a larger number of positive detects than either routine sampling or wet weather sampling meeting other criteria. Remotely sensed soil moisture data indicating saturated or near saturated conditions may therefore be an important criterion for flagging events where sampling for *Cryptosporidium* and *Giardia* is warranted. Regardless of whether such sampling is conducted, utilities should note that high efficiency water treatment is likely especially important during these periods.

The potential for (1) failure of beaver impoundments or (2) the occurrence of events causing scour of sediment stores within impoundments are of significant concern in terms of beaver effects on downstream water quality. It is thus suggested that known beaver impoundments in drinking water supply watersheds be either removed or monitored. In-situ depth, temperature, pH, and conductivity probes strategically placed near beaver impoundments have the potential to provide information on when they are spilling, thus perhaps warranting monitoring for pathogen presence downstream. Additional study of the importance of bed sediments is needed. Passive bed samplers may be useful in this regard.

Lastly, future monitoring schemes would benefit by including a number of contingency samples for specifically evaluating the effect of filter clogging on detection. Contingency samples could be used to filter a predetermined, large volume of water by running the filters in sequence (changing when one becomes clogged) and then reporting the results determined across the sample set. Alternatively, spiking individual samples with inactivated cysts to provide a percent recovery value per sample holds promise for identifying the rate of false negative environmental samples.

CHAPTER 1
PROJECT OVERVIEW

BACKGROUND

The drinking water community is faced with the challenge of ensuring the safety of their drinking water supplies. However, waterborne disease outbreaks in the U.S. over the last two decades have shown that threats from pathogens such as *Cryptosporidium* and *Giardia* are difficult to protect against, particularly without a clear understanding of the variability of pathogen occurrence in both space and time. A majority of the data collected to date on the occurrence of *Cryptosporidium* and *Giardia* in surface waters has concentrated on characterization of pathogen concentrations under average conditions. Data are lacking on pathogen variability in sub-basins characterized by different pathogen sources, during and following events such as rainfall, elevated turbidity or fecal coliform counts, or following land use change. An additional noted deficiency is the lack of multi-year databases to assess year-to-year variability. As a result, current watershed monitoring methodologies do not provide a means to evaluate the comparative contributions of pathogens from different sources, under varying hydrologic conditions and watershed characteristics, and land use activities. This study is thus a critically needed research effort to advance our understanding of pathogen variability in sub-basins characterized by different pathogen sources and watershed characteristics, during and following events. The results of this research may be combined with the results from other studies in order to assist public water suppliers and regulators to identify monitoring program requirements necessary to reduce vulnerability of drinking water supplies to pathogens.

PROJECT OBJECTIVES

Sampling, analytical, and statistical methods were used to attain several research objectives. These objectives centered on advancing current understanding on the variability of pathogen occurrence and transport through watersheds, particularly during and following events. The results of this research were used together with information from other studies to develop methodologies aimed at minimizing the risk of waterborne disease outbreaks in drinking water supplies. Specifically, the research objectives include:

- To obtain data on the transport of *Giardia* cysts, *Cryptosporidium* oocysts, and other microorganisms of fecal origin from selected land use areas during wet weather events.
- To obtain data on the seasonal and year-to-year variability of pathogen occurrence during wet weather.
- To determine both the relative magnitude and timing of specific water-quality parameters from various land use areas.
- To examine factors which affect the spatial, seasonal, and annual variability of pathogen occurrence for a range of watershed conditions, including meteorological (precipitation depth, duration, intensity, time since previous precipitation), hydrologic (stream order), basin (size, average slope, shape), and/or land use characteristics.

- To evaluate relationships between protozoa and standard water quality parameters, standard indicator organisms, and alternative indicator organisms.
- To identify the best sampling techniques and number of samples for adequate characterization of pollutant concentrations and loads, particularly human pathogens, during storm events within reasonable economic and technical constraints.
- To evaluate the viability of using source-specific indicator organisms to identify pollutant source-areas during stormflow when combined with an optimized sampling plan and GIS data.
- To facilitate more informed watershed management decisions and the development of appropriate watershed protection plans.

APPROACH

Microbial and physical water-quality data were collected at six sampling locations over a period of two to three years and under a variety of meteorological conditions. Three distinct land uses (wildlife, dairy farm/agricultural, and residential) were isolated by pairing the sampling locations. An upstream – downstream sampling strategy was used to isolate export from the dairy and residential study sites. For the wildlife study site, a paired watershed approach was used, with sampling transects established at the outlets of two distinct watersheds, one influenced solely by ranging wildlife and one influenced by both ranging wildlife and an established beaver colony. Data collection frequency included monthly, storm-event based, and continuous sampling. The objectives of the project are to develop and validate a strategy for selection of sampling locations, frequencies and methods to accurately depict pathogen occurrence and variability in relation to various sources within watersheds during and after weather, hydrologic, or land use events. Of particular interest is monitoring for occurrence of *Cryptosporidium* and *Giardia*.

The analyzed constituents include several emerging pathogens and alternative source-specific indicator organisms in addition to traditional water-quality and microbial analyses. Traditional water-quality analyses included pH, temperature, turbidity, conductivity, total suspended solids, total phosphorus, total Kjeldahl-nitrogen, and ammonia nitrogen. Total coliforms and fecal coliforms were measured at all sites as industry standard microbial analyses. Monitoring for *Giardia* cysts and *Cryptosporidium* oocysts was included at all sites to provide insight into the transport of these emerging human pathogens. Sorbitol-fermenting *Bifidobacteria* was measured as an indicator of human sewage contamination for samples from sites affected by residential development and septic systems. *Rhodococcus coprophilus* was measured as an indicator of grazing animal manure contamination for sites impacted by animal rearing operations and free-ranging wildlife. *Clostridium perfringens* spores were measured as an indicator of the presence of cyst and spore type pathogens for all sample sites. Somatic (F^+) and CN (F^-) specific coliphages were measured as an indicator of fecal contamination in general and as an indicator of the presence of viruses for all sample sites. The final database consists of occurrence and loading data for the above parameters during twenty-one distinct storm events and three dry weather events, in addition to monthly standard water quality data and continuous physical data.

Storm period sampling and analysis for several of the microbial analyses is unique to this study. This study provides new information about the variability and rate of occurrence of these organisms over the course of storm events, their transport characteristics through the watersheds under study, and their viability for distinguishing sources of contamination in mixed land use areas such as New England. The collected data may thus be used to propose sampling strategies for accurate identification of contamination potential and determination of constituent loads, including the relative importance of wet weather sampling, appropriate sampling locations, the number of samples to be collected, the timing of sample collection both during an event and seasonally, and the constituents to be measure to ensure the integrity of surface-water drinking supplies.

This work is of particular interest in Massachusetts, where the Quabbin-Ware-Wachusett reservoir system provides unfiltered drinking water to 2.5 million people in Boston and surrounding communities, Figure 1-1. The evaluation of storm-event loads and the detection of other pathogens and indicator organisms within the water supply watersheds are active areas of research for the Department of Conservation and Recreation (DCR), Water Supply Division, the agency charged with maintaining source water-quality of the drinking water reservoirs. Current monitoring for human pathogens in watersheds surrounding the reservoirs relies mainly on fecal coliform and standard water quality data collected on a bi-weekly basis, supplemented by more detailed alternative indicator and disinfection byproduct studies carried out with the assistance of the University of Massachusetts (UMass). Historically three wet weather watershed events were monitored for standard water quality parameters.

In-reservoir monitoring for *Cryptosporidium* and *Giardia* at the Quabbin and Wachusett Chicopee Valley Aqueduct (CVA) intakes has been conducted as part of the Interim Enhanced Surface Water Treatment Rule (IESWTR) and to prepare for the upcoming Long-Term 2 Enhanced Surface Water Treatment Rule (LT2ESWTR). A complete overview of the Massachusetts Water Resources Authority (MWRA) pathogen monitoring program and results may be found on the web (MWRA, 2005) and is summarized herein. At Wachusett, monthly samples were collected from 1994 to mid-January 1999 at Shaft 4 and later at the Cosgrove Intake. Since mid-January 1999, sampling for *Cryptosporidium* and *Giardia* at Cosgrove Intake has occurred weekly. At Quabbin Reservoir, the CVA intake is monitored bi-weekly for *Cryptosporidium* and *Giardia*. EPA Method 1623 has been used since April 2004. Based on the data collected by the MWRA, almost all values at Quabbin (CVA Intake) and Wachusett (Cosgrove Intake) have been less than the detection limit, with the goal of 50L filtered achieved each time. In calendar year 2004, only one of twenty-six samples tested positive for *Giardia* at the CVA Intake, resulting in an average of 0.1 cysts/100 L; no *Cryptosporidium* was found. At Cosgrove Intake, one sample in fifty-two tested positive for *Cryptosporidium*, resulting in an average of 0.03 oocysts/100L, while five samples out of fifty-two tested positive for *Giardia*, resulting in an average of 0.2 cysts/100 L. All detections were less than 5 cysts or oocysts per 100L. In addition, MWRA has voluntarily partnered with Tufts University to conduct research monitoring on weekly composite samples from one location. The large volume sample method used is much more sensitive. Results from this study found average *Cryptosporidium* levels of 0.04 oocysts/100L. (since 2001) and average *Giardia* levels of 0.01 cysts/100 L (since 2002) (MWRA, 2005).

Based on previous studies, the DCR has several areas of concern in terms of potential sources of contamination from its watersheds to the drinking water reservoirs. Runoff from farms, generally located in "mixed-use" land use areas in Massachusetts, is of particular interest due to the high association of *Cryptosporidium* infections with dairy cattle (calves and heifers). Contamination from residential areas is of particular interest as significant resources have been invested in extending sewer lines and replacing older systems; pet waste is also a concern in these areas. The influence of beaver impoundments on water quality is of interest due to their ubiquitous presence throughout New England and their association with *Giardia*. Study areas meeting these specifications were thus selected for this research.

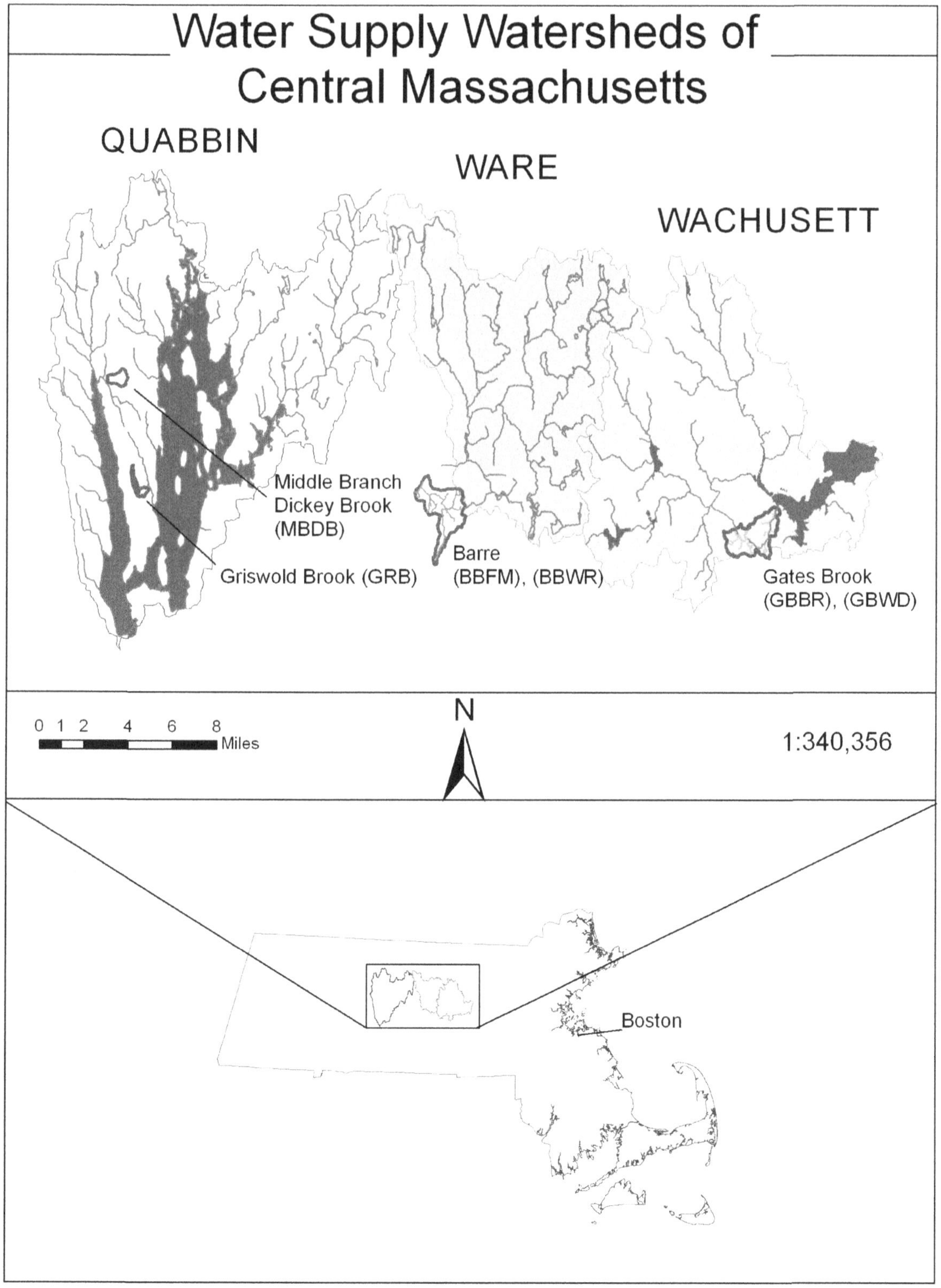

Figure 1-1 The drinking water supply system for Boston and surrounding communities

REPORT ORGANIZATION

This report is organized as follows. Chapter 2 provides an overview of watershed management and microbial indicators, including factors that affect levels of microorganisms in surface waters. Chapter 3 provides a more detailed overview of results from other watershed monitoring studies of strong relevance to this research. Chapter 4 presents the materials and methods, including study site characterization, used in this research. Chapter 5 presents a broad overview of the event results, including a comparison of monthly and event data, event mean concentrations, and event loads. Chapter 6 presents the findings with respect to *Cryptosporidium* and *Giardia* occurrence. Chapter 7 summarizes the conclusions and recommendations from the research.

CHAPTER 2
WATERSHED MANAGEMENT AND MICROBIAL INDICATORS

WATERSHED MANAGEMENT

History

Prior to the advent of modern water treatment technologies, the protection of water supply sources relied on the concept of source isolation. Before the 1900s, source isolation was used throughout the United State to guard against contamination of raw water. This was predominantly accomplished through acquisition of watershed lands. With the development of water treatment technologies and the increased demand for agricultural and urban land uses resulting from population growth in the 20th Century, it was no longer cost effective to acquire an entire watershed to ensure the safety of the source water.

Before World War I, all of the basic elements of conventional drinking water treatment including coagulation, sedimentation, filtration and disinfection had been developed. In 1925, the Treasury Department announced that water treatment could be a substitute for source isolation and protection. As a result, source isolation fell into general disuse. The neglect of source isolation as a method of water quality assurance had the effect of increasing public health related issues such as cancer-causing chemicals found in drinking water, outbreaks of waterborne diseases, and in some cases failure for the drinking water to meet government safety standards. Several drinking water surveys conducted by different agencies during the 1970s publicly questioned the effectiveness of relying strictly on water treatment technology as a method to assure drinking water safety. As a result of these surveys, the interests in drinking water source protection reemerged (Burby *et al*, 1983).

Microbial Monitoring

Monitoring for organisms in drinking water sources to detect the potential presence of pathogens and fecal contamination is the most widely used method to assure the microbial safety of drinking water in the United States. Traditional indicators like total and fecal coliforms are considered a reliable indicator of water quality and a measure of the adequacy of water treatment (APHA, 1998). However, since the indicator organisms can be natural inhabitants of soil and plant material, their detection in water does not provide absolute evidence of the presence of pathogen-bearing fecal contamination (Collins *et al.*, 1995). Furthermore, the different environmental attenuation and survival characteristics between coliforms and pathogens such as viruses, protozoa, and other parasites may account for the existence of pathogens without the presence of coliforms (or the presence of coliforms without the presence of pathogens) (Ferguson *et al.*, 1996). In addition, traditional microbial indicators provide no information as to microbial sources. For these reasons, the traditional indicators could be inadequate for indicating the presence of potential pathogens and fecal contamination. Research into alternative indicators that are specific to the source of contamination (microbial source tracking) would be of greater use to watershed management agencies as they would allow the agencies to modify or implement management strategies in order to reduce sources of contamination.

The source-tracking function of alternative indicators is also attractive because any classically trained bacteriologist employed by water utilities should be capable of performing the monitoring tests. Methods based on traditional microbiological techniques are economically desired, because molecular and serological methods (such as amplification of targeted genetic material using polymerase chain reaction (PCR) and serotyping of specific organisms) require specialized training and/or expensive instrumentation not typically available in water utility laboratories (Brock *et al.*, 1994; Sobsey *et al.*, 1995).

One approach to microbial source tracking (MST) is to analyze for fecal organisms specifically shed by different host species. Several microorganisms have been proposed as indicators of specific water pollutant sources. *Rhodococcus coprophilus* and sorbitol-fermenting *Bifidobacteria* are promising alternative bacterial indicators of human fecal or domestic wastewater (Mara and Oragui, 1983; Jagals *et al.*, 1995) and animal fecal contamination (Mara and Oragui and, 1983; Mara and Oragui, 1985; Jagals *et al.*, 1995), respectively. Studies of F-specific Ribonucleic Acid (RNA) coliphages have also shown that the four serogroups of these viruses can be used to determine if the source of fecal contamination as human, non-human, or both (Sobsey *et al.*, 1995).

Management Strategies

Protection of the source water through a watershed protection program can be achieved through any of the following: ownership of the land by the water utility (either the total watershed area or buffer strips around the reservoir and its tributaries); land use agreements with owners to manage or limit the activities in the watershed; special water supply watershed ordinances and land use regulations (Burby *et al.*, 1983).

Ownership of the land within the watershed is the most effective measure of protecting the source water, but it does not always offer a cost-effective solution. Land use agreements offer a less costly alternative to land ownership, but may be less effective. Implementation of special ordinances and land use regulations may be less costly than land acquisition, but require on-going effort to assure good compliance on the part of land owners.

In order for watershed managers to determine which combination of watershed protection approaches to use, sound scientific data is needed regarding the nature and magnitude of point and non-point sources in question. The more specific the identification of on-going or future problems (as related to source) become, the better management can implement strategies which target the source of the contamination. This concept of "specific" introduced the need for source-specific indicators.

USEPA Surface Water Regulations

The United States Environmental Protection Agency (USEPA) developed the Surface Water Treatment Rule (SWTR) in 1989. The rule governed all public water systems (PWSs) using surface water sources or groundwater sources under the direct influence of surface water. The goal of the SWTR was to provide multiple barriers of treatment to control pathogens in drinking water.

The SWTR required all PWSs to conform to four major criteria. One, disinfection residual was to be maintained within the distribution system. Two, three log (99.9%) removal or inactivation of *Giardia* and four log (99.99%) removal of viruses was to be attained by treatment. Three and four, all PWSs had to meet the specified combined filter effluent turbidity performance standards and *un*filtered PWSs were required to implement a watershed control program. The watershed control program included a characterization study of the watershed's hydrology characteristics, land ownership, and activities, which could have an adverse effect on the source water quality. The SWTR established a maximum contaminant level goal (MCLG) of zero for *Giardia lamblia*, viruses, and *Legionella* sp. The SWTR did not specifically control for *Cryptosporidium* sp. which has been shown to be fairly common in the environment.

The Interim Enhanced Surface Water Treatment Rule (IESWTR) was later developed by the EPA in 1998 to build upon the treatment requirements in the SWTR and improve the control of microbial pathogens, specifically *Cryptosporidium* sp. in drinking water. The rule applied to all PWSs which served 10,000 people or more. Two major components of the IESWTR were an established MCLG of zero for the genus *Cryptosporidium* and for *un*filtered water supplies to include *Cryptosporidium* monitoring in watershed control programs. The Long Term 1 Enhanced Surface Water Treatment Rule (LT1ESWTR), established in 2002, built upon the framework established in the 1998 IESWTR. The rule applied the same set of regulations in the IESWTR to PWSs serving less than 10,000 customers.

The USEPA proposed the LT2ESWTR in August 2003 (USEPA, 2004). The proposed rule builds upon the framework of the IESWTR and LT1ESWTR to improve control of microbial pathogens in drinking water, specifically *Cryptosporidium* sp., and to address trade offs between microbial risk and disinfection by-product risks. The key provisions of the proposed rule include source water monitoring of *Cryptosporidium* sp.; additional inactivation of *Cryptosporidium* sp. by all filtered systems based on source water quality; and complete inactivation of *Cryptosporidium* sp. by all *un*filtered systems. To meet new *Cryptosporidium* sp. treatment requirements, the rule develops a number of treatment and management options (i.e. microbial toolbox) available to PWSs to implement.

Source water quality monitoring of *Cryptosporidium* sp. requires filtered PWSs serving 10,000 or more customers to sample source water for *Cryptosporidium*, *E. coli*, and turbidity either on a monthly or bi-weekly basis for twenty-four months, starting within six months of the LT2ESWTR finalization. Source water monitoring in small systems, or PWSs serving less than 10,000 people, will not occur until after the required large system monitoring is finished. Small filtered systems will first conduct one year of bi-weekly monitoring for *E. coli* and turbidity. Systems with sufficiently high levels of *E. coli* (annual mean concentration exceeds 10 *E. coli* per 100 mL for lakes or exceeds 50 *E. coli* per 100 mL for rivers) will be required to subsequently sample for *Cryptosporidium* oocysts, *E. coli*, and turbidity on a bi-weekly basis for one year. Large and small unfiltered (meeting all the filtration avoidance criteria of 40 CFR 141.71) systems must monitor for *Cryptosporidium* unless they provide 3 log inactivation of *Cryptosporidium* by the time treatment is required. It should be noted that utilities may request that previously gathered, equivalent Quality Assurance/Quality Control (QA/QC) standard data be reviewed by the USEPA and accepted in lieu of new monitoring. In addition, systems providing the maximum level of treatment are not subject to the monitoring requirements. The proposed regulations do not require sampling schedules to account for hydrological conditions.

The LT2ESWTR also provides guidance on sampling locations. Sampling should be conducted prior to any treatment but at a point such that the source water is not subject to further runoff. In addition, for systems where it applies, samples should be taken after presedimentation, after the off-stream storage reservoir, after the terminal reservoir, or after riverbank filtration. Plants with multiple sources may sample after sources have been blended, use composite samples, or use weighted results of individual samples.

The monitoring results will be used to classify filtered systems in one of four risk bins based on an average concentration. Large PWSs may calculate their average *Cryptosporidium* level either as an overall average (if the plant collected at least 48 samples, or used bi-weekly sampling), as the highest average for any consecutive 12-month period if the plant collected 24 to 47 samples, or as the highest value if the plant failed to collect all of the required 24 monthly samples. Small PWSs may calculate their average level either as an overall level if they collected 24 samples (at least two per month) or use the highest value if the plant failed to collect all of the required 24 bi-weekly samples. A second round of monitoring is required beginning six years after the initial treatment requirement classification.

MICROBIAL INDICATORS

Standard Microbial Analysis Indicators

Total Coliform

Since 1912, total coliform (TC) have been used as an indicator of fecal contamination (Montgomery, 1985). Total coliforms are currently defined as "all aerobic and facultative anaerobic, Gram negative, nonsporulating, rod-shaped bacteria that ferment lactose with gas formation at 35°C" (or "produce a red colony with a metallic (golden) sheen within 24 h incubation at 35°C on an Endo-type medium") (APHA, 1998). This is an operational definition and depends on the detection/enumeration method rather than bacterial taxonomy. It includes several species of bacteria: *Enterobacter* spp.; *Citrobacter* spp.; *Klebsiella* spp. ; *Serratia* spp.; *Hafnia* spp.; and *E. coli* (Arango, 2000). Although initially total coliforms were used as indicators of human fecal contamination, correlations between the presence of TC and contamination from warm-blooded animals have been demonstrated on several occasions (Whitman *et al.*, 1995). Thus, TC are used as indicators of fecal contamination despite the fact that they have been shown to replicate in some environmental waters (Hazen and Toranzos, 1990) and many of the species comprising the coliform group have non-fecal origins (Hardina and Fujioka, 1991). Non-fecal coliforms can originate from soil, vegetation, and wastes of industrial processes such as pulp and paper processing. Their presence can be associated with high organic carbon levels in the water (Kuhn *et al.*, 1991).

Fecal Coliform

Fecal coliforms (FC) are thermotolerant and associated almost exclusively with the feces of warm-blooded animals (APHA, 1998). The functional definition of fecal coliforms is the same as that of TC with the exception that the organisms meet the criteria when grown at 44.5°C. Because these organisms are able to grow and ferment lactose at 44.5°C, their origin is associated more frequently with the feces of warm-blooded animals than TCs (Toranzos and McFeters, 1997). However, some bacteria, such as *Enterobacte*r and *Klebsiella spp,* fit in this operational definition, but do not have fecal origins (Baudisova, 1997). In addition, some true fecal coliform substrains do not ferment lactose (Baudisova, 1997). *E. coli* is the most representative species of fecal coliforms, and in some cases it may be measured in place of fecal coliforms as an indicator (Krumperman, 1983).

Total Heterotrophic Plate Counts

Total heterotrophic plate counts (THPC) (anaerobic or aerobic) determine the total number of bacteria able to grow on a wide variety of carbon/energy sources. Heterotrophic bacteria are naturally occurring, and their presence in drinking water is not indicative of a health risk. As such, heterotrophic bacteria are not regulated by the Safe Drinking Water Act (SDWA). The SDWA does recognize that heterotrophic bacteria, in excessive numbers (greater than 500 CFU/mL) as measured by the heterotrophic plate count (HPC) technique, can inhibit the growth of TC and the development of the diagnostic green-to-golden metallic sheen on m-endo plates. In order to better protect drinking water, the SDWA requires that any total coliform membrane filtration plate that has greater than 200 non-coliform CFU/100 mL be re-sampled and analyzed by another technique (e.g., multiple tube fermentation or presence-absence). The SDWA also assumes that an adequate disinfection residual is maintained in areas of the distribution system where THPC counts are consistently less than 500 CFU/mL.

Heterotrophic bacteria, which include the aerobic and facultative-anaerobic heterotrophs, are more resistant to both the chemicals and processes used to remove TC , FC, and *E. coli* from water. Since some heterotrophic bacteria are more resistant to disinfection than other indicators, they are useful as an indicator of treatment efficiency or deteriorating water quality in the distribution system.

The relationships between THPC and TC were studied by Edberg and Smith (1989). Statistical analysis based on the results of 26,158 samples from different sources such as raw sewage, treated effluent, and distribution water from a public supply, indicated that the presence of THPC had a low correlation coefficient with TC. In other words, enumeration of one indicator was not found to be indicative of the other; however, as indicated above, the measurement of each may provide different information about the water in question.

Microbial Source Tracking Indicators

In addition to the existence of non-fecal originated coliforms and their replication in certain environmental waters, the most commonly used fecal indicators, TC and FC, cannot provide information as to the specific source of the contaminant, as they are found in both human and animal feces (Sinton *et al.*, 1998). This has prompted investigations into alternative indicators capable of providing source specific information (Sinton *et al.*, 1998). Despite the shortcomings of using coliforms as a water quality indicator, the information provided by microbial source tracking indicators would only supplement coliform tests for watershed management agencies and are typically not used to indicate pathogen risk.

Sorbitol-fermenting Bifidobacteria

Sorbitol-fermenting *Bifidobacteria* have been investigated as potential specific indicators of human fecal contamination because sorbitol is a food additive used only in food intended for humans. *Bifidobacteria* are strict anaerobes from the family Actinomycetaceae. They are Gram-positive or Gram-variable anaerobic rods having bifurcating morphology, and do not produce spores (Holt *et al.*, 1994). As common inhabitants of the digestive track of warm-blooded animals, *Bifidobacteria* can ferment lactose without gas production. They can produce lactic and acetic acid in a ratio of 3:2, a characteristic that lowers the pH of their surrounding media (Payne *et al.*, 1999).

Studies of sorbitol-fermenting *Bifidobacteria* sources have been conducted by many researchers. The species capable of fermenting sorbitol have been isolated almost exclusively from human feces and water samples subject only to human contamination (Mara and Oragui, 1983). Experiments conducted by Jagals and Grabow (1996) have shown that about 58 to 93 percent of the total number of *Bifidobacteria* isolated from human sources were able to ferment sorbitol. *B. breve*, *B. adolescentis*, *B. catenulatum* and *B. pseudocatenulatum* have been reported as sorbitol-fermenting species (Scardovi, 1981; Mara and Oragui, 1983). They have been isolated from the digestive tract and feces of human infants and adults, the human vagina, sewage and infrequently from the feces of suckling calves. In raw sewage, septic systems and cesspools, the concentration of sorbitol-fermenting *Bifidobacteria* and *E. coli* are approximately equal, between 10^6 and 10^7 per 100 mL (Mara and Oragui, 1983). Since sorbitol-fermenting *Bifidobacteria* have been isolated in bovine rumen and suckling cow feces infrequently (Scardovi, 1981), their utilization as a human fecal contamination indicator in the vicinity to cattle farms should be scrutinized.

As anaerobes, *Bifidobacteria* species' environmental survival characteristics are important in order to ascertain their use as microbial source tracking indicators. Anaerobes typically do not continue to multiply after they have been excreted into the environment, but they may be transported through soils and in water.

Along with the physiological states of the microorganisms, the physical and chemical nature of the soil is an important factor in determining their survival. Soil is likely to contain more anaerobic zones than water (Long *et al.*, 2002). Sorbitol-fermenting *Bifidobacteria* have been shown to persist in sewage, septic tanks and cesspools with densities of 7×10^6/100 mL, and their numbers appear to be relatively unaffected until chlorination through conventional wastewater treatment processes (Resnick and Levin, 1981). Previous studies have recorded survival times for *Bifidobacteria* of up to 12 days in lake waters at ambient temperatures (Gyllenberg *et al.*, 1960) and as short as 5 hours in sterile fresh water (Resnick and Levin, 1981). Survival times between 10 and 24 hours were estimated from a study at UMass when *B. breve* were subjected to simulated typical surface water environments. A longer survival time (48 hours) has been recorded from the same study conducted at UMass when the organisms were exposed to a simulated septic system environment (Mahar, 2002). Considering different factors which can affect *Bifidobateria*, their estimated surface water survival period is 48 hours or less, and therefore they can only be considered as an indication of recent contamination of human origin (Mahar, 2002).

Rhodococcus coprophilus

Rhodococcus coprophilus have been considered a good candidate for an indicator of animal fecal contamination (Mara and Oragui, 1981 and 1983; Jagals *et al.*, 1995). It is an actinomycete of the family Nocardiaceae and its optimal growth environment is the dung of herbivores. In addition, studies by Rowbotham and Cross (1997a and b) found that grasses and soils are suitable environments for its survival but are not amenable for growth.

R. coprophilus has been isolated from the feces of grass eating farm and wild animals (dogs, horses, sheep, cows, deer, pigs, ducks, rabbits) in addition to samples of water subject to fecal contamination from these animals. Long *et al.* (2002) analyzed feces from two beaver and found no *R. coprophulus*. In addition, *R. coprophilus* has not been isolated from human feces or water samples subject mainly to human contamination (Mara and Oragui, 1985, Jagals *et al.*, 1995). Long (1999) found *R. coprophilus* was only detected in samples collected downstream from water contaminated by grazing animal feces in the Wachusett Reservoir watershed in Massachusetts. Thus, it has been considered a good candidate for an indicator of animal fecal contamination (Mara and Oragui, 1983; Mara and Oragui, 1985; Jagals *et al.*, 1995). The same researchers also suggested some limitations to using *R. coprophilus* as an indicator organism, including the long time required for its isolation and identification (18 days), the cumbersome method of isolation, and the unsuitability of growth medium when organisms are present in very low numbers.

The environmental survival study conducted by Arango (2000) indicated that *R. coprophilus* survived better in lower TOC concentration and at lower environmental temperatures. A reduction in numbers of *R. coprophilus,* on the order of 75 to 90 percent, was observed after two weeks at 27 ºC to 29 ºC (typical of summer temperatures), while the reduction of *R. coprophilus* at 4 ºC (typical of winter temperature) was not significant. Therefore, small loads of fecal contamination might not be detectable after about 2 weeks during the summertime, when water temperatures are elevated (Long *et al.*, 2002). Because of the long environmental survival characteristics of *R. coprophilus* in environmental waters, up to 120 days (Mara and Oragui, 1983), care must be taken to implement an appropriate sampling design to hydrologically isolate potential animal sources.

Somatic and F-specific coliphages

A number of researchers have considered bacteriophages, (viruses that infect bacteria), as good indicators of fecal contamination (Sobsey *et al.*, 1995). Somatic coliphages are viruses that attach to receptors on the cell wall of coliform bacteria and may contain single or double stranded DNA as genetic material (Brock *et al.*, 1994). These phages are commonly detected in feces of human, cattle, pigs, chickens and other animals. They are found in especially high numbers in sewage and wastewaters from slaughterhouses and hospitals. Morinigo *et al.* (1992) consistently detected somatic phages in waters where enteroviruses were detected, even when the latter were in very low concentrations (1-10 enteroviruses/10 L). This suggests that coliphages have the potential to indicate the presence of enteric viruses. Morinigo *et al.* concluded that somatic coliphages should be included in regular monitoring as supplementary indicators to coliforms.

F-specific RNA coliphages, together with F-specific DNA coliphages, comprise the F-specific or male-specific coliphages which infect bacteria via F-Pili. F-Pili are hair-like structures encoded by a plasmid of the incF incompatibility group and are only produced at temperatures higher than 30°C (Sobsey *et al.*, 1995). F-specific coliphages can contain single stranded RNA or DNA as their genetic material. Of particular interest are RNA phages. F+RNA phages are present in human feces and other animal samples when the pertinent hosts exist (Haavelar *et al.*, 1990). Studies on feces and sewage have shown that F+RNA coliphages can be divided into the following four antigenic groups (I-IV) based on serotyping (Gerba and McLeod, 1976; Furuse *et al.*, 1978; and Haavelar *et al.*, 1990):

- Group 1 – Non-human animal.
- Group 2 – Primarily found in human feces, but sometimes pig feces.

- Group 3 – Exclusively human in origin.
- Group 4 – Primary of non-human origin, with rare human association.

The recognition of four serological groups suggested that they could be useful as indicators of both human and non-human fecal contamination.

A number of watershed problems have been investigated using typing of F+RNA coliphages for source tracking. Long *et al.* (2003) used F+RNA coliphage typing to investigate fecal pollution sources in a Massachusetts source water watershed. Water samples were collected for 13 months from a highly protected area, a wildlife impact area, and a residential impact area, which were determined based on land use data. The major contamination for all three sites was from non-human sources, as indicated by F+RNA coliphage results. Coupled with land use analysis, birds were indicated as the probable non-human fecal source. Using F+RNA coliphage genotyping, Griffin *et al.* (2000) determined the source of fecal contamination in Hommosassa Springs, Florida. Results indicated that the origins of fecal contamination in those areas were predominantly animal sources (Group I and Group IV) from a wild animal park operated in the area.

Clostridium perfringens

Clostridium perfringens is an anaerobic, Gram-positive, sporeforming rod. It is widely distributed in the environment and frequently occurs in the intestines of humans and many domestic and feral animals. Spores of the organism persist in soil, sediments, and areas subject to human or animal fecal pollution. *Clostridium perfringens* is associated with human fecal waste, where it is present in relatively high numbers, although in a number that is two orders of magnitude lower than coliforms (Bisson and Cabelli, 1980).

Clostridium perfringens have a survival time which is longer than that of oocysts for *Cryptosporidium* because it is very resistant to environmental and disinfection stresses in its spore form. Spores of *C. perfringens* have been proposed as a surrogate indicator for cysts and oocysts of both *Giardia* and *Cryptosporidium*. Its vegetative cell is very sensitive and has a short survival time in the environment.

Based on the characteristic of the organism exsiting in two forms that have different survivals in the environment, Bisson and Cabelli (1980) proposed their use as indicators of waste water contamination in conjunction with coliforms. Specifically, *C. perfringens* can serve as an indicator of the presence of chlorinated or unchlorinated sources of fecal pollution, but advised against its use as indicator of water quality for recreational waters. Bison and Cabelli also recommended the use of the spores as conservative tracers in survival studies of pathogens.

Fujioka and Shizumura (1985) found a correlation between the degree of pollution in water and the presence of *C. perfringens* in stream waters in Hawaii. They proposed the utilization of this organism as indicative of the degree of treatment and age of wastewater. Fujioka and Shizumura concluded that *C. perfringens* was more useful as an indicator of treated vs. non-treated sewage than a microbial source tracking indicator.

Pathogens

Cryptosporidium

Cryptosporidium has long been recognized as an important pathogen of farm and domestic animals, but human cryptosporidiosis (disease caused by *Cryptosporidium*) was not recognized until 1976 (Atherton *et al*, 1995). Symptoms of this illness include typical gastrointestinal symptoms, ranging from mild gastroenteritis to abundant and prolonged watery diarrhea (Craun *et al.*, 1998). The severity and duration of the illness is often greater in immunocompromised persons (i.e. infants, pregnant women, the elderly, and those with severely weakened immune systems) than in healthy individuals and may be fatal among this population. *Cryptosporidium* sp. can infect a wide variety of mammals and birds (Craun *et al.*, 1998; Donnelly and Stentiford, 1997; AWWA, 1999a). Domestic livestock, pets, and wild animals, such as beaver, are potential hosts of the *C. parvum* species (Smith *et al.*, 1995). Cattle feces appear to be the primary source of *Cryptosporidium* in the environment (Donnelly and Stentiford, 1997; AWWA, 1999a).

Cryptosporidium is a single celled parasitic protozoan. There are at least four recognized species of *Cryptosporidium*. *C. parvum* and *C. muris* are known to infect mammals (Smith *et al.*, 1995). *C. baileyi* and *C. meleagridis* infect birds (Smith *et al.*, 1995). The genus species *C. parvum* is the major species responsible for clinical disease in humans and infected mammals (Smith and Rose, 1998). However, immunocompromised individuals may become infected with animal-specific species of *Cryptosporidium*.

Transmission has been demonstrated to occur directly from person to person through the fecal-oral route and by the consumption of contaminated food and drink (Donnelly and Stentiford, 1997; Smith *et al.*, 1995). The potential exists for transmission by the waterborne route (Smith *et al.*, 1995). Infected wildlife within a watershed can contribute to the pool of waterborne oocysts through endemic infection of indigenous mammals (Smith and Rose, 1998). Infected hosts can excrete between 10^9 and 10^{10} oocysts within one defecation and both domestic and feral animals can be reservoirs of infection (Smith and Rose, 1998). Smith *et al.* (1995) found that environmentally robust oocysts can survive for six months suspended in water while Mancini and Plummer (1995) found slightly longer survival times of up to 54 weeks. As with any organism, environmental survival can depend on strain and physiological condition when released into the environment in addition to environmental parameters. A broad host range and a high output of oocysts cause a high level of contamination in the environment (Smith and Rose, 1998).

Cryptosporidium exhibits an extremely complex life cycle, most of which occurs in the small intestine of its host. The organism is initially ingested as an oocyst (egg containing cyst), excreted in the feces on the infected host (Smith *et al.*, 1995). Oocyst are typically spherical with a diameter of 4-6 µm (although some report diameters up to 8 µm) and are highly resistant to most environmental conditions as noted above (with the exception of dehydration) (LeChevallier and Norton, 1995; Levine, 1984; Fayer *et al.*, 1998). Once ingested, the tough protective outer layer of the oocyst is softened by the host's digestive system. The oocyst splits to release four sporozoites into the small intestine. These sporozoites enter the epithelial cells of the small intestine, to parasitise the cell contents. After the first round of infection is complete the sporozoites undergo asexual reproduction. Using this process, the organisms can develop into a large population in a comparatively short time. Following numerous reproductive cycles, the parasite population may begin sexual reproduction. Two types of oocysts are produced, with thick resistant walls for excretion, the second with thin walls that may be excreted or may rupture within a lower portion of the infected host's intestine to repeat the cycle of infection.

None of these cycles of infection and reproduction render *Cryptosporidia* open to medical intervention and control (Fayer *et al.*, 1998). The oocyst normal infective dose (ID_{50}), if ingested in water or food, is between 30 and 132 oocysts although one report places the minimum infectious dose as low as one oocyst for immunocompromised individuals (internet sources: http://www.cdc.gov/ncidod/publication/ brochures/cryptos.htm).

Municipal drinking water treatment providing filtration and disinfection with chlorine can reduce the risk of contracting cryptosporidiosis. Chlorine by itself however, is not effective against *Cryptosporidium* (AWWA, 1999b). This fact and data collected as part of the Information Collection Rule (ICR) has served to catalyze the implementation of the proposed LT2ESWTR. As noted above, the LT2ESWTR has set out monitoring requirements for systems serving greater than 10,000 (large systems) and those communities serving 10,000 or less (small systems) (DiGiovanni, 2003). In addition, USEPA has set out guidance regarding source water protection and treatment technologies that can be used to satisfy pathogen reduction credits specifically aimed at *Cryptosporidium* oocysts (DiGiovanni, 2003; USEPA, 2004).

Giardia cysts

Giardia sp. is a ubiquitous flagellated protozoan parasite that inhabits the gastrointestinal tract of their host species (Dunlap *et al.*, 2002). *Giardia* cysts have been isolated from avian (great blue herons and parrots) as well as mammalian species (AWWA, 1999a). *Giardia* is commonly found in human, beaver, muskrat, and dog feces, and in lakes and rivers. There are three type species of *Giardia*, *G. agilis, G. muris*, and *G. duodenalis* (Smith *et al.*, 1995). *Giardia* of the *duodenalis* species, also recognized as *Giardia lamblia,* causes infection and disease in humans (Smith *et al.*, 1995). Cross-species infections have also been demonstrated to occur in studies that infected animals with cysts isolated from humans (AWWA, 1999 a). Species affected included rats, gerbils, guinea pigs, beaver, dogs, cats, raccoons, and three varieties of sheep.

The life cycle of the *Giardia lamblia* (*Giardia duodenalis*) is simple: the parasite exists in two distinct morphological forms, the reproductive trophozoites (non-invasive) and the environmentally resistant cyst (Smith *et al.*, 1995). *Giardia* cysts are excreted in feces and are typically 8-14 μm in length by 7-10 μm in diameter (Sexton *et al,* 1995). In its cyst form, an egg-shell-like covering protects the microorganism from environmental hazards such as chlorine, which can act as a disinfectant. Cysts excreted in the feces of infected hosts may remain viable for several months (Dunlap *et al.*, 2002); when viable cysts are ingested, exposure to digestive processes initiates excystation to the trophozoite stage. In the trophozoite stage, *Giardia* grow and multiply in the intestine of the host causing the symptoms of giardiasis (Dunlap *et al.*, 2002). During the trophozoite stage, cysts are also produced and excreted, completing the life cycle.

FACTORS THAT AFFECT LEVELS OF MICROORGANISMS IN SURFACE WATERS

Land Use Effects

Since the 1970s, it has become apparent that the freshwater resources in the United States (US) were receiving significant pollutant loads from non-point sources (NPS); (Potter and Gilliland, 1988). Compared to point source pollution, which has been for the most part successfully controlled, NPS pollution is much more difficult to control. This is due to the difficulty in isolating the contribution from individually dispersed sources in a scientific and legally defendable manner (Berka *et al.*, 2001).

Residential Development

Microbial pollution can result from urban land use. Both urban and suburban sources of FC include feces from birds, rodents, wild animals, domestic animals, and especially domestic pets (Weiskel *et al.*, 1996). In urbanized watersheds, NPS runoff is considered to be a major source of many pollutants, including microorganisms (Klein, 1979; Bannerman *et al.*, 1993; Weiskel *et al.*, 1996). Urban impoundments commonly receive NPS pollution of many types and from a variety of sources, such as lawn and garden chemicals, pet and wildlife feces, petrochemicals and heavy metals from parking lots and streets. Storm and other surface water runoff events may deliver large amounts of contaminants into urban watershed systems. According to the characteristics of an urban area, NPS impacts are likely to vary according to demographic and land use factors. Kocasoy (1995) found population density to be a significant estimator of microbial pollution. Mallin *et al.* (2000) also reported a significant correlation between watershed human populations and both mean FC and *E. coli* abundance.

Regarding land use factors, Mallin *et al.* (2000) reported that the combination of quantity (amount of developed land) and quality (impervious surface area) of the land use is important in determining receiving water bacteriological quality. For this study, development was defined as land utilized for human benefit including agricultural fields, residential and commercial development. Lands remaining in forest or as unmanaged open fields were considered not developed. Further analysis demonstrated that the quality rather than quantity of land development was the most important influence on urban and suburban NPS fecal coliform bacterial pollution. Their data showed that when an area contained a large percentage (27.3%) of impervious surface and was in a less developed watershed (less overall improved land), higher average fecal coliform counts resulted than for an area with a smaller percentage (12.5%) of impervious surface located in a more highly developed watershed (more overall improved land). They concluded that NPS runoff from areas with 20% impervious surface or more can highly degrade receiving water quality. Maiolo and Tschetter (1981) previously reported a similar relationship.

In contrast, the study conducted by Moorhead *et al.* (1998), which was an effort to identify factors that correlate with bacterial populations in urban waters, found no positive relationships detected between TC or *E. coli* and the water quality parameters measured (chemical oxygen demand (COD), biological oxygen demand (BOD), nitrogen or phosphorus) or landscape use measures (total watershed area, impervious surface area, area adjacent to open land). The authors investigated 20 playa (permanent urban impoundments due to the runoff) lakes located within the city of Lubbock, Texas. Ten playas were sampled on two consecutive days from three locations in each playa. The lack of relationships between TC and water quality parameters, as well as TC and landscape use measures, indicates that in order to assess the microbial quality of water, microbial testing is required.

Agricultural Activities

Agriculture is rapidly emerging as one of the largest contributors of NPS pollution to surface waters in North America (Kellogg *et al.*, 1994; USEPA, 1996). Weidner *et al.* (1969), Robbins *et al.* (1972), Harms *et al.* (1975a, 1975b) and other researchers, all reported that TC and FC densities present in runoff from agricultural lands were frequently in excess of the respective state water quality standards for primary contact recreation. Levels in stormwater runoff from rural areas ranged from 220 to 180,000 colony forming units (cfu) per 100 mL for TC, from below detection limits to 56,000 cfu/100 mL for FC, and from 24 to 150,000 cfu/100 mL for Fecal Streptococci (FS), (Weidner *et al.*, 1969).

Other Land Use Types

The extent and types of NPS pollution can vary with type of land use. Concentrations of TC and FC in forest and recreational areas were compared with other land use types by Faust (1982) and Bolstad and Swank (1997). FC and TC concentrations and proportions in agricultural soils and forests were investigated by Faust (1982). The soil concentrations of FC were highest in pastures and averaged from 3 to 396 most probable number (mpn) per gram (g). In cornfield and forest soils, FC were present in comparatively low numbers, averaging from 3 to 26 and 6 to 27 mpn/g, respectively. *E. coli* was dominant in forest soil (67% of isolates were confirmed as *E. coli.*), while they were a minor soil community member at 8% in cornfield soils.

Data also indicated that the proportion of *E. coli* was directly related to the degree of soil fecal pollution. Land use type would thus impact the types of organisms ultimately reaching surface waters. Bolstad and Swank (1997) found consistent, cumulative, downstream changes in TC, FC, and FS when the stream flowed through forest, agricultural, recreational, and residential land. They reported downstream increases in microbial indicator concentrations, which were coincident with downstream human-caused changes in land use. However, the magnitude of increase was variable.

Seasonal Effects and Environmental Factors

Seasonal effects and environmental factors affecting microbial loading to environmental waters (*i.e.* FC and TC concentrations) appear to be related to microbial occurrence, survival and transport variables. Numerous studies (Skinner *et al.*, 1974; Doran and Linn, 1979; Doran *et al.*, 1981; Jawson *et al.*, 1982; Howell *et al.*, 1995; Anderson *et al.*, 1997; Hunter *et al.*, 1999) found that surface water FC concentrations were generally higher during warmer months than during cooler months. This seasonal variation could be related to environmental temperature, incident solar radiation, livestock management practices, manure handling practices, and land store recovery (Hunter *et al.*, 1999).

Temperature

Temperature of water is a very important factor for aquatic life. It influences the rate at which metabolic and reproductive activities occur, and determines which fish species can survive. Temperature also affects the concentration of dissolved oxygen as well as influences the activity of bacteria and toxic chemicals in water (Murphy, 2002).

Temperature variations among seasons can affect the levels of microorganisms in surface waters. The observable effect of temperature on microorganism survival in surface waters is a complex interaction of accelerated die-off as temperature increases between 5°C and 15°C for coliforms (McFeters *et al.*, 1974) and temperatures able to sustain growth (above 7.8°C) for *E. coli* (Ingrahm, 1987). It has been demonstrated that coliforms can grow in pristine tropical waters at ambient temperatures (Hazen and Toranzos, 1990).

Incident Solar Radiation

Numerous studies have reported that sunlight can inactivate enteric bacteria (Fujioka *et al.*, 1981; Curtis *et al.*, 1992; Auer and Niehaus, 1993). A combination of elevated temperature and solar irradiance was cited as the cause of microbial disinfection in some studies (Conroy *et al.*, 1996; Joyce *et al.*, 1996; Rijal and Fujioka, 2001). Solar UV radiation (200 to 400 nm) accompanied by thermal contributions (up to 55°C) can completely disinfect a water sample that is highly contaminated (20×10^5 cfu/100 mL *E. coli.*) within 7 hours (Joyce *et al.*, 1996). If a stream is shallow, sunlight can penetrate deeply enough to reach the bottom layers and may inactivate enteric bacteria (Baudart *et al.*, 2000). In particular, the decay coefficient for FC from sewage was demonstrated to be linearly related to average irradiance at depth (Auer and Niehaus, 1993). Therefore, concentrations of coliforms during summer when the solar radiation is strong could be reduced by solar disinfection effects.

The impact of solar radiation varies among microorganisms. In a study evaluating the effectiveness of combining solar radiation and solar heating for disinfecting contaminated water (Rijal and Fujioka, 2001), the log reduction of several indicators in an increased order are: total heterotrophic bacteria < *C. perfringens* < F-specific RNA virus < enterococci < *E. coli.* < Fecal coliform.

Stream Flow Rates

Stream flow could affect the levels of microorganisms in the water indirectly through factors such as DO concentration, solar inactivation, and sediment resuspension. The volume and velocity of water flowing in the water body can affect the DO concentrations. In fast-moving streams, rushing water is aerated by bubbles as it churns over rocks and falls down hundreds of tiny waterfalls. These streams, if unpolluted, are usually saturated with oxygen. In slow moving waters, oxygen only enters the top layer of water while deeper water is often low in DO due to decomposition of organic matter by bacteria that live either near the bottom of the water column or in the sediments (Murphy, 2002). Lower flow rates also allow the water to warm up quickly due to the longer exposure time to the sunlight (Baudart *et al.*, 2000). Increased stream flow and turbulence can resuspend bacteria adsorbed to the sediment (Sayler, *et al*, 1975; Kay and McDonald, 1980; Ferguson, *et al*, 1996; Weiskel, *et al*, 1996; Atherholt, *et al*, 1998; Baudart, *et al*, 2000).

Organic Matter in Ambient Waters

Organic matter plays a major role in aquatic systems. It affects biogeochemical processes, nutrient cycling, chemical transport, and microbial interactions (Hem, 1985). Organic matter in water consists of thousands of components, including macroscopic particles, colloids, dissolved macromolecules, and specific compounds (Sawyer *et al.*, 1994), which may enter water bodies from rainfall runoff. Domestic and industrial wastewaters also contribute organic contaminants in various amounts.

The larger the carbon or organic content of a body of water is, the greater the potential for oxygen depletion. A high organic content can lead to an increase in the growth of microorganisms that contribute to the depletion of dissolved oxygen supplies (Hem, 1985). Higher levels of organic matter in water could enhance surface water survival of microorganisms. This phenomenon was reported for the survival of *E. coli* in marine sediments (Gerba and McLeod, 1976). These results are consistent with results of a study conducted by Ferguson *et al.* (1996), where the presence of total organic matter was associated with a significant increase in the density of FC. Lim and Flint (1989) demonstrated that the addition of simple carbon molecules to sterilized lake water prolonged the survival of spiked *E. coli* at both 15°C and 37°C. Klock (1971) demonstrated that different mechanisms might be required for enteric organisms to utilize dissolved and particulate organic materials. Since the nature of organic matter is highly variable, the quantitative effects of total organic matter cannot be generalized.

Water Quality Parameters

Other water quality parameters like DO and turbidity can also affect levels of microorganisms in the water. DO concentrations in the water were very important to the levels of sorbitol-fermenting *Bifidobacteria* owing to their strict anaerobic living habits. Research has found that increased turbidity is often associated with an increase in bacterial pollution (Scherer *et al.*, 1992; Irvine and Pettiboine, 1996). Mallin *et al.* (2000) noted turbidity was positively correlated to enteric bacteria abundance. A highly significant proportion of fecal bacterial indicators can be associated with particulate matter in water (Sayler *et al.*, 1975). Scherer *et al.* (1992) noted that FC and FS half-lives in surface waters doubled in the presence of sediments. Faust *et al.* (1975) demonstrated that the half-life of *E. coli* in estuary water could be extended by 40% by the addition of clay particles. However, quantitative relationships between turbidity type and level, and microbial concentrations appear to be site-specific.

Sediment Qualities

Densities of coliforms in sediments were significantly greater than those in the overlaying waters in a bay area (Erkenbrecher, 1981). Erkenbrecher (1981) also found the overall densities of bacteria were lower in the region with higher-salinity water and coarser sediment than sites where the salinity of water was low and sediments were effected by tidal action. A study conducted by Gerba and McLeod (1976) found that *E. coli* demonstrated a longer survival time in the seawater when sediment material was present. The longer survival of *E. coli* in sediment was attributed to the greater content of organic matter present in the sediment over that of the seawater alone.

Livestock Management Practices

Seasonal variations in land use practices, such as presence/absence of animals in grazing land, also contributes to the seasonal variations noted in surface water microorganism levels (Hunter *et al.*, 1999). As might be expected, increased FC and FS enumerations were noted in runoff from a grazed watershed in the Pacific Northwest after warm, spring weather but not in an adjacent ungrazed watershed (Jawson *et al.*, 1982). Researchers reported that the highest average levels of TC, FC, and FS in grazing land runoff occurred when animals were present (Jawson *et al.*, 1982). Similarly, a comparison of TC, FC, and FS levels in runoff between grazed and ungrazed land showed that the levels were 5 to 10 times higher from the grazed land (Doran and Linn, 1979). Thus, the practice of barn housing grazing animals in regions with severe winters can contribute to decreased microbial levels during cold months with the concomitant increase in microbial loading to surface waters during warm months when the animals are again pastured. Studies such as these confirm that detailed land use practice information may be necessary to elucidate the underlying cause of observed microbial seasonal variations.

Precipitation

Precipitation can affect the population of bacteria in environmental waters in two general ways. First, a variety of surface materials from different sources may be washed from the terrestrial environment into receiving waters. Second, increased stream flow and turbulence can resuspend bacteria adsorbed to the sediment (Grimes, 1975; Sayler *et al.*, 1975; Erkenbrecher, 1981; McDonald and Kay, 1981; Ferguson *et al.*, 1996; Weiskel *et al.*, 1996; Atherholt *et al.*, 1998; Baudart *et al.*, 2000).

A number of studies involving various drinking water sources found that water body coliform concentrations were higher immediately after a rainfall (McDonald and Kay, 1981; Jagals *et al.*, 1995; Baudart *et al.*, 2000). Baudart *et al.* (2000) reported river bacteria loads to coastal waters varied greatly during the year and were highly related to storm events (95% of the annual loads for FC). Stukel *et al.* (1990) investigated 15 small community drinking water systems in rural New England and found a strong relationship between the duration and amount of rainfall in the days preceding sampling, and the coliform contamination levels. McDonald and Kay (1981) demonstrated that the geometric mean level of TC and *E. coli* more than doubled after a rainfall event when compared to average dry weather levels. Lamka *et al.* (1980) noted that higher levels of coliforms were enumerated during sampling periods with rainfall in the preceding 24 hours than during periods with no preceding rain. Kelch and Lee (1978) found that FC counts in Tillamook Bay, OR, were correlated with the average precipitation in the six days preceding sampling. Bolstad and Swank (1997) reported that the cumulative impacts attributed to land use alteration were much greater during storm events. While all these researchers found increases in indicator concentrations in surface waters after precipitation events, the magnitude of those changes were highly variable.

Several studies have highlighted the importance of sediments as potential reservoirs for microorganisms in the aquatic environment (Erkenbrecher, 1981; Grimes, 1975). Bacteria adsorbed to sediment survive longer because they may be protected from the influence of inactivation factors, and the presence of organic matter could stimulate growth (McFeters *et al.*, 1974; Scherer *et al.*, 1992; Davies *et al.*, 1995). Ashbolt *et al.* (1993) reported that sediments often contain 100 to 1000 times as many fecal indicator bacteria than the overlying water. These sediment-trapped bacteria represent an important source of fecal pollution, since contaminated particles can be resuspended during bottom disturbances by rainfall events (Howell *et al.*, 1995; Wilkinson *et al.*, 1995).

Ferguson *et al.* (1996) reported that the effects of precipitation may not affect all microorganisms in pools similarly. In an estuarine system, FC, FS, *C. perfringens* spores and *Aeromonas spp.* increased in water samples after precipitation events. In contrast, only FC sediment concentrations were significantly increased by rainfall while sediment concentrations of other indicators, including FS, *C. perfringens* spores, F-RNA coliphages, *Aeromonas spp.*, *Giardia spp.* and *Cryptosporidium spp.* did not demonstrate this trend. Studies indicate that precipitation events can increase microbial inputs to surface waters; however, the magnitude and composition of those changes vary on a site-specific basis.

Watershed Characteristics

Soil Texture

Hunter and McDonald (1991) reported that FC can survive in a semi-permanent bacterial land store in surface soils. The size of this store depends upon a number of factors, one of which is the fecal input rate (which is influenced by agricultural land use practices). In the study conducted by Hunter *et al.* (1999), the agricultural land use areas investigated included rough, semi-improved and improved pastures for sheep grazing. The term "improved" was used to describe the degree of land use intensification. During its passage through the study area, the study stream became significantly contaminated by fecal bacteria. These results suggest the presence of a semi-permanent store of fecal bacteria in the study area soils as well as a hydrological transport mechanism capable of moving bacteria from the soil to the receiving water. The highest FC concentrations did not appear in the samples collected adjacent to the most intensive agricultural land. Peak FC concentrations occurred at the end of a wooded stream where immediate grazing was poor and animal access was restricted. It was hypothesized that observed influence on FC concentrations reflected a balance between land use intensification (higher inputs of fecal material) and a reduction in the potential for bacterial survival in the "improved" soils. Antagonistic effects between land use intensification and decreasing efficiency of hydrological transport pathways have also been hypothesized to influence the observed FC results.

Hunter's group also found fecal coliform concentrations in stream water were significantly higher during the summer (mid-June to end of August) than during other times of year. The watershed land store size of surface soil enteric bacteria was cited as the main reason for this trend in their study. Land store size is the absolute quantity of bacteria present in a given volume of land. The land store size depends on the balance between fecal input rate, bacterial die off rate, and the rate of removal of bacteria in runoff. During the summer, evapotranspiration was at a maximum and rainfall was at a minimum, which resulted in substantial accumulation in land store size since losses through transport out of the land system were minimized. When summer storms occur, increased bacterial loading to receiving waters can also occur (Kay and McDonald, 1980).

Drainage Characteristics

A study conducted by Barbe *et al.* (1998) demonstrated that drainage characteristics can affect the maximum coliform counts after storm events. Fecal coliform concentrations after storms were analyzed within two adjacent watersheds in southeast Louisiana. They concluded that maximum coliform counts were observed two days earlier in the smaller watershed (with 192 sq. miles of drainage area) than in the bigger watershed (with 450 sq. miles drainage area).

Organism Size, Shape and Other Factors

The transport characteristics of the microbial organisms may vary due to differences in size, shape, surface characteristics, and environmental robustness. Additional information on these characteristics not included in the previous sections is briefly reviewed here.

Cryptosporidium vary from approximately 3 to 6 or 8 μm in diameter depending on their life stages – schizonts and thin or thick walled oocysts (Fayer *et al.*, 1998; Levine, 1984). *C. parvum* oocysts are the genus species of *Cryptosporidium* most responsible for clinical disease in humans and infected animals and have a specific gravity of 1.0 to 1.1 (Chapra, 1997; Walker *et al.*, 1998). Some sedimentation velocity studies have been conducted. Sediment velocity for a single *Cryptosporidium* oocysts are small, estimated at 0.5 $\mu m\ s^{-1}$ and even less, 0.3 $\mu m\ s^{-1}$ at cooler temperatures (Medema *et al.*, 1998; Graczyk *et al.*, 1998; Lisle and Rose, 1995; Walker and Stedinger, 1999; Badenoch, 1990). Electrostatic surface charges of *Cryptosporidium* oocysts are impacted by pH, ionic strength and ionic make-up of the water and organic matter content because they are somewhat hydrophobic, physiochemical adsorption processes an be important mechanisms affecting transport, particularly in soils (Roberston and Edberg, 1997; Harvey and Garabedian, 1991; Bales *et a*l., 1995).

Most assessments of the rate of oocysts inactivation have been carried out in laboratory settings focusing on single stressors; little field research has been conducted that examines inactivation in response to multiple environmental stressors (Walker et al., 1998). In one study (Robertson *et al.*, 1992), 89% to 99% of oocysts in river water were dead at 176 days. Sattar *et al.* (1999) noted that pH's in the range of 5 to 9 had little influence on oocysts survival and that anaerobic conditions only slightly affected survivability. However oocysts were sensitive to natural sunlight and desiccation. This study also found greater die-off in natural waters when compared to water that was filtered or irradiated. Sattar *et al.* (1999) suggest that this was due to a combination of microbial antagonisms, water chemistry, and other factors. Addition of humic acid decreased oocysts die-off. Temperatures greater than 37° C have been noted to cause greater excystation of C. parvum oocysts than lower temperatures (Fayer and Leek, 1984). In addition, Olson *et al.* (2000) found faster inactivation at 25° than 4° C. Olson *et al.* (2000) also compared survival times of *C. parvum* and *Giardia* in water, soil, and feces. *C. parvum* survival time was greater in water (10 weeks) than in soil or cattle feces (4 weeks). *Cryptosporidium* was more environmentally robust than *Giardia*.

Giardia cysts tend to be slightly larger than *Cryptosporidium* oocysts, typically ranging from 8 to 10 μm in diameter (Sexton, 1995). F^--specific RNA coliphages tend to be icosahedrons with a diameter of approximately 24 ηm while F^--specific DNA coliphages are filamentous in nature with a size of 810 by 6 ηm (Sobsey *et al.*, 1995). Coliforms are short, approximately 1 μm, rods. Sorbitol-fermenting *Bifidobacteria* tend to be non-autoagglutinating and do not adhere to surfaces well. They are pleomorphic rods that can vary from small (0.8 μm) to relatively large (2 μm) lengths (Perez *et al.*, 1998). *Rhodococcus coprophilus* are norcardioform actinomycetes. They can exist in a coccoid (sphere), rod, or filamentous life form, with sizes varying from <1 μm to over 2 μm (Holt *et al.* 1994). *Clostridium perfringens* vegetative cells tend to be rods ranging from 2-4 μm while *C. perfringens* endospores are typically <1 μm and spherical or ovoid in shape (Woodruff, 1992). *E. coli* can range from 1.1 – 1.5 μm by 2.0 – 6.0 μm depending on growth conditions (Holt *et al.*, 1994). Based on their sizes, it is anticipated that most of the above organisms will move as wash load - typically particles less than 4 to 8 micrometers (Fisher, 1995). Their shapes, sizes, and surface characteristics, however, may result in different transport mechanisms.

CHAPTER 3
WATERSHED MONITORING STUDIES OF PARTICULAR RELEVANCE

OVERVIEW

In the past decade, the number of watershed monitoring studies has increased dramatically. Many of these studies have moved from focusing only on standard water-quality constituent data (nitrogen, phosphorus, dissolved oxygen, temperature, suspended sediment, etc.) to include microbial measurements. In addition, watershed monitoring studies involving microbial measurements have moved from traditional parameters (such as fecal coliform) to include organisms such as *Giardia*, *Cryptosporidium*, *Clostridium perfringens*, and others. Most of these studies have focused on fixed-interval sampling, such as bi-weekly or monthly, thus providing a mixture of base- and "single value" event-flow data. While focused storm-event collection of water-quality constituent data has become more common, few studies collecting storm-event water samples for microbial analyses exist. Storm-event sampling and analysis of *Giardia*, *Cryptosporidium*, and source-specific indicator organisms are particularly lacking due to the complexity of the analyses, labor and other cost constraints, and holding time issues. Particularly relevant work conducted by other researchers is summarized below. Inclusion of all salient studies is beyond the scope of this discussion.

WATERBORNE OUTBREAKS

Cryptosporidium and *Giardia* infections have occurred worldwide, indicating the ability of these parasites to adapt to numerous environments (Smith *et al.*, 1995). The relative robustness of oocysts and cysts and the low infectious dose have contributed to the occurrence of these parasitic protozoa infections (Smith *et al.*, 1995). While such infections occur frequently in the developed countries of the world, they are almost ubiquitous in developing countries where sanitation is often primitive (Smith *et al.*, 1995).

Infection with *Cryptosporidium* and *Giardia* may be transmitted either directly (person to person), from animals, or from contamination of the environment (Smith *et al.*, 1995). Transmission by the waterborne route can result in large numbers of consumers being infected by ingesting contaminated water, resulting in outbreaks of waterborne disease (Smith *et al.*, 1995). Outbreaks of waterborne Giardiasis have occurred with relatively high frequency in PWSs drawing from pristine surface water sources. This suggests that contributions of cysts and oocysts from wildlife may be significant (Smith *et al.*, 1995). Waterborne protozoa, including *Giardia lamblia,* have also been a significant cause of outbreaks for the past 25 years (Craun *et al.*, 1998). In British Columbia, a waterborne outbreak of Giardiasis was linked to an infected beaver (Smith *et al.*, 1995).

TARGETED WET WEATHER SAMPLING

Work in Wisconsin focused on using upstream-downstream monitoring to evaluate the effectiveness of barnyard best management practices on storm-runoff constituent concentrations (Stuntebeck 1995; Stuntebeck and Bannerman, 1998). In this work, intensive automated stream water sampling at upstream-downstream locations was successfully used to isolate pollutant loads coming from specific source areas during storm-runoff periods. While samples were analyzed for fecal coliform bacteria, emphasis was placed on determining storm-runoff period microbial loading (of fecal coliform) rather than optimization of the event period sampling procedures. Work in Arkansas focused on using automated stream water sampling to collect storm-runoff period and baseflow pollutant load and concentration data in two watersheds (Nelson *et al.*, 1998). Statistical methods were then used to determine the optimum sample number and timing for determining pollutant loads with high confidence and low bias. In New York, monitoring of the Delaware, Catskill, and Croton watersheds has been quite extensive since 1987, with pathogen monitoring data available since 1992 (Ashendorff *et al.* 1997).

In Massachusetts, several projects have focused on determining stormwater and mainstem loading of bacteria and other constituents to the Lower Charles River Watershed (Parker, 2000), the Merrimack and Neponset watersheds (Camp Dresser McKee, 2002), and the Blackstone River (Wright *et al.*, 2001). Sample analyses have included fecal coliform, total coliform, and, for some of the studies, additional indicator organisms (*E. coli*, enterococci, *Clostridium perfringens* and coliphages). However, collection of several of the indicator organisms and human pathogens of interest to drinking water utilities is limited in the number of stormflow samples. Studies have also been conducted within the Wachusett Reservoir watershed to collect storm-event data for traditional microbial and water-quality parameters (Patricia Austin, personal communication), and to test for alternative indicator organism presence (Long, 2002).

Studies focusing on wet weather transport of *Cryptosporidium* and *Giardia* are fewer. Atherholt *et al.* (1998), investigated the variability of *Giardia* cysts and *Cryptosporidium* oocysts levels in the Delaware River by collecting daily (Monday – Friday) water samples for analysis from within the Trenton (N.J.) Water Works treatment plant for several periods over the course of a year. Monthly data were also collected. Significant relationships between concentrations in *Giardia* and *Cryptosporidium* with precipitation and turbidity were found. Investigations by other researchers, however, have resulted in a mixture of findings (see Atherholt *et al.*, 1998 for a brief review).

Kistemann et al. (2002) studied microbial loads during extreme rainfall and runoff in three small German watersheds. Routine sampling was also conducted. Sampling during events consisted of one to five samples representing the rising limb, peak, or falling limb. A total of about 500 L were pumped through each filter. They found variations in concentrations over the flood hydrograph that were site/event specific. Sometimes *Cryptosporidium* and *Giardia* concentrations peaked immediately before flow, sometimes they peaked with flow, and sometimes they decreased over the course of the event. Increased levels of microorganisms in response to precipitation occurred even within the least anthropogenically effected watershed. Analyses found significant differences between routine and event samples. In particular, loads contributed during heavy rainfall and runoff events were significant. Resuspension of bed sediments or erosion of soil and aged fecal matter were identified as a major source of *Giardia*. Combined sewage overflows and sewage treatment plan outfalls were also identified as sources.

LeChevallier et al. (1991) studied event and routine transport of *Giardia* and *Cryptosporidium* in six large (35 to 2700 mi^2) rivers located in the continental United States. During routine sampling, *Giardia* and *Cryptosporidium* were detected in on average 16% and 10% of the samples, respectively. The average *Giardia* level across sites was 5.2 cysts/100L during routine sampling and 24.1 cysts/100 L during event sampling. The average *Cryptosporidium* level across sites was 1.5 oocysts /100L during routine sampling and 3.3 oocysts/100 L during event sampling. The range of individual samples at sampling locations was significantly larger. Correlations with hardness and *E. coli* were identified for *Giardia*, but no significant correlations were found for *Cryptosporidium*.

The various studies have had a range of success in terms of identifying correlations between *Giardia*, *Cryptosporidium*, and other parameters. Further storm-event sampling is necessary to help (1) determine the concentrations of *Giardia*, *Cryptosporidium*, and other microbial organisms during the first flush and (2) identify important factors affecting these loadings. In particular, because the current methods of detecting and quantifying these organisms are time consuming and expensive, there is a need to develop storm-runoff sampling schemes that optimize microbial and water-quality parameter sample collection analysis type, timing, and number.

BED SEDIMENTS

Several researchers have noted that bed sediments may act as a reservoir, and thus a source, for fecal contamination (see, for instance, Crabill *et al.*, 1999; Howell *et al.*, 1996; Kebabjian, 1994; McDonald *et al.*, 1982; Sherer, B.M., 1988; Tunnicliff and Brickler, 1984). Multiple site along-stream grab sampling, including upstream and downstream of suspected sources, of water and sediment has been conducted along Oak Creek in Arizona (Crabill *et al.*, 1999). This study found that both bed sediment and additional sources (grazing cattle, wildlife, septic tanks, etc.) of fecal pollution were important, but was unable to clearly distinguish these additional sources due to lack of further microbial analyses and/or hydrologic transport data. Similar studies have also been conducted along the lower Buffalo River in NY (Irvine and Pettibone, 1996) and, with the inclusion of indicator bacteria analyses, in Kaelepulu Stream, Hawaii (Roll and Fujioka, 1997).

CRYPTOSPORIDIUM AND *GIARDIA* OCCURRENCE IN SURFACE WATERS

Cryptosporidium oocysts have been detected in all types of surface waters, including pristine waters, or those not receiving community wastewater discharges (Craun *et al.*, 1998). Surveys show that over 80 percent of the tested U.S. surface water supplies are contaminated with *Cryptosporidium* sp. with densities ranging from 0.001 to 484 oocysts per liter (Smith and Rose, 1998). Rose (1988) showed that 83 percent of rivers within protected watersheds in Western US were contaminated with oocysts ranging from non detect (ND) to 13 oocysts per 100 Liters with an average of 2 oocysts per 100 Liters. Ong et al. (1996) studied *Giardia* and *Cryptosporidium* occurrence in two adjacent watersheds located in British Columbia, Canada. Samples were collected at numerous locations throughout the watersheds, including downstream sites where irrigation intakes were located. Levels of both *Giardia* cysts and *Cryptosporidium* oocysts increased from the headwaters to downstream. Higher concentrations of *Giardia* cysts were observed in March – May while higher concentrations of *Cryptosporidium* oocyts were observed in February – May, with February the peak in calving. The observed occurrence of *Cryptosporidium* oocysts and associated risk of cryptosporidiosis have been recognized by regulatory agencies as a significant threat to public health. Tables 3-1 and 3-2 summarize some of the available literature with regard to *Cryptosporidium* and *Giardia* levels in the environment. This list is by no means exhaustive.

MODELING WORK

Work in the Saw Kill Watershed, a tributary of the Hudson River, New York, combined weekly water-quality sampling in twelve sub-watersheds with geographical information system (GIS) data in order to develop an geographical information system-based delivery model for pathogen loading (Fraser *et al.*, 1998). Sampling for pathogens, however, was limited to fecal coliform and was conducted only over an eight week period in a single year. The resulting model proved useful for ranking potential pollution impacts from livestock but was limited due to lack of information on stream processes influencing bacterial transport, potential background contamination from non-livestock related sources of contamination (wildlife and septic systems), and additional microbiological measurements.

Modeling for *Cryptosporidium* in Ontario incorporated probability density functions representing pathogen availability in the watershed with a distributed hydrologic model driven by radar based rainfall inputs (Dorner et al., 2004). In addition, modifications of the Guelp All-Weather Sequential-Events Runoff (GAWSER) model have been proposed to account for *Cryptosporidium* delivery and transport (LeChevallier et al., 2002).

Table 3-1
***Cryptosporidium* in the environment**

Water Type	% Positive	Average oocysts/100 L	Range oocysts/100 L	Notes
Irrigation canals – farm drainage (AZ) (Madore et al, 1987)	100	555,000	530,000 – 580,000	2 samples
Lakes (Rose et al, 1991)	53	9.3	ND – 307	34 samples
Rivers (pristine) (Rose et al, 1991)	32	29	ND – 24,000	59 samples
Source waters of surface water treatment plants (LeChevallier et al, 1991)	87	270	ND – 48,400	66 water treatement plants in 14 states and 1 Canadian province – 85 samples
Surface water supplies for drinking water plants (LeChevallier and Norton, 1995)	51.5	240	ND – 6,510	1991-1993, 262 samples at 72 water plants
Rivers in a protected watershed (Western USA) (Rose, 1988)	83	2	ND – 13	6 samples
Catskill Watershed Delaware Malcolm Brook (Okun et al, 1997)	46 37 52	1.4 0.8 1.0	ND – 17.3 ND – 15 ND - 48	3 protected watersheds that serve as drinking water source for NYC
Phillippi Creek (FL) (Rose, and Lipp, 1997)	13	16	ND – 158	An urban stream in Sarasota, 16 samples
Streams (FL) (Rose, and Lipp, 1997)	4	6.6	ND – 157	In the vicinity of Sarasota, 24 samples
Tampa Bypass Canal (FL) (Rose, 1993)	43	3.1	ND – 11	7 samples

(continued)

Table 3.1 (Continued)

Water Type	% Postive	Average oocysts/100L	Range oocysts/100L	Notes
German drinking water reservoir main tributaries (Kistemann *et al.*, 2002)				
Nauholzbach (3.27 km^2)				98% forest
Kall (19.7 km^2)		22.32	Max 65.6 SD 17.75	56.3% dairy Ag and 19% forest, sewered. 11 event samples.
Wahnback (45 km^2)		27.48	Max 147.1 SD 37.69	63% Ag, 21% woodland, sewered and unsewered. 14 event samples
Delaware R. at Trenton, NJ (Atherholt *et al.*, 1998)		20	0 - 280	
British Columbia (Ong *et al.*, 1996)				
BMID intake				
Phase 3	52	3.5	1.7 – 44.3	27, weekly, slight peak in May
BMID upstream of cattle	47	5.6	0.5-34.4	32, weekly
BMID downstream of cattle	47	13.3	1.4-300	30, weekly, peak in Feb, when calving & May
VID intake				
Phase 1				3, monthly
Phase 2	29	9.2	4.8-51.4	19 over 6 months
VID headwaters				3 samples, lake source, monthly
Cattle area (reference to Rose et al., 1988)		2,850		
Protected area (reference to Rose et al., 1988)		8		
Various source waters (LeChevallier *et al.*, 1991)				
Routine sampling	9.7	1.5	Nd – 117	
"Event" sampling CT, GA, IA, ON, OR, TN	12.7	3.3		Events not necessarily flow based

Table 3.2
***Giardia* in the environment**

Water Type	% Positive	Average cysts/100 L	Range cysts/100 L	Notes
British Columbia (Ong *et al*., 1996)				
BMID headwaters	67	9	1 - 200	10, once per site
BMID intake				
Phase 1	100	229	7 – 2,215	70, twice a week
Phase 2	100	202	179 - 261	4, weekly
Phase 3	89	60.4	4.6 – 1,880	27, weekly
BMID upstream of cattle	72	5.2	0.5-34.4	32, weekly
BMID downstream of cattle	73	9.0	0.6-42.9	30, weekly, Ap & May peaks
VID intake Phase 1	100	3.8	.6-13.2	3, monthly
Phase 2	97	26	2-114	70, weekly
VID headwaters	0			3 samples, lake source, monthly
Agricultural protected area	44		1.2 – 24.7	
Forested protected area	12		1.2 – 2.1	
Cattle area (reference to Rose et al., 1988)		120		
Protected area (reference to Rose et al., 1988)		0.9		
German drinking water reservoir main tributaries (Kistemann *et al*., 2002)				
Nauholzbach (3.27 km^2)				98% forest
Kallbach (19.7 km^2)		3.08	Max 21 SD 6.27	56.3% dairy Ag and 19% forest, sewered, 11 event samples
Wahnbach (45 km^2)		10.6	Max 21.6 SD 7.25	63% Ag, 21% woodland, sewered and unsewered. 14 event samples
Delaware R. at Trenton, NJ, 1996 (wet year) (Atherholt *et al*., 1998)	--	74	0 – 800	72 samples including 12 monthly and 60 seasonal (three consecutive 5-day week samples per season)
Various source waters (LeChevallier *et al*., 1991)				
Routine sampling	16.3	5.2	ND –1333	
"Event" sampling	22.8	24.1		

CHAPTER 4
MATERIALS AND METHODS

STUDY SITE LOCATIONS

Data were collected in order to isolate the influence of three land use types on water quality, particularly during storm events, Table 4-1. The three distinct land uses studied included an agricultural (dairy cow) site, a high density residential (septic/sewer) - commercial site, and a wildlife (ranging wildlife versus ranging wildlife and beaver) site. At both the agricultural and residential study sites, two along-stream sampling transects were established. One transect was located immediately upstream of the area of interest while the second transect was located a short distance (~ 1 km) downstream. Upstream-downstream sampling allowed determination of load contributions from the study area. At the wildlife site, a paired watershed sampling strategy was employed due to the small order of the streams studied and a resulting inability to sample upstream. Sampling transects were established at the outlets of two distinct watersheds, one influenced solely by ranging wildlife and one influenced by ranging wildlife in addition to an established beaver colony. At all sampling transects, either a weir was installed or a culvert was used in order to accurately determine discharge from stage measurements. The three study sites and the sampling locations at each are described in detail within this section. All watersheds are located in the Commonwealth of Massachusetts, Figure 1-1.

Table 4-1
Summary of sampling locations

Influence	Sampling Location	Description	Code
Wildlife*	Middle Branch Dickey Brook	Ranging wildlife	MBDB
	Griswold Brook	Beaver influenced	GRB
Agricultural†	Burrow Brook Woods Road	Upstream of cow pasture	BBWR
	Burrow Brook Farm	Downstream of cow pasture	BBFM
Residential‡	Gates Brook Woodland Street	Upstream of residential and commercial area	GBWD
	Gates Brook Brayton Road	Downstream of residential and commercial area	GBBR

Notes: * MBDB and GRB are located at the outlet of two distinct, although similar, watersheds. †BBWR and BBFM are up- and downstream monitoring locations along the stream of a third watershed, used to isolate the affect of a cow pasture. ‡GBWD/GBBR are up- and downstream monitoring locations along the stream of a fourth watershed, used to isolate the affect of intense residential/commercial development.

Source: Rees *et al.*, 2005. Transport of *Cryptosporidium* and *Giardia* in Massachusetts Watersheds During Storm Events. In *Managing Watersheds for Human and Natural Impacts: Engineering, Ecological, and Economic Challenges*. Copyright ©2005 ASCE. Reproduced with permission.

Table 4-2
GIS data available from the Commonwealth of Massachusetts

Data Layer Name			
USGS topographic quadrangle images	Land Use	Major drainage basins	Water-quality monitoring stations
1:5000 black & white digital orthophotos	Surficial Geology	Major watersheds	Stream-gaging stations
State outline	Soils (pending)	Major ponds and major streams	USGS data-collection stations
Town boundaries	Hydrography	Drainage sub-basins	Massachusetts Highway Department Roads

The Commonwealth of Massachusetts maintains an extensive GIS database. These data were invaluable for assistance in study site selection, site delineation, and site characterization. Several of the available data layers of use to this project are listed in Table 4-2 (MassGIS, 2001). Additional soils GIS data are available from Penn State (Miller and White, 1998). Digital Elevation Model (DEM) data were obtained from the United States Geological Survey (USGS) Earth Science Information Center (ESIC) through the Earth Resources Observation Systems (EROS) Data Center website. The USGS National Elevation Dataset (NED) data were used.

Wildlife Study Sites

Beaver (*Castor canadensis*) cut trees and build dams, activities that alter stream morphology, hydrology, organic matter transport, nutrient cycling and decomposition dynamics and may impact downstream water quality. Beaver inputs to a stream can be a significant source of nitrogen and phosphorus and may impact the trophic state of the receiving water body. Studies have shown that beaver are a source of *Giardia* and potentially of *Cryptosporidium* (Smith et al., 1995). Beaver were nearly brought to extinction from the New England landscape at the beginning of the 20th century, but reintroduction programs in the early 1900's have led to extensive beaver habitat throughout the Northeast (Moulton, 2001). In addition, strict laws regulating the capture and destruction of beavers exist in the Commonwealth. The effect of beaver on water quality was a focus of this study.

Windsor Dam, the downstream end of Quabbin Reservoir, is located approximately 20 minutes from the UMass Amherst campus and is the site of one of the two DCR field headquarters. Quabbin Reservoir consists of two branches which are separated by the Prescott Peninsula. The Prescott Peninsula is an almost entirely forested, protected water supply watershed area free from anthropogenic influence except for research activities and limited logging. Prescott Peninsula lies within the Appalachian Highlands Province. Soils throughout the peninsula are typically sandy-loam and are relatively shallow, with bedrock at approximately 1.25 m beneath the surface (Miller and White, 1998). Direct access is restricted to DCR personnel. Access to Prescott Peninsula can be difficult in the winter months, after significant snowpack accumulation, and during the melting season when the access roads become mired in mud and are closed by the DCR; a four wheel drive vehicle is typically essential for safe access.

The Prescott Peninsula of the Quabbin Reservoir is inhabited by a variety of free ranging wildlife such as white-tailed deer (*Odocoileus virginianus)*, black bear *(Ursus americanus),* moose *(Alces alces),* and wild turkey *(Meleagris gallopavo).* Moose and deer affect widespread areas of the forest with their grazing habits, which over time can alter everything from the type of tree cover to the undergrowth of the forest (Senus, 2002). Whitetail deer densities within Massachusetts range from 10 per square mile in northwestern Massachusetts to 50 to 60 per square mile on Nantucket Island (Mass Wildlife website, 2004). The statewide population ranges between 85,000 to 95,000 animals (Mass Wildlife website, 2004). The Massachusetts Division of Fisheries and Wildlife do not monitor the population densities of moose or black bear; however, Mass Wildlife (2004) reports a statewide population of 2000 black bear in 2002 and 500 to 700 moose in 2003. Population management strategies are currently implemented in Massachusetts for deer and bear by hunting. The DCR conducts controlled hunts for whitetail deer on the Quabbin reservation, including the Prescott Peninsula. Mass Wildlife (2004) reports 153 animals were harvested during the Quabbin controlled hunt in 2002. This was a 51 percent increase in harvested animals compared to the 2001 Quabbin hunt.

Beaver *(Castor canadensis)* also inhabit Prescott Peninsula. Beaver can significantly alter the hydraulics and hydrology of a watershed by building dams and creating large ponds. The effects of beaver tend to be more localized compared to the widespread effects of the free ranging wildlife. Beaver populations and densities are particularly important as these animals can directly influence water quality. Beaver populations are managed in Massachusetts through regulating trapping from November through April. The Commonwealth of Massachusetts also permits trapping of beaver outside these dates for the purpose of protection from threats to human health and safety. Permits are issued for these cases on an individual basis through the Massachusetts Wildlife Department, which also monitors beaver population throughout the state.

Monitoring for a variety of water-quality parameters (temperature, dissolved oxygen, conductivity, pH, ammonia, nitrate, total Kjeldahl nitrogen, total phosphorus, silicon, total organic carbon, dissolved organic carbon, particulate organic carbon, UV254, and trihalomethane formation potential) has occurred since 1997 at approximately twelve locations throughout the Prescott Peninsula (Paul Reyes, personal communication). Limited Department of Environmental Projection (DEP) sampling was also conducted at five locations in 1996. Historical USGS stream gauging data also exist for several locations, although all sites directly on the peninsula have been inactive for several years. In addition, Prescott Peninsula has been the location of several detailed studies, including studies on the management of a forested watershed for source water protection (Barten et al., 1998) and the effect of beaver impoundments on stream biogeochemistry (Moulton, 2001).

Beaver activity throughout the peninsula has been documented since 1952, when beaver population on the peninsula was at a historic low. During the period for which data are available, beaver activity peaked in the late 1970's to early 1980's, Figure 4-1. Following an approximate decadal decline in population, the number of beaver on the peninsula increased from 1996 to 2001 and has slightly decreased since. The quality of food supply for beaver varies throughout the peninsula, but young (1 to 2-inch diameter) hardwood saplings that beaver prefer are not overly plentiful. Beaver thus typically occupy a site only for three to seven years, at which point their food sources have been extinguished, and then migrate to a nearby location (Dan Clark, personal communication). Beaver location data indicate that there have been over 50 distinct locations where beaver have been active on the peninsula since 1988, the earliest year for which location data are available from the DCR. The locations of active beaver sites on the Prescott Peninsula in 2000 - 2003 are shown on Figure 4-2.

Close proximity to the UMass campus and DCR headquarters, a history of extensive research on Prescott Peninsula, and site safety made Prescott Peninsula an optimal location for the wildlife (beaver influenced) study site. Several watersheds where past beaver activity had been noted were scrutinized as potential wildlife study sites. The watersheds were initially visited to determine hydrologic and hydraulic suitability and similarity to watersheds included in the forest management study. Several of the sites were eliminated due to a lack of long-term food supplies for the beaver and evidence of only a modest level of current activity.

Griswold Brook (GRB) was determined to be the best potential site in terms of geomorphic similarity to watersheds included in the forest hydrology study, site accessibility, current beaver activity, and long-term food supplies. Historical data suggest beaver have inhabited the watershed since 1988 and possibly earlier although geographical locations prior to 1988 were not available. During 2000 – 2003 (the latest for which data are available), beaver inhabited the headwater impoundments of GRB. The active beaver impoundment includes a series of dams and several lodges. The outlet of the impoundment is approximately 1 km upstream from the GRB weir station. While this is a longer period of time than beaver typically occupy sites on the peninsula, the food sources at this location appeared adequate for the beaver to remain over the course of this study.

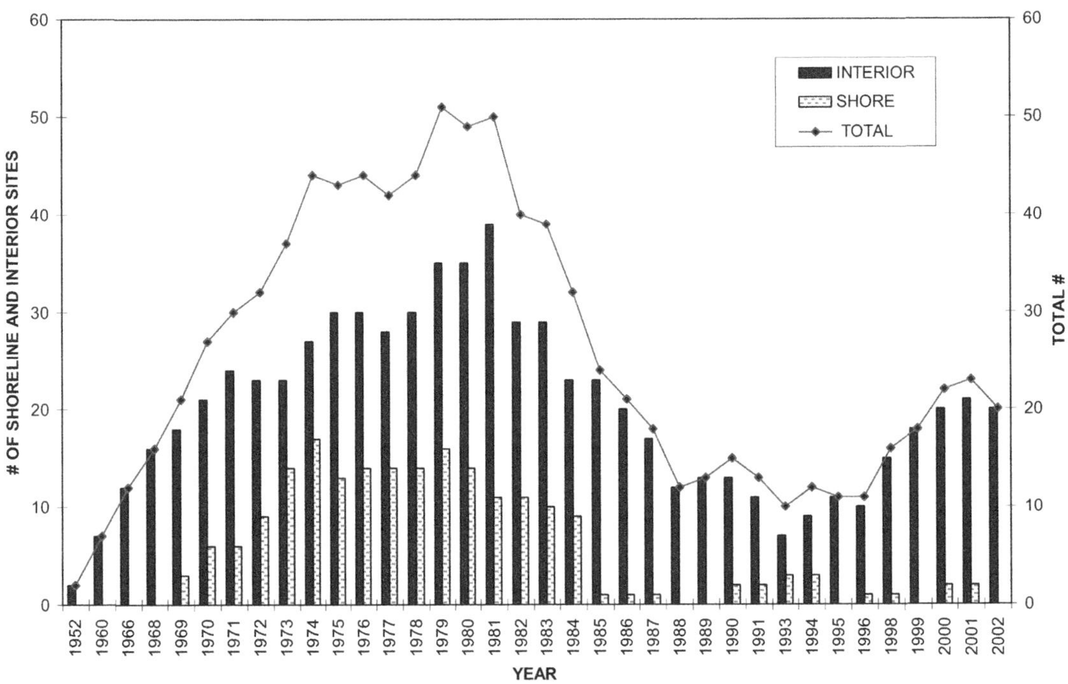

Figure 4-1 Number of active beaver sites on Prescott Peninsula, Quabbin Reservation, 1952-2000. Data provided by the DCR.

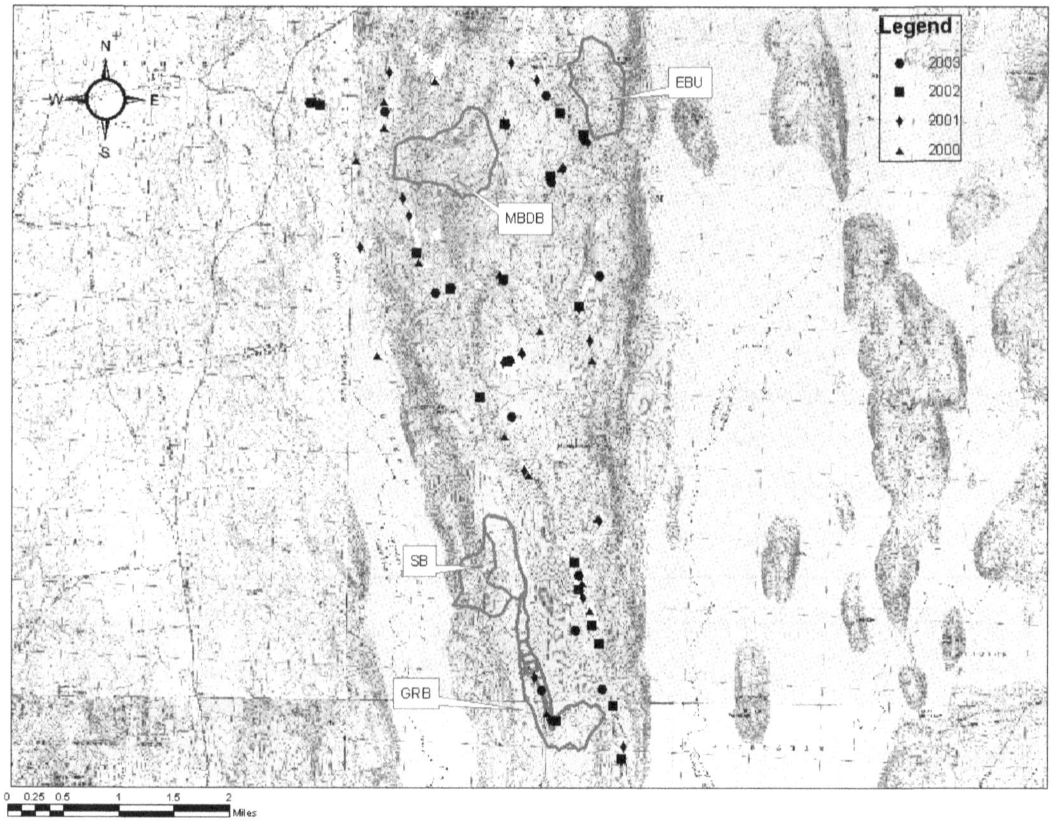

Figure 4-2 Location of active beaver colonies on Prescott Peninsula, 2000 - 2003

The Middle Branch Dickey Brook (MBDB) watershed was chosen as the control watershed unaffected by beaver. This decision was based on similarities between Griswold Brook (GRB) and MBDB and the expectation that MBDB would remain pristine, except for the influence of free-ranging wildlife, over the course of the study due to lack of food sources for beaver. Historical data (1988 to 2001) suggests beaver inhabited an area on the eastern edge of the MBDB watershed in 1995 and 1998 but have not been present in the watershed since then. MBDB had the added advantage of having an existing weir for discharge calculations.

A paired watershed study of the MBDB and GRB watersheds, located on Prescott, was thus used to study the relative effects of ranging wildlife and beaver on water quality, Figure 4-3. The two watersheds are similar in watershed characteristics, Table 4-3. GRB, however, is smaller than MBDB. While the total watershed average slope for MBDB is slightly larger (Table 4-3), the bottom "L" portion of GRB (see Figure 4-3) is actually more steeply sloped (data not provided). Additional open and water land use percentages in GRB are due to the beaver activity in its headwaters. Land use details for MBDB and GRB are provided in Figure 4-4 and Figure 4-5.

Table 4-3
Summary of watershed characteristics for the wildlife study watersheds

Site Name	Total Area (km^2)	Percent Area per Land Use Type: Forest	Open Land	Water	Stream Length (km)	Slope
MBDB (ranging wildlife)	1.15	95%	4%	1%	1.6	0.057
GRB (beaver)	0.79	87%	10%	3%	1.8	0.047

Source: Rees *et al.*, 2005. Transport of *Cryptosporidium* and *Giardia* in Massachusetts Watersheds During Storm Events. In *Managing Watersheds for Human and Natural Impacts: Engineering, Ecological, and Economic Challenges*. Copyright ©2005 ASCE. Reproduced with permission.

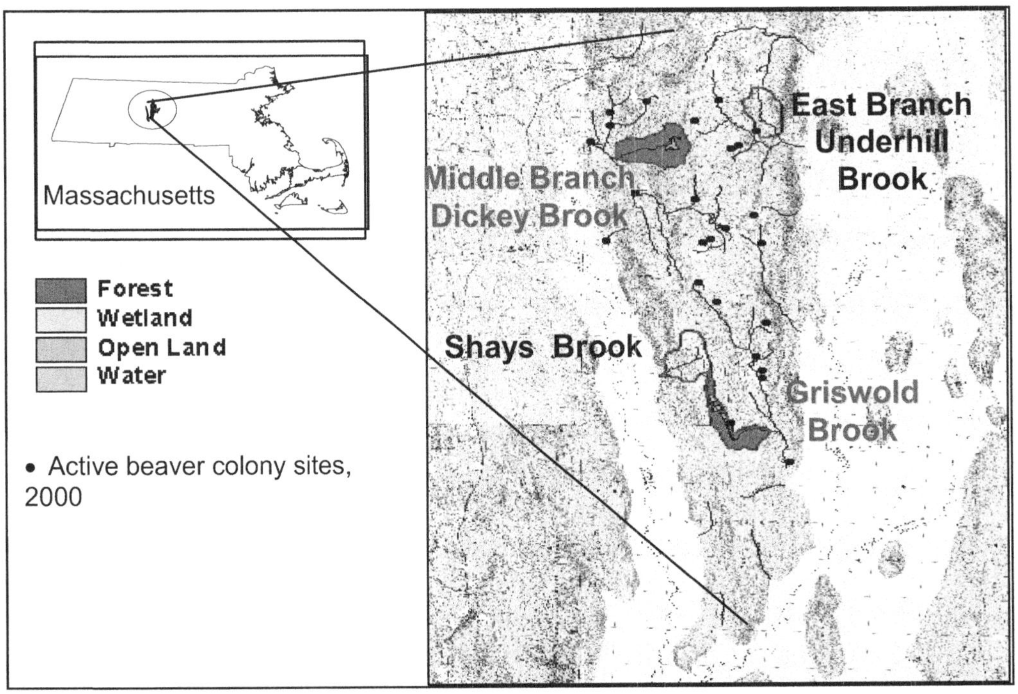

Figure 4-3 Location of MBDB and GRB study watersheds. Both watersheds are almost fully forested. Two additional watersheds are instrumented as part of a forest management project. Active beaver colonies as of 2000 are shown (data from DCR).

Source: Rees *et al.*, 2005. Transport of *Cryptosporidium* and *Giardia* in Massachusetts Watersheds During Storm Events. In *Managing Watersheds for Human and Natural Impacts: Engineering, Ecological, and Economic Challenges*. Copyright ©2005 ASCE. Reproduced with permission.

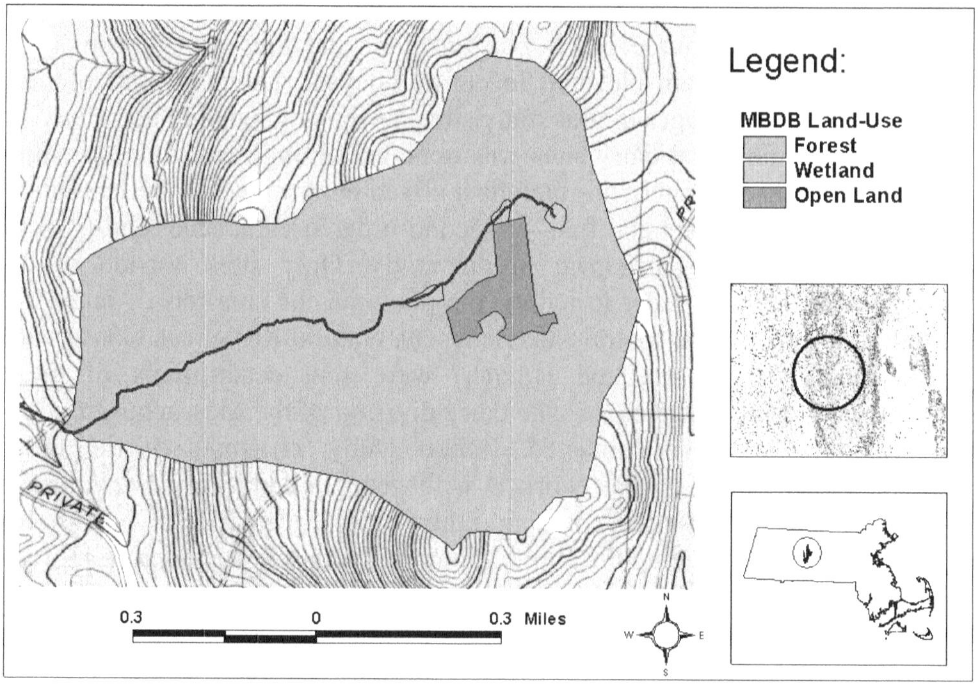

Figure 4-4 Details of MBDB land use

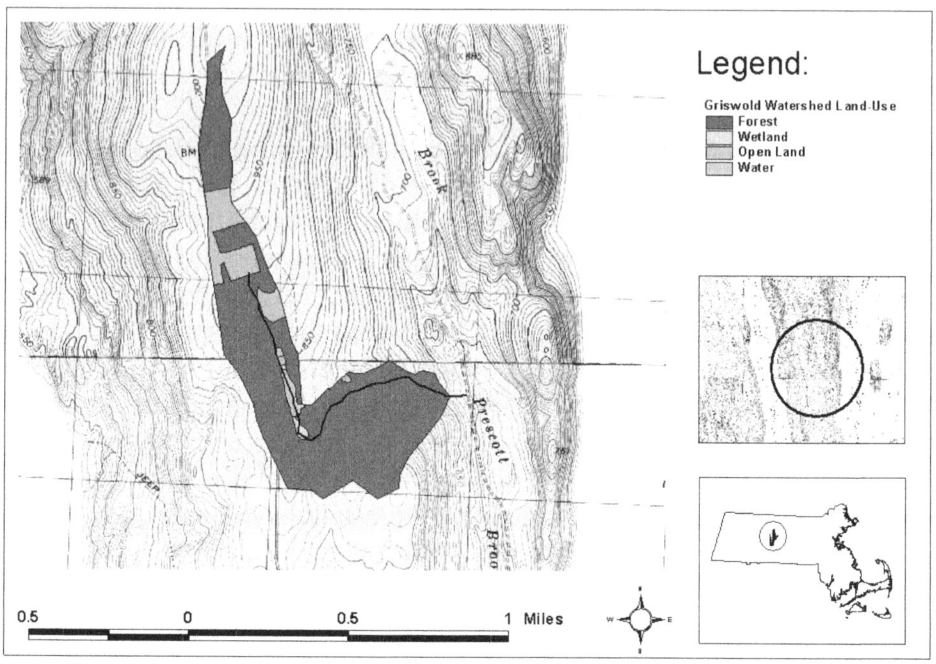

Figure 4-5 Details of GRB land use

Agricultural Study Site

Heifers are a known source of *Cryptosporidium* and thus runoff from cattle pastures during storm events may contain elevated levels of oocysts (Donnelly and Stentiford, 1997; AWWA, 1999a). A dairy farm/agricultural study site was thus included in this study.

An initial list of potential dairy sites was obtained through consultation with DCR staff scientists. Several key site characteristics include a mixed herd (calves or heifers intermixed with older cows), cow access to a perennial first- or second-order stream, accessibility of the site, and willingness of the farmer to participate in the study. Only sites outside of direct MDC jurisdiction were considered in order to reduce the potential for conflicts of interest and farmer liability. In addition, three of the central Massachusetts Natural Resources Conservation Service offices (Greenfield, Northampton, and Holden) were sent descriptions of the study and personally contacted for site suggestions. The dairy division of the Massachusetts Department of Food and Agriculture was also contacted. Robert Duby, chairman of the Department of Veterinary and Animal Sciences, has worked with several dairy farmers in the region and provided additional names and helped contacting the farmers. A total of 18 potential sites within an hour drive of UMass and meeting the above site requirements were identified. Fifteen of these sites were dairy cattle and three were beef. The farms were typically small (less then 50 head) operations.

Most of the dairy farms were contacted after assessing their hydrology and hydraulics either first hand or via topographic and digital orthophoto data. A confidential in-depth report on the quality of streamflow upstream and downstream of their property, identification of best management practices (BMPs) for site improvements, assistance in applying for BMP implementation funds, and help implementing BMPs were promised as incentives to participate in the study. However, only one dairy farm expressed a willingness to participate in the study. In general, dairy farmers in the Commonwealth of Massachusetts are in a precarious financial position. The fear of being forced to initiate changes at their site was typically too great to overcome, despite assurances of confidentiality.

The farm selected for the study is located along Burrow Brook in the Ware River watershed in central Massachusetts, approximately 45 minutes from the UMass campus. An upstream – downstream monitoring approach was used to isolate the effects of cattle on water quality, Table 4-4 and Figure 4-6. While MassGIS land use layers list the fields as "cropland", they are in reality pastures. At the beginning of this study, the cow herd consisted of sixty head of dairy cattle, twenty of which were typically heifers. In the late spring and summer, the herd was typically allowed to graze in a steeply sloped pasture (the northern pasture) every two to three days. Direct access to the stream is restricted from this pasture, but runoff directly enters a Burrow Brook tributary. The herd was also allowed to graze in an abutting field (the southern pasture) on a rotating basis. From this field, the cows have direct access to Burrow Brook, which they use as a drinking water source. Typically by the middle of June, 12 to 15 heifers were placed in this field full time. Visual inspection of the stream indicated several areas where the cows preferentially access the stream. The two fields and their position relative to Burrow Brook are shown in Figure 4-6. Over the course of the project, the dynamics of the herd changed and it currently consists of a varying number of beef cattle with the same access patterns to the pastures. This confounding issue will be discussed in more detail in Chapter 6. Beaver also inhabit the stretch of Burrow Brook between the upstream and downstream sampling locations.

They have used the remnants of a road to build a dam, creating a large swampy area upstream of the cow pasture.

Land use upstream of the site is a mainly a mixture of forest and residential with some agriculture, Table 4-4. A new suburban area is currently under development and contributes significant sediment loads to the brook during storm events. No water quality data from other sources for the Burrow Brook watershed is known to exist.

Table 4-4
Summary of watershed characteristics for the agricultural sampling locations

Site Name	Total Area (km^2)	Percent Area per Land Use Type					Stream Length (km)
		Agriculture	Forest	Urban	Water	Other	
BBFM	8.96	9.7%	74.3%	10.6%	2.0%	3.4%	17.75
BBWR	7.82	6.7%	76.4%	11.5%	2.2%	3.2%	15.76

Source: Rees *et al.*, 2005. Transport of *Cryptosporidium* and *Giardia* in Massachusetts Watersheds During Storm Events. In *Managing Watersheds for Human and Natural Impacts: Engineering, Ecological, and Economic Challenges*. Copyright ©2005 ASCE. Reproduced with permission.

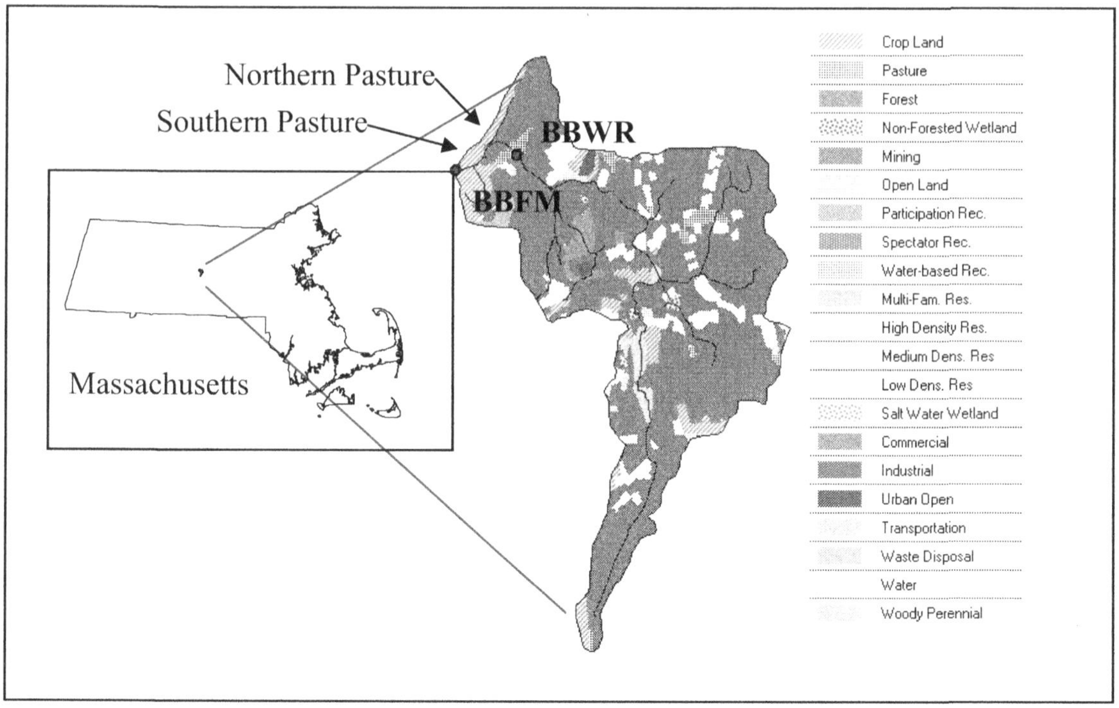

Figure 4-6 Location of the upstream (BBWR) and downstream (BBFM) monitoring locations (dots) within the Burrow Brook watershed. The two sites isolate a dairy/beef cattle pasture and crop land.

Source: Rees *et al.*, 2005. Transport of *Cryptosporidium* and *Giardia* in Massachusetts Watersheds During Storm Events. In *Managing Watersheds for Human and Natural Impacts: Engineering, Ecological, and Economic Challenges*. Copyright ©2005 ASCE. Reproduced with permission.

Residential Study Site

Priority was placed on selecting a residential study site with known leaky septic or sewer systems within the Wachusett Reservoir watershed based on stated DCR needs. Several key site characteristics included semi-dense residential development, an area distinct enough that it could be isolated from upstream land use, availability of long-term fecal coliform and/or discharge data, and willingness of landowners to participate in the study. DCR staff scientists identified the main-stem reach of the Gates Brook as a potential study site fitting the key site characteristics.

The Gates Brook sub-watershed of the Wachusett Reservoir Basin is located approximately 1 hour and 15 minutes from the UMass campus. The Wachusett DCR office and lab, however, are located approximately five minutes from the watershed. The DCR has collected depth, discharge, temperature, dissolved oxygen, percent saturation, conductivity, pH, total dissolved solids, fecal coliform, ammonia, nitrate, total phosphorus, iron, silicon, turbidity, standard alkalinity and EPA alkalinity at ten sites within the Gates Brook watershed, identified by squares in Figure 4-7. The exact parameters measured and period of record vary from site to site, but the most extensive database extends from 1998 to the present. In addition, environmental testing for alternative indicator organisms occurred at seven sites within the Gates Brook sub-watershed during Spring/Summer 2000 (Long et al., 2002), identified by triangles in Figure 4-7. The Gates Brook watershed was thus selected to study the influence of residential properties (septic and sewered), commercial properties, and transportation systems on water quality, Table 4-5 and Figure 4-7. There are no combined sewer outfalls in Gates Brook. The upstream sampling transect for this study was co-located with historical DCR sampling location Gates 9 and the downstream sampling transect was located just upstream of Gates 4. Culverts at these locations were used to accurately determine discharge from stage measurements rather than through the installation of weirs. Both human and pet wastes were considered potential sources of pathogen contamination in the watershed. In addition, limited grazing animal (horses) and ranging wildlife live within the watershed. The sampling locations used for this study are identified with circles.

Table 4-5
Summary of watershed characteristics for the residential sampling study site

Site Name	Total Area (km^2)	Percent Area per Land Use Type: Agriculture	Forest	Urban	Other	Stream Length (km)
GBBR	5.9	7.6%	36.3%	52.0%	4.1%	10.0
GBWD	4.3	10.4%	41.0%	43.3%	5.3%	6.7
Intervening	1.6	0%	23.6%	75.6%	0%	3.3

Source: Rees *et al.*, 2005. Transport of *Cryptosporidium* and *Giardia* in Massachusetts Watersheds During Storm Events. In *Managing Watersheds for Human and Natural Impacts: Engineering, Ecological, and Economic Challenges*. Copyright ©2005 ASCE. Reproduced with permission.

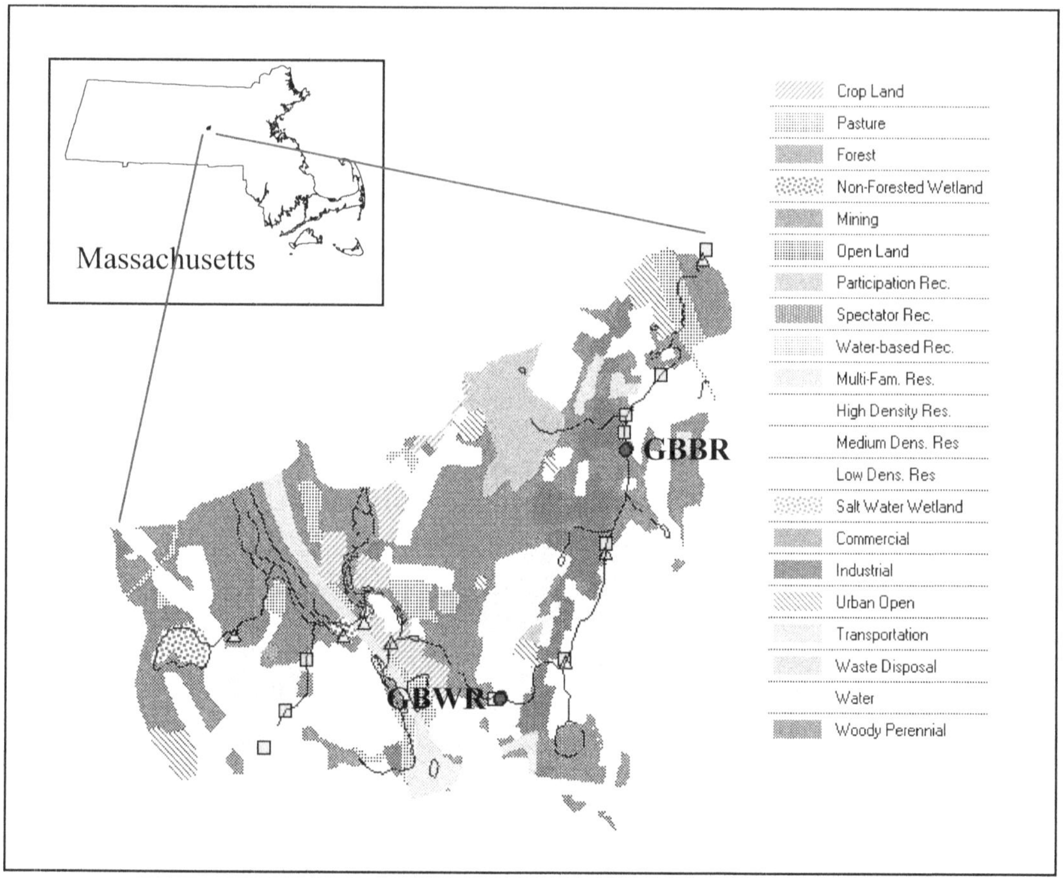

Figure 4-7 Location of the upstream (GBWR) and downstream (GBBR) monitoring locations within the Gates Brook watershed. The two sites isolate a heavily urbanized area.

Source: Rees *et al.*, 2005. Transport of *Cryptosporidium* and *Giardia* in Massachusetts Watersheds During Storm Events. In *Managing Watersheds for Human and Natural Impacts: Engineering, Ecological, and Economic Challenges*. Copyright ©2005 ASCE. Reproduced with permission.

MONITORED PARAMETERS AND SAMPLE TIMING

Continuous Monitoring

Stage, conductivity, pH, temperature, and turbidity, Table 4-6, were continuously monitored at each sampling transect per study site. Precipitation was monitored at only one of the sampling transects per study site. During *non-event* periods, continuous data were collected at 15-minute intervals; during *event* periods, continuous data were collected at 5-minute intervals. The start of continuous data collection at the sites varied, Table 4-7, due to personnel limitations and the time required for obtaining landowner and Conservation Commission permission to install equipment and conduct sampling. While data collection often started prior to these dates, these dates denote the time when the most reliable data collection began.

Table 4-6
Continuously monitored physical and water quality parameters and the equipment used for the measurements

Equipment	Purpose
Stevens/Greenspan CS4-1200	Conductivity pH Temperature
Stevens/Greenspan TS1200	Turbidity
Stevens/Greenspan PS-1200	Pressure (depth)
Hydrolgic Inc. TB-3 Tipping Bucket Rain Gauge	Precipitation
Solarex 20 Watt Solar Panel	Backup power
AxSys MPU Datalogger	Data storage

Source: Rees *et al.*, 2005. Transport of *Cryptosporidium* and *Giardia* in Massachusetts Watersheds During Storm Events. In *Managing Watersheds for Human and Natural Impacts: Engineering, Ecological, and Economic Challenges*. Copyright ©2005 ASCE. Reproduced with permission.

Table 4-7
Summary of the start date of continuous data collection at each sampling location

Study Site	Sampling Location	Date
Wildlife	MBDB	October 25, 2001
	GRB	October 22, 2001
Agricultural	BBWR	May 19, 2003
	BBFM	May 20, 2003
Residential	GBWD	September 24, 2002
	GBBR	October 8, 2002

Source: Rees *et al.*, 2005. Transport of *Cryptosporidium* and *Giardia* in Massachusetts Watersheds During Storm Events. In *Managing Watersheds for Human and Natural Impacts: Engineering, Ecological, and Economic Challenges*. Copyright ©2005 ASCE. Reproduced with permission.

The equipment selected to facilitate the continuous monitoring is listed in Table 4-6. Operating ranges were selected to best match the expected conditions of each site. The range and associated accuracy of the equipment deployed at each site is summarized in Table 4-8. Accuracy of the Steven's water quality monitoring equipment is similar to that of the equivalent products manufactured by other vendors.

The continuous in-situ water quality sensors were deployed at each site in fabricated holders. The electronics and programming features of the sensors are designed so that they may be powered and recorded via shared equipment in a streamlined fashion, reducing cost and maintenance challenges while improving efficiency and system reliability. Steven's electronic data loggers were deployed in weatherproof housings to record data at each study site. Further details about the in-situ water quality sensors and deployment are provided below in succeeding paragraphs.

Data Storage

Data was stored via an AXSYS MPU data logger SDI-12 input (Stevens, Beaverton, Oregon) and downloaded monthly in the field to a laptop computer. AXSYS systems were factory installed into weatherproof Hoffman fiberglass housings. However, during sampling events the housings were often open to access the system for data observation. For this reason, pressure treated plywood housings were constructed to house the AXSYS systems. The base of the housing was constructed using pressure treated 2" x 4" lumber secured together using 2 ½" galvanized decking screws. The housing was constructed of ½" pressure treated exterior plywood. A linseed oil based brown stain was applied to the plywood structure. The joints were sealed with construction grade silicone to provide additional weatherproofing. The access door was hinged using a 1" stainless steel piano hinge and opened to provide a work platform. Screened vent ports were installed in the sides of the housing to provide air circulation throughout. A galvanized hasp was fastened to the outside of the door to provide a place to lock the housing.

The AXSYS system operated on a 12-volt sealed gel battery wired in parallel to a 20-watt solar panel (Solarex, Federick, Maryland). The solar panel recharged the battery and operated the system during daylight hours. The battery provided power to the system during cloudy days and throughout the nights. Steven's SDI-12 data cables with bulkhead connectors were deployed to connect the AXSYS to the probes. The cables were inserted through 5/8" garden hose to provide a protective outer casing. The SDI cables, rain gauge, and power supply were connected to the AXSYS data collector. The AXSYS systems were setup to log data to an electronic memory bank in 15-minute intervals during non-event periods. Five of the 12 channels were assigned and setup for each parameter monitored. At sites including a tipping bucket raingage, channel nine was used as the rain gauge (pulse) input from the TB-3. The procedures used to setup the AXSYS data logger channels and parameters are presented Bourdeau (2004).

Table 4-8
Range and associated accuracy of equipment for the continuously monitored parameters

Parameter	Range	Accuracy
Turbidity		
MBDB, GRB		
12/01 – 5/02	0 to 250 NTU	± 2 %
Since 5/02	0 to 50 NTU	± 2 %
BBWR, BBFM	0 to 40 NTU	± 2 %
GBWD, GBBR	0 to 250 NTU	± 2 %
pH (all sites)	0 to 14 units	± 0.2 units
Temperature (all sites)	0 to 50 °C	± 0.2 °C
Conductivity		
MBDB, GRB		
12/01 – 5/02	0 to 200 μS	± 0.5 % full scale
Since 5/02	0 to 1000 μS	± 0.5 % full scale
BBWR		
5/03 to 4/04	0 to 1000 μS	± 0.5 % full scale
Since 4/04	0 to 500 μS	± 0.5 % full scale
BBFM		
Entire Preiod	0 to 500 μS	± 0.5 % full scale
GBWD		± 0.5 % full scale
Entire Period	0 to 1000 μS	± 0.5 % full scale
GBBR		
5/03 to 4/04	0 to 500 μS	± 0.5 % full scale
Since 4/04	0 to 1000 μS	± 0.5 % full scale
Precipitation (all sites)	at 0.5 in/hr	0.5 %
	at 1 to 3 in/hr	1.0 %
Pressure (Depth) (all sites)	0.0 to 2.5 m	± 0.05 % full scale

Solar Panel

The AXSYS system operated on a 12-volt battery wired in parallel to a 12" x 12" 20-watt solar panel (Solarex, Federick, Maryland). The panel was attached to a 4"x4"x8' pressure treated post using the adjustable aluminum bracket supplied by the manufacturer. The post was secured in the ground using a Post-Up bracket, which was driven flush to grade with a sledgehammer. To prevent the panel and post from tipping in extreme winds, three ¼" aircraft grade aluminum guy wires were installed. The cables were secured to the post with two-inch galvanized eyebolts and to the ground with ½" diameter x 18" rebar guy anchors. The angle of the panel with respect to the post was set to optimize the time at which the panel was at a ninety-degree angle with the sun. Excessively high voltage output in late fall and early spring caused fuses to blow and disrupt power early in the study. This was fixed by using larger fuses.

Pressure Transducer

Since the USGS is the premier flow-monitoring agency in the United States, meeting USGS flow-monitoring criteria was a major consideration in equipment selection. Stevens/Greenspan pressure sensors are standard equipment for the USGS, meeting their QA/QC acceptance requirements. The Stevens/Greenspan pressure sensor has also proven reliable in a wide range of field conditions including ice-covered streams, high flows, low and high temperatures, and high suspended-sediment loads. This reliability has been documented both by the USGS and by Co-PI Paul Barten, who has used this product for over 10 years in salt marshes, wetlands, and beaver ponds in addition to stream channels. As the equipment was to be deployed year round, it needed to function with high accuracy under a wide range of field conditions.

For this study, the PS-1200 pressure sensor (Stevens/Greenspan, Beaverton, Oregon) with a 0.0 to 2.5 meter range was selected for each sampling location. The sensor measured the depth of water above the bottom of the sensor in meters to three decimal places. The pressure transducers were positioned on the upstream face of the weirs in a fabricated metal bracket at an elevation below that of the V-notch. The depth of water above the sensor was converted to depth of water above the V-notch using periodic offset measurements. The depth of water above the V-notch was used to generate discharge data using the empirical equations for a sharp crested V-notch weir (see Equation 4.1). At the two residential sampling locations, depth of flow through a culvert rather than over a weir was monitored and rating curves were used to generate discharge data. At GBBR, the pressure sensor was attached to the upstream face of the box culvert. At GBWD, a fabricated metal bracket secured in the stream immediately upstream of the culvert was used to hold the pressure sensor.

Steven's Greenspan recommends the PS1200 be calibrated to local atmospheric conditions. The calibration was performed prior to installing the probe in the bracket. The PS1200 was also equipped with a closed venting system, CVS-001, which compensated for changes in local atmospheric pressure. The CVS-001 was setup prior to attaching it to the PS1200 vent tube using procedures described in the Steven's PS1200 User Manual, 1.4th edition.

Conductivity, pH, and Temperature

The CS4-1200 multi-parameter sensor (Stevens/Greenspan, Beaverton, Oregon) was selected to monitor pH, conductivity, and temperature. The recording range of the CS4-1200 varied between study sites due to underlying physical water quality differences, Table 4-8. Each sensor was secured to the fabricated metal support structure positioned in the stream. Temperature and conductivity were reported in Celsius and micro Siemens per cm, respectively. The metal support bracket was fabricated and installed into the streambed at a location upstream of the pooling caused by the weir. The bracket was constructed of 1 ¼" x 1 ¼" x ¼" angle iron and 1 ½" x ¼" steel pipe. Posts constructed from number 3 and 7 reinforcing bar supported the bracket. The posts were driven into the streambed and guyed with ¼" diameter cable to provide rigidity at high flow conditions. Sensors were attached to the support structure using two stainless steel hose clamps and positioned on the stand so the water level was above the water quality sensors. The probe was aligned with the pH sensor on the downstream side to prevent debris from damaging the sensor.

Turbidity

Steven's Greenspan model TS1200 turbidity sensors were deployed to monitor turbidity. The recording range of the TS1200 varied between study sites due to underlying physical water quality differences, as listed in Table 4.8. Turbidity was reported in nephelometric turbidity units (NTU) within the probe's recording range. Each TS1200 probe was originally attached to a fabricated holder in a vertical orientation similar to the CTDP 1200 sensor. However, ambient light interfered with the TS1200 sensor resulting in inaccurate turbidity data. A modified holder was deployed to prevent interference of ambient light. Each probe was inserted into a 24" long, 6" diameter black PVC tube as recommended by Steven's. The tube was attached to the rebar probe stand using a stainless steel hose clamp. The tube was oriented parallel with the stream flow positioned on the streambed.

Precipitation

Precipitation was monitored using Hydrologic Inc. TB-3 tipping bucket rain gauges. At the GRB study site, the gauge was deployed in a logged clearing which provided a 30 – 45 degree clear sight angle from the top of the gauge to the adjacent tree line. At the BBFM site, the gauge was deployed in the southern pasture near the housing for the AXSYS. The pasture provided a 30 – 45 degree clear sight angle from the top of the gauge to the adjacent tree line for 270°. The desirable open radius was slightly affected over approximately 90° due to proximity to trees along the stream edge. The nearest trees, however, were approximately 10 feet away and lost rain catch effects should be small. This placement was necessitated by landowner requests over placement and concerns over the cows knocking over equipment. For both the GRB and BBFM rain gauges, a wooden platform was constructed to secure the gauge into position approximately two feet above grade. The base was constructed from a 4" x 4" x 2' pressure treated post with a 12" x 12" x ¾" pressure treated plywood deck. Due to extensive scrub brush and small trees at the residential site, GBBR, the rain gauge at that site was secured into position approximately eight feet above grade. This more elevated position provided a 30 – 45 degree clear sight angle from the top of the gauge to the adjacent tree line. The base was modified over that at the other sites by using a 4" x 4" x 8' pressure treated post. At all sites, the post was secured in the ground using a Post-Up bracket. The gauge was fastened to the plywood deck using three 2" x 3/8" galvanized hex-bolts and leveled using a fish eye bubble level.

Maintenance

Data recorded to the AXSYS data log were downloaded on a monthly schedule. In addition, data were downloaded prior to monitoring of a storm event at a site and then again at the end of each storm event to avoid potential data loss. A laptop computer with MODTERM software was used to extract the data from the internal data log of the AXSYS logger and write the data to a local text file generated by the program. The procedure to download data is available from Bourdeau (2004).

Calibrations of the Steven's water quality monitoring equipment were performed on a monthly basis. Steven's Greenspan CTDP 1200 pH, conductivity, and temperature and TS1200 turbidity probes were calibrated in accordance to Steven's recommendations as found in Steven's user guides. The conductivity wire loop method was used to calibrate the conductivity sensor of the CTDP1200. This method was preferred because of the consistent accuracy of the Steven's wire loops. In addition, no waste was generated using this method. The Formazin method was used to calibrate the TS1200 turbidity probes. The pressure transducers and rain gauge were not calibrated on a routine basis as recommended by Steven's. Detailed calibration procedures are available in Bourdeau (2004) and are summarized below.

At the time of calibration, the probes were removed from the field and returned to the laboratory. Each probe was thoroughly washed with a Liquinox solution and then rinsed twice with Super Q water. The laptop computer was connected to the SDI 400 calibration box and then powered on. The probe to be calibrated was then connected to the SDI 400 Calibration Box with a SDI-12 bulkhead connector cable. A 12-volt sealed gel cell battery was then connected to the SDI 400 Calibration Box. Once the hardware setup was complete, the CALTECH program was opened to perform the calibration.

The CALTECH program is a simple user-friendly software package developed by Steven's Greenspan to calibrate Steven's Water Quality Monitoring equipment. The program opens to a main page with a pull down menu, which allows the user to select the parameter to be calibrated. The parameter available in the pull down menu will be specific to the probe connected to the CALTECH box. For example, if a TS1200 turbidity sensor is attached, only turbidity will be shown as an option in the pull down menu. Standard calibrations were performed using the "offset and gain correction" option. This option clears the probe of a previous calibration and uses a two-point data set to calibrate.

Problems Encountered

Several problems were encountered with the data collection systems through the period of study. These problems included sediment and debris clogs in the turbidity tube, ice, power failures and probe failures. In addition, low flow conditions were sometimes encountered. At BBFM, the cows knocked over the AXSYS housing and ripped out cables, resulting in a stretch of lost data.

The Steven's Greenspan TS1200 performed in nearly all flow conditions. However, problems were encountered which affected the data integrity. Most commonly, a biofilm grew on the sensor face resulting in extended turbidity recordings of 40.0 NTU. Throughout the fall, the PVC turbidity tube would collect leaves, woody debris, and sediment, which resulted in increased non-representative turbidity readings. Low flow conditions common in the summer would also result in sediment clogged PVC turbidity tubes, which would result in non-representative records. Roughly 40 percent of continuous turbidity data was lost due to these problems. Drift in the readings was also common. As a result, turbidity data is generally considered to be unreliable.

During periods of extremely low temperatures, typical of New England's winters, ice would cover the streams. To prevent damage to the probes, they were removed during these periods. Such conditions inhibited the ability to sample and collect data.

The initial setup of the AXSYS system was experimental and required a trial and error period. One of the major problems throughout the trial and error period was power failure of the 12-volt system. The sealed gel 12-volt batteries often drained low in power (less than 10.0 volts) resulting in a "low power, not logging data" condition. Solar panels, which received excess durations of insolation, distributed greater than 14.0 volts to the system and caused the in-line fuse to blow resulting in a similar "no power, not logging data" condition. The power supply problems were alleviated by installing a series of fuses (4 Amp followed by a ¾ Amp) in-line between the solar panel and sealed gel 12-volt battery.

Periodic failures of equipment resulted in lost data or the inability to collect data. The majority of problems encountered required sending equipment to Steven's for repair resulting in long periods of time without the ability to collect data, or the ability to collect only partial data. Missing stage (discharge) data due to equipment failures was typically filled in based on data from the paired sampling site. An AXSYS troubleshooting guide was developed to help solve and repair other problems encountered with the AXSYS systems. Failures of the AXSYS data collector included a broken internal clock, and a blown internal fuse. In the event that a data collector was exposed to water, the data collector would turn to a "not logging data" condition. This was caused by fluid entering the data card slot on the AXSYS face resulting in a temporary internal short circuit. The problem was relieved after the fluid was removed using desiccant pillow packets.

Access onto the Prescott Peninsula was granted by the DCR in the form of a research access permit. The permit was issued with a key to access the Prescott by service road 17 through security gate Number 17. Service road 17 is a three-season road that spans the entire Prescott from the north to the south. Travel on service road Number 17 was limited to the use of an off road vehicle throughout the winter and spring months. Snow and ice resulted in impassible road conditions and the inability to access the study sites to service and sample. The spring season brought thawing conditions which resulted in impassible muddied roads. The terrorist activity of September 11, 2001 resulted in a national heightened security level at all water supply municipalities, including the Quabbin Reservoir. All entry onto Quabbin property and the Prescott Peninsula was limited for the fall of 2001. Thus, access to the Prescott study sites for event sampling was temporarily discontinued.

Monthly Monitoring

Monthly data were collected on either the second Wednesday (wildlife and agricultural sites) or second Thursday (residential sites) of each month. Monthly data collection at the two wildlife sites began in November 2001; monthly data collection at the agriculture and residential sties began in August 2002. Water quality parameters monitored on a monthly basis are shown in Table 4-9. In June 2002, monitoring switched from ammonia to nitrite and nitrate/nitrite due to consistent non-detectable levels of ammonia in the samples. Monthly monitoring for alternative indicator organisms, *Cryptosporidium* oocysts and *Giardia* cysts was cost prohibitive and thus not included.

Table 4-9
Water quality parameters monitored on a monthly basis

Monthly Water Quality Parameters	
Fecal Coliform Enumeration	Total Phosphorus
Total Coliform Enumeration	Total Kjeldahl Nitrogen
Suspended solids	Nitrate/Nitrite*

Notes: * Beginning in May 2002.

Storm-Event Monitoring

Parameters Monitored and Timing

Monitoring for both the monthly water quality parameters and several additional parameters was conducted during storm events, Table 4-10. The number of samples analyzed per sampling location during events varied by parameter. Only one study site, and thus two sampling locations, was active during each event. Bulk samples were typically collected at 30-minute intervals and stored in coolers on ice. However, during periods of low rainfall or during long events, 60-minute sampling intervals were used. Similarly, during periods of extreme rain and sharp rises in the hydrograph, samples were collected at 15-minute intervals. Bulk sample collection was initiated by the start of precipitation in order to enable detection of the first flush from both the isolated land use area and upstream sources (Stuntebeck and Bannerman, 1998).

Once the hydrograph shape and timing were known, bulk samples were identified for analysis; aliquots were then prepared and preserved as required by analysis method. The identification of bulk samples appropriate for analysis during events was set so that the rising-limb, peak, and falling-limb of the hydrograph were well represented for each parameter, with a larger emphasis on capturing the rising-limb and peak. Due to the variable number of analyses per event based on parameter type (Table 4-10), not all collected samples were analyzed nor were aliquots for all parameters prepared for all bulk samples analyzed; rather, personal judgment was used to determine the appropriate bulk samples to analyze for each parameter, based on the available number of analyses and hydrograph response.

Throughout the beginning stages of the study at the Prescott sites, bulk samples collected at the same time at the paired watershed sampling locations were selected for analysis. However, preliminary analysis of the data suggested the hydrological response of the paired watersheds was significantly different. The difference in response time was due mainly to the influence of the beaver impoundments, with additional differences in temporal response caused by spatial variability of rainfall rates and timing. As a result, identification of bulk samples for analysis was completed independently at each site. Simultaneous sampling at the agricultural and residential sites was similarly relaxed to facilitate sampling with smaller field crews. At these sites, identification of bulk samples for analysis was also completed independently, but tended to correspond more closely due to correlation between the up- and down-stream sampling locations.

Over the course of the project, two storm events were monitored at each study site during spring, summer, and fall (Apr – Jun, Jul – Sep, and Oct – Dec, respectively). During the winter (Jan – Mar), a single storm event was monitored at each study site. Winter storm-event monitoring was limited due to safety concerns resulting from icy conditions, severe cold weather which resulted in stream freezing, and limited winter month rainfall events. In addition, a dry weather event was monitored at each study site. Dry weather events consisted of collecting spatially distributed bulk samples throughout the study site watershed for analysis as per storm events, Table 4-10

Table 4-10
Microbial and standard water quality paramaters monitored during storm events, including the frequency of sampling per sampling location

Standard Analyses	No. per Event per location
Fecal Coliform Enumeration	8
Total Coliform Enumeration	8
Suspended solids	16
Total Phosphorus	16
Nitrate and Nitrite*	16
UV254	8
Total Heterotrophic Plate Counts	3 or 8

Special Analyses	No. per Event per location
F^- Coliphage Enumeration with CN13 host	3
F^+ Coliphage Enumeration with Famp host	3
R. coprophilus Enumeration and Confirmation	3
Clostridium perfringens	3
Sorbitol-fermenting *Bifidobacteria* Enumeration and Confirmation	3
Giardia	3
Cryptosporidium	3

Notes: *Ammonia was originally included in the study rather than nitrate/nitrite. However, due to low detections, nitrate/nitrite was substituted beginning in June 2002. *Source*: Rees *et al.*, 2005. Transport of *Cryptosporidium* and *Giardia* in Massachusetts Watersheds During Storm Events. In *Managing Watersheds for Human and Natural Impacts: Engineering, Ecological, and Economic Challenges*. Copyright ©2005 ASCE. Reproduced with permission.

Parameter Inclusion Justification

A brief description of the purpose for including each indicator organism or alternative indicator organisms during storm events is provided herein. A detailed literature review was provided in Chapter 2. Laboratory methods are described in a following section. Emerging pathogens *Cryptosporidium* and *Giardia* were included as a focus area of the study.

During events, TC and FC were included as traditional indicators for determining the presence and general extent of microbial contamination. However, research has demonstrated that coliform bacteria may not be ideal indicators of the presence of pathogenic viruses, protozoa, or other emerging pathogens (Long *et al.*, 2002; Rose *et al.*, 1986; DeNileaon, 1998). Total heterotrophic plate counts were originally conducted on eight samples per sampling location per event. This frequency was dropped to three per event as the counts were high and it was not felt that the higher temporal resolution data for THPC's provided additional insight over the TC and FC data.

Clostridium perfringens has been associated with human fecal material in field monitoring studies and is known to be a good measure of the degree of pollution in a stream (Fujioka and Shizumura, 1985). In treatment processes, the behavior of these organisms mimics that of cyst and spore type pathogens. Detection of *Rhodococcus coprophilus* is considered a good indication of grazing animal fecal contamination (Jagals *et al.*, 1995). Detection of sorbitol-fermenting *Bifidobacteria* is considered a good indication of human sewage contamination (Long *et al.*, 2002). Studies suggest that F+ and F- coliphages are good indicators of fecal contamination in general and have the potential to indicate the presence of enteric viruses. F+ coliphages have the potential to discriminate between human (groups 2 and 3) and animal (groups 1 and 4) fecal pollution (Long *et al.*, 2002).

Source specific indicators *Clostridium perfringens*, F+ and F- coliphages, sorbitol-fermenting *Bifidobacteria*, and *Rhodococcus coprophilus* were included at select sites based on literature indicating their potential usefulness for identifying contamination and its likely sources. Analyses for *Clostridium perfringens* were conducted at all sites. Analyses for *R. coprophilus* were restricted to the wildlife and agricultural sites and analyses for sorbitol-fermenting *Bifidobacteria* were included only at the residential sites. Samples were not analyzed for coliphages after summer 2003 due to low counts during previous events and the length of time required to process the samples for the UMass special analyses. The UMass special analyses were dropped twice (agriculture site events June 2003 and December 2003) to facilitate event collection when personnel were not available or the turn around time between events was too short.

LABORATORY ANALYSIS METHODS

The laboratory Standard Method (SM) or USEPA method used for water quality parameters analyzed for on a monthly basis are provided in Table 4-11. The laboratory methods used for parameters monitored only during events are provided in Table 4-12. Nutrient and total suspended solids analyses were performed by the Massacusetts Water Resources Authority (MWRA) laboratory at Deer Island. *Cryptosporidium* and *Giardia* analyses were conducted by the Environmental Associates Lab in Ithaca, NY, using EPA method 1623 (USEPA, 1999). Alternative indicator and THPC analyses were performed at UMass.

Total and fecal coliform analyses for Prescott Peninsula (wildlife) and Barre (agricultural) were conducted by the DCR lab at Quabbin while those for Gates (residential) were conducted by the DCR lab at Wachusett. For some events at Barre, the Wachusett lab performed the coliform analyses due to personnel constraints at Quabbin. In addition, beginning in July 2004, the Quabbin and Wachusett DCR laboratory staffs were replaced by MWRA personnel. For Wachusett, the lab was closed and analyses were conducted at the MWRA Southborough facility. Total coliform analyses utilized m-Endo LES agar (Difco, Detroit, MI) plates with incubation at 35°C for 22 to 24 hours. Fecal coliform analyses utilized m-FC agar (Difco, Detroit, MI) plates with rosolic acid addition and waterbath incubation at 44.5°C for 24 ± 2 hours.

Table 4-11
Summary of laboratory methods for monthly data

Parameter and Method	Sample Volume	Preservation	Holding Time	Detection Limit (Reporting Limit RL)
Total Phosphorus EPA - 365.1	1 Liter	Sulfuric Acid	28 days	0.002 mg/L (RL <0.005 mg/L)
Total Suspended Solids EPA - 160.2	500 ml	Refrigeration	7 days	1.7 mg/L (RL < 5.0 mg/L)
Total Kjeldahl-Nitrogen‡ EPA – 351.2	1 Liter	Sulfuric Acid	28 days	0.040 mg/L (RL <0.05 mg/L)
Total Kjeldahl-Nitrogen‡ SM 4500-N_{org}-B	1 Liter	Sulfuric Acid	28 days	0.200 mg/L (RL <0.2 mg/L)
Ammonia-Nitrogen*† EPA - 350.1	60 ml	Filtered and Frozen	180 days	0.001 mg/L (RL <0.005 mg/L)
Nitrite*† EPA – 353.2	60 ml	Filtered and Frozen	90 days	0.0002 mg/L (RL <0.005 mg/L)
Nitrite/Nitrate*† EPA – 353.2	60 ml	Filtered and Frozen	90 days	0.003 mg/L (RL <0.005 mg/L)
Nitrate*† EPA – 353.2	60 ml	Filtered and Frozen	90 days	NO3 measured as subtraction of NO2 from NO3/NO2
Total Coliform§ SM 9222 B, membrane filtration	Various dilutions	NA**	Max 24 hr, 6 hr goal	Variable**
Fecal Coliform§ SM 9222 D, membrane filtration	Various dilutions	NA**	Max 24 hr, 6 hr goal	Variable**

Notes: *In order to reduce interference in the lab analysis, the samples are filtered in the field using a glass fiber Acrodisc (pore size = 1 micron; Gelman Sciences, Inc.) attached to a syringe. †We discontinued Ammonia starting May 2002 and substituted Nitrate and Nitrite. ‡ Total Kjeldahl Nitrogen method was switched to EPA 351.2 effective May 1, 2003. § – DCR labs report presumptive results, as verifications are not performed on every plate. ** – dependent on the volume filtered, at least 1 cfu/mL..

Table 4-12
Summary of methods for parameters monitored only during storm events

Parameter	Method	Instrumentation/Comments
Total Heterotrophic Plate Counts	Modified SM9215b	Millipore total count water testers
Sorbitol-fermenting *Bifidobacteria*	Membrane filtration presumptive, biochemical reaction confirmation	Human bifido sorbitol agar (HBSA), reinforced clostridial agar (RCA-S), Gram stain reaction, F6PPK enzyme reaction, sorbitol fermentation, nitrate reduction
Rhodococcus coprophilus	Membrane concentration-resuspension, confirmation	MM3 media and Bennett's agar
F^- Coliphage Enumeration	Single agar layer method, EPA 1602	*E. coli* CN13 host
F^+ Coliphage Enumeration	Single agar layer method, EPA 1602	*E. coli* Famp host
Clostridium perfringens	Membrane filtration with growth on m-CP agar.	
Cyrptosporidium and *Giardia*	EPA 1623	Aim to filter 20 L
UV254	SM 5910 B	UV-VIS Spectrophotometer

Sorbitol-fermenting *Bifidobacteria* was enumerated using a membrane filtration presumptive, biochemical reaction confirmation procedure adapted from Beerens (1991) and Mara and Oragui (1985) for use in the UMass laboratories (Long and Arango, 1998; Long, 1998). A sample volume is measured and filtered onto prewetted 0.45 μm (Type HA) gridded membranes (Fisher Scientific, Fair Lawn, NJ). The filter tower is rinsed with sterile PBS, and vacuum aspirated. The membranes are then transferred onto human bifid sorbitol agar (HBSA). The plates are allowed to remain exposed to ambient air conditions for only up to two hours. Inoculated plates are incubated inverted anaerobically (BBL GasPak Jar Systems; Fisher Scientific, Fair Lawn, NJ) at 35-37°C for 48 ± 2 hours. After incubation, the plates are counted for yellow or mustard colored, domed and mucoid colonies which are considered sorbitol-fermenting *Bifidobacteria* presumptive positives. A representative number of presumptive positive colonies are transferred within two hours onto reinforced clostridial agar (RCA-S) with 10 g/L sorbitol added for confirmation testing. Appropriate confirmation reactions are as follows. Isolates must ferment glucose, lactose, and sorbitol to produce acid without producing gas. The isolates must be strict anaerobes that do not reduce nitrate or peroxide (catalase negative). The isolates must be non-motile and demonstrate variable Gram stain reaction. This test requires six days for completion of the confirmed tests.

R. coprophilus was enumerated using a membrane concentration-resuspension, spread-plating method developed in the UMass laboratories (Arango and Long, 1998; Long, 1998). A volume of sample is measured and filtered through a 0.2 μm (Type GTTP) membrane. The membrane is transferred to a tube containing a 0.1% sodium pyrophosphate Bennett's broth solution to release sorbed bacteria. The suspension is heat-treated by immersing for 6 minutes in a 55°C waterbath to kill a majority of background bacteria. Replicate 100 μL aliquots of each appropriate dilution are spread-plated onto MM3 media, selective for actinomycetes. Ten replicate spread-plates for each volume filtered are used. Once all plates are spread, the plates are incubated inverted at 30°C for 10 to 14 days in a humidified incubator. The plates are then removed and placed upright at room temperature (20 to 28°C depending on the season) and exposed to light for at least 3 days. The plates are observed daily, and when positive controls inoculated with a type strain of *R. coprophilus* (NRRL-B 16537) demonstrate appropriate colony development and pigmentation, the plates are enumerated. After exposure to light, orange and stellate colonies without aerial hyphae are determined to be *R. coprophilus*. *R. coprophilus* colonies tend to develop less pigment when present on crowded plates. For equivocal colony identification, the colony will be transferred onto Bennett's agar, four quadrant streaked, incubated, and observed for confirmation as *R. coprophilus*. In order to alleviate the burden due to the number of analyses that must be completed in a short period of time, only a finite number of representative presumptive colonies will be confirmed rather than all colonies on the presumptive plate during events.

Coliphages were enumerated using membrane concentration-elution method, and plated using the double agar layer method developed by Sobsey *et al.* (1990 and 1995) and adapted for use in the UMass laboratories (Long, 1998). A 4M magnesium chloride solution is added to a volume of water sample. The sample is gravity filtered through a prewetted 0.45 μm (Type HA) membrane, and vacuum aspirated to remove any remaining water. The filter is eluted into a 50 mL centrifuge tube using a beef extract-glycine-Tween 80 eluent. The eluent is then mixed with an aliquot of log phase *E. coli* host (CN13 for somatic phages and Famp for F-specific phages) for 15 to 20 minute to facilitate infection. The eluent/host solution is then plated using the double agar layer method and the plates are incubated inverted at 37°C for 22 to 24 hours. Infected host cells are lysed by the phages and create a clearing or plaque in the host lawn, and results are reported as plaque forming units.

Clostridium perfringens spores were analyzed using membrane filtration with growth on m-CP agar. Although there is no standard method for analysis of C. *perfringens* spores, the accepted method is as follows (Bisson and Cabelli, 1979). The samples will be heat shocked to stimulate spore germination, the sample will be filtered and incubated on m-CP agar anaerobically at 45°C for 18 to 24 hours. After incubation, the plates are exposed to ammonium hydroxide fumes and appropriately colored colonies, red to dark pink, are enumerated.

STORM-EVENT PREPARATION AND MOBILIZATION

Event Criteria

A variety of conditions had to be met prior to an event in order to sample it successfully. Previous dry periods, properly functioning field equipment, successful preparation of sampling equipment, and coordination of lab facilities and sampling teams were all required. Security authorities, such as DCR rangers, local and state police in addition to landowners also required notification prior to an event.

Rainfall Depth

The target rainfall depth for sampling events was between 6 and 19 mm (0.25 and 0.75 inches). Senus (2002) defined the median event depths for events in central Massachusetts with 1 to 2 hours of intermittent no precipitation as 15 and 18 mm (0.6 and 0.7 inches), respectively. These accumulations were less than the 2-year return period, 1-hour precipitation accumulation of 27 mm (1.05 inches) for central Massachusetts (McKay and Wilks, 1995). Predictions from numerous weather forecasters were monitored as the initial steps in planning an event. Typically, the Hydrologic Prediction Center was consulted to estimate the amount of rain expected. If the amount was greater than 6mm (0.25 inches), further preparations were performed. However, target events that were sampled often resulted in less precipitation than predicted. As the predicted storm systems approached, NEXRAD radar data and satellite images were consulted routinely to determine probability of rain over the sampling sites. In some cases, storm systems broke up or materialized in greater strength a short distance from the sampling sites. This resulted in some events with low rainfall accumulations; analyses were conducted for these events due to the already committed investment by field and laboratory personnel

Length of Storm

Event sampling was physically and mentally demanding and storm duration was an important consideration. Four hour rain events were preferable to sample. A minimum of sixteen grab samples were required to be collected, yielding a 15 minute sample schedule over a four hour event and 15 minute and thirty minute schedule over a six hour event. As the duration of the storm increased, the event became increasingly complicated to manage. The samplers' discretion was used to collect fifteen, thirty, sixty, ninety, or two hour samples depending on the rain intensity. The overall goal was to sample over the rise, peak, and fall of the storm hydrograph with a minimum of sixteen samples collected. Often more than sixteen samples were collected and stored in coolers until the end of the event. The goal of sixteen samples per event was met by selecting from the collected samples to include the entire hydrograph. This method was preferred to sampling the minimum number and risk missing sampling over the peak of the hydrograph. Samples were selected to capture periods of intense precipitation and hydrograph rise and the hydrograph peak. In addition, samples were selected to provide a fairly uniform temporal distribution. Coverage of the falling limb of the hydrograph was often limited as duration of the falling limb often exceeded feasible sampling time. Feasibility was based on the limitations of sample hold times collected at the beginning of the event and sampling personnel coverage. For long events where coverage was available, bulk samples were split into aliquots at intermediate points of the event.

Storm length varied seasonally. Summer thunderstorms were short and intense, however, they were also very unpredictable. Frontal systems, common in the fall and spring, were easier to predict but were typically longer in length. Winter storms typically resulted in snow or mixtures of snow and rain, which often occurred over one or more days resulting in an excessively long sampling plan. Winter conditions were also more extreme, and required more personnel in shorter work shifts.

Antecedent Conditions

A dry period of three to five days was optimal prior to sampling an event. This ensured nutrients, solids, and microbial concentrations were not influenced by a previous event and likely represented baseline conditions. To ensure overland flow occurred, the ground either needed to be partially saturated or the storm predicted to be of a sufficiently high intensity or large accumulation to saturate the ground and produce significant runoff. In addition, it was necessary for the monitoring equipment to be submerged in the stream and to have been calibrated within the previous three weeks.

Event Mobilization

Equipment Preparation

Prior to an event, bottles were cleaned and autoclaved and sampling supplies were restocked. A pre event protocol was developed to ensure proper equipment preparation. In addition, a pre event equipment checklist was completed to ensure all necessary items were available. A minimum of thirty-two bottles, 16-2 Liter and 16-4.3 Liter, were washed, autoclaved, and distributed equally into two sets of sampling coolers. In addition, sixteen 250 ml bottles were washed and autoclaved and evenly distributed into two small coolers designated for microbial analysis aliquots. Two splitting kits were stocked with the necessary equipment to successfully split event samples and ensure quality data.

The MWRA supplied aliquot bottles for nutrient and total suspended solids (TSS) analyses, coolers, ice packs, and chains designated to each study site by sample ID. Table 4-13 lists the splitting bottles with corresponding id's, volumes, and preservatives for the nutrient and SS analyses. In addition, Prescott sampling sites were identified as MD61 (for MBDB) and MD62 (for GRB) to conform to other MWRA sampling location identifiers at Prescott. The designations used herein were used by MWRA to identify the agricultural (BBFM and BBWR) and residential (GBBR and GBWD) sampling locations.

The *Cryptosporidium-Giardia* (CG) filter sample kits required to perform the CG analysis included a SHURflo Model 8002-793-238 self-priming 1.7 GPM pump, a discharge hose, a screened intake hose, filter housing, Filta-Max® filters, and a 12-volt sealed Marine deep-cycle battery. The battery was required to be fully charged to 12.8 volts before an event to ensure proper flow rate was achieved when pumping. In addition to the kits, two clean 25 L Nalgene carboys were required. One of these was filled with at least 10 L of Super Q water which was used to rinse the filter apparatus after testing. The second was used to discharge the filter discharge to monitor the volume filtered (volume passed through filter). The rinse water was filtered through the pump, lines, and housing before a CG filter run was performed. Site-specific sampling totes, containing auxiliary gear for sampling personnel, were restocked prior to every event.

Sampling Personnel Preparation

Event sampling required personnel to work long hours in adverse conditions and was physically demanding. For these reasons, personnel were often rotated in shifts. However, rotating personnel required contacting samplers in advance to set up schedules and transportation plans. Samplers with the greatest degree of experience and expertise were often required to remain at the site through the entire event to supervise.

Organization of sampling teams and rotating personnel required good communication. Typically, a detailed forecast compiled from numerous web sites was sent out Monday morning to all sampling personnel. If a tentative sampling event was in the forecast, instructions were sent to the samplers to promptly respond with their availability and transportation needs. A detailed list of personnel and their available times was compiled and a rough sample plan was prepared identifying personnel, coverage times, and assigned locations.

Transportation was often limited due to sampler's reluctance (mud, water, and reliance issues) to use personal vehicles. Thus transportation issues often dictated the format of the sampling plan. Transportation was most restrictive at the Prescott (wildlife) sampling locations. The GRB study site was located on a three season access road and required the use of an off road vehicle throughout much of the year due to extreme road conditions. Sampling personnel were able to access MBDB using the University cargo van or personnel non-off road vehicle, however, only in the summer and early fall. Typically, a DCR vehicle was supplied or a cargo van was rented. This enabled some flexibility in the sampling plan and aided in the communication problems associated with sampling the paired watersheds (no cell phone or two-way radio coverage). When the University cargo van was not available, one was rented. For some events multiple vehicles were rented to facilitate transportation.

Table 4-13
Aliquot bottle ID and parameter matrix

Container ID	Sample Volume	Preservative	MWRA Test Code	Description
P1L-A	1000 mL	H_2SO_4	TKNLAQBDA	Total Kjeldahl Nitrogen
		H_2SO_4	TPL-AQADA	Total Phosphorus
P500	500 mL	NONE	TSS-AQADA	Total Suspended Solids
P60F	60 mL	Filter and Freeze	NO2FAQAAN	Nitrite
		Filter and Freeze	NO3FAQAAN	Nitrate
		Filter and Freeze	NOCFAQAAN	Nitrite/Nitrate

Laboratory Preparation

The laboratories included in post event analysis required advanced notification of event sampling. The feasibility of sampling an event often was limited by laboratory schedules and sample hold times.

The UMass and DCR laboratories performed microbial analyses and required forty-eight and thirty-six hours advanced notification, respectively, to anticipate the delivery of samples and prepare media used in the analyses. An event could be cancelled after notifying UMass or DCR; however, this could result in materials being wasted. The MWRA Deer Island lab performed nutrient analyses and required twenty-four hours of notification to ensure sample delivery and preservation. Environmental Associates Laboratory (EA labs) located in Ithaca, New York performed analyses on the CG filters and required 24 hours of advanced notification.

Sample hold times and shipment procedures also limited the feasibility of sampling an event. Table 4-14 lists samples with their corresponding hold and shipment times. Also listed in Table 4-14 are the days that were thus feasible to sample. The microbial hold times of twenty-four hours were within USEPA guidelines. However, USEPA preferred a six-hour hold time, which was often not feasible due to the duration of sampling events.

The microbial and CG samples analyzed by DCR and EA labs, respectively, were the most limiting of the post event analyses with respect to sample hold times. This constraint was occasionally eased when laboratory personnel worked on Saturday or Sunday. This allowed samples to be collected at any time throughout the week (Sunday 8AM through Sunday 8AM). This condition required previous approval from lab managers of both facilities.

Table 4-14
Sample hold and shipment times and sampling feasibility

Sample ID	Description	Hold Time (incl. Ship.)	Shipment Time	Feasible Sample collection Times
CG	6 (3 per site)	4 days	24 hours	Friday 8 AM – Thursday 12 PM
MWRA Nutrient	32 (16 per site)	4 days	24 hours	Sunday 8AM – Friday 7 AM
DCR Microbial	16 (8 per site)	24 hours	< 8 hours	Sunday 8 AM – Thursday 10 AM
UMASS Microbial	6 (3 per site)	24 hours	< 8 hours	Sunday 8 AM – Sunday 8 AM

Site to Mobilize and Feasibility

The final decision to field an event was made after all the preliminary conditions were met and sampling personnel, laboratory staff, and security organizations were organized and notified. Upon said decision, the equipment was transported to the selected sampling sites. The event was executed with the goal to collect the required number of samples over the complete storm hydrograph.

A preliminary sampling plan of sample collection times was created among the samplers prior to departing to the sampling sites. Included in this plan were meeting times and locations when samplers would communicate and tailor the sampling scheme to the actual storm. Infield forecast updates were limited to radio news broadcasts and by cellular phone communication, when available, to contact personnel with online weather radar and forecast access. The forecasts were incorporated into the sampling plan as part of the tailoring process.

SAMPLE COLLECTION AND HANDLING

As noted previously, prior to each sampling event (dry or wet), bulk sample bottles, fecal/total coliform aliquot bottles, and filtering equipment were washed thoroughly with suitable non-phosphate detergent and hot water, rinsed thoroughly with hot water, and autoclaved at 121°C for 15 minutes for sterilization. Plastic screw-cap closures were loosely fastened when autoclaving and fully tightened following cooling to prevent bottle deformation. Autoclave tape was used to attach the lids/bottles and thus identify properly cleaned bottles.

Due to the narrow/shallow nature of the streams included in this study, composite depth-integrated sampling was neither feasible nor necessary. Bulk samples were thus collected as manual grab samples in pre-sterilized wide-mouth polypropylene large volume (2 or 4.3 L) sampling bottles without pre-rinsing. Open bottle sampling is considered to be representative of average cross-section conditions for shallow streams with velocity less than 2.0 ft/s and uniform distribution of the constituent in the water column (Shelton, 1994). At each sampling transect, a site appropriate for the collection of open bottle grab samples was thus identified. This location was typically located a short distance upstream of the effects of the weir pool. A stream section where flow was consistently well mixed and characterized by uniform flow was selected in order to minimize problems caused by poorly mixed flows (Shelton, 1994). The selected sections were also deep enough to allow for immersion of the bottle or consisted of a free fall under which the sample bottle could be held. Automated, flow-weighted sample collection was not feasible due to monetary constraints, concerns over sample integrity (microbial analyses), the need for large volumes (microbial analyses and CG), and the desire to filter water pulled directly from the stream (CG analyses) to avoid settling issues.

Bulk sampling bottles were kept closed until they were to be filled. Originally samples were to be collected using a fresh pair of sterile gloves treated with rubbing alcohol to "sterilize" after donning and immediately prior to sampling. This proved to be infeasible given the field conditions during wet weather. Instead, particular care was taken to avoid contamination of the bottle lip, inner lid, and inner bottle. To fill, bottles were grasped near the base (2L) or by the molded handle (4.3L) and placed in the stream with the bottle mouth pointing upstream so that any hand or bottle contamination was washed away from the bottle. Keeping the mouth of the bottle under the surface, the cap was removed but not set down, the bottle allowed to fill, and the bottle was recapped. Approximately one-inch of airspace was left at the top to facilitate mixing by shaking. After filling, the bulk sample bottles were labeled with the site, date, and time then stored in coolers with ice bottles (not loose ice) and/or blue packs. At the end of the event or during shift changes, the coolers were transferred to an appropriate facility to enable splitting of the aliquots for analysis.

A SHURflo Model 8002-793-238 self-priming 1.7 GPM pump, a discharge hose, a screened intake hose, a filter housing fitted with a Filter-Max® filter, and a 12-volt sealed Marine deep-cycle battery were used for collecting *Cryptosporidium* and *Giardia* samples. Prior to sample collection, the empty hose-filter housing setup was rinsed by pumping 10L of Super Q through it. The filter housing was then opened and a clean Filter-Max® filter was inserted. The screened intake hose was then placed in the stream (but not on the stream bottom), the outlet hose was placed in an empty carboy, and the appropriate volume of water was pumped through the filter. Originally a goal of 50 L was set, as this is the volume of water that each filter can sample. However, we encountered filter clogging issues due to sediment loads in our tributary watersheds. We were able to filter 50 L during one event at the wildlife sampling sites, but were never able to filter this large of a volume at either the agricultural or residential sites. Based on initial field experience, we modified our sampling strategy to filter 20 L of water at the wildlife sampling locations and 10 L of water at the agricultural and residential sampling locations.

Originally flow regulators were used to insure that the filter rate did not exceed 2 L per minute. In storm-event field conditions, the regulators resulted in too much head loss to facilitate filtering within a reasonable amount of time. The initiation and completion of sample filtering was instead recorded, and these were checked to make sure they were not less than five (10 L sample) or ten (20 L sample) minutes.

Filters to be analyzed were removed from the filter holders, placed in zip-lock bags with a small volume of stream water to prevent drying, and then placed in individual Tupperware containers labeled with the filter serial number, date, time, and location of sample collection. These were placed on ice in small coolers and shipped overnight to the Environmental Associates Laboratory in Ithaca, NY.

Matrix spikes were conducted at the rate of 1 in 20 samples. Water for these samples was collected from the stream in disposable carboys and shipped overnight on ice to the lab, where it was spiked and filtered.

Pre-sterilized and prepared (in terms of preservative) aliquot bottles for total phosphorus, nitrate-nitrite, total Kjeldahl-nitrogen, and total suspended solids were provided by the Massachusetts Water Resources Authority (MWRA) as previously noted. Acid washed bottles for UV254 aliquots and pre-sterilized bottles for microbial aliquots were prepared by UMass. Bulk samples were split into appropriate aliquot bottles after transfer to an appropriate facility, typically the laboratory. During some events, streamside splitting was necessary. When streamside splitting occurred, splitting was done under a rain tarp to avoid potential contamination from rain or atmospheric deposition. Prior to splitting, the bulk samples were vigorously mixed by inverting ten times. Upon splitting, labels were affixed to all aliquot bottles denoting the sample location, date, time, preservative method, and analyses to be performed. Aliquots were then placed in clean coolers with blue packs, or frozen as necessary, and transferred to the appropriate laboratory for analysis. Laboratory analysis and preservation methods were provided above.

DISCHARGE

Weir Design and Installation

Sharp crested weirs were installed at four of the sampling sites to calculate discharge from continuously monitored stage data. A sharp crested weir is a simple device used for discharge measurement (Roberson *et al*, 1998). A detailed survey of the channel cross-section formed the basis of the weir design process. At GRB, the weir was installed approximately 1 km downstream from the beaver colony. This weir site was selected for several reasons: (1) reasonable access from a three-season road; (2) a confined channel with sufficient depth for a weir and stilling basin; and (3) enough distance from the beaver colony to avoid damage or disturbance. Similar reasoning was used to determine the placement of the BBWR and BBFM weirs; the best hydraulically available cross-sections with close proximity to the road while maximizing distance between the two sites were selected. Each weir was designed according to the stream's physical characteristics and a preliminary peak discharge estimate.

Hydraulic and structural guidelines from King and Brater (1963) and Gordon et al. (1992) were used. Marine plywood panels were cut to match the stream cross-section and weir design. A steel plate was cut and sharpened to a knife edge for the main overflow; steel channel was fabricated to top the overflow section. In the field, the plywood was anchored to the streambed with 5/8" reinforcing rod and fiberglass bracing. A 45-mil rubber pond liner was affixed to the plywood, covered with the steel plate and channel, and formed to the streambed. Sandbags were used to weigh the liner down and eliminate leakage beneath the weir. Additional weir construction details are available in Bourdeau (2004).

Weirs at MBDB and GRB were designed as 120° V-notches with rectangular overflows. The cross section was used to set the configuration of the bottom edge of the weir section. The V-notch configuration was designed to accommodate a portion of the estimated peak discharge while a rectangular weir section above the V-notch section was designed to pass the remainder of the peak discharge. The peak discharge was estimated using a 24-hour, 50 year recurrence interval event assuming 100 percent runoff generated from the watershed area. Total precipitation accumulation for such an event in the Quabbin area is 5.5 inches (Wilks and Cember, 1993). The estimated peak discharge for the MBDB and GRB weirs were 65 cfs and 45 cfs, respectively (Bourdeau, 2004). The GRB weir was configured as a 120-degree sharp crested V-notch with a maximum flow (Q) of 4.33 cfs. A rectangular (1' x 24') cross section above the V-notch weir was designed to handle the remaining portion (90 percent) of the peak discharge. The configuration of the MBDB weir was also a 120-degree sharp crested V-notch weir with a maximum depth of flow over the weir crest of 1.0 foot and a corresponding maximum discharge of 4.33 cfs. The rectangular overflow section was designed to accommodate the remaining 94 percent of the peak discharge.

BBFM and BBWR were designed as trapezoidal weirs and overflow sections. The trapezoidal configuration was used to accommodate sufficient runoff through the cross-section without overtopping during high flows. Significantly higher runoff response was estimated for these sites compared to the Prescott Peninsula sites. As a result, sensitivity at low flows was sacrificed to insure that neither flooding of the road immediately downstream of the BBWR site nor of the cow pasture at the BBFM site occurred due to installation of the weirs.

The capacity of the upstream weir (BBWR) is sufficient to accommodate 2.5 – 3.4 inches of precipitation or snowmelt over a 24-hour period, equivalent to stream discharge on the order of 80 to 180 cfs. This capacity is sufficient for two to five year return interval events characteristic of the region. The capacity of the downstream weir (BBFM) is sufficient to accommodate 2.5 inches of precipitation of snowmelt over a 24-hour period, equivalent to stream discharges on the order of 85 cfs. This capacity is sufficient for 2-year return interval events characteristic of the region. Considerable attenuation of flow peaks between the sites is expected due to the beaver ponds.

Discharge Calculations

Pre-calibrated weir rating equations were used to convert depth of flow over the weir to discharge. A sharp crested triangular or V-notch weir is used for low flow conditions and has an opening (θ°) ranging between 40 and 120 degrees (French, 1985). Fluid depth above the V-notch, H, can be used to generate discharge by Equation 4.1.

$$Q = \frac{8}{15} K \sqrt{2g} \tan(\frac{\theta}{2}) H^{\frac{5}{2}} \qquad (4.1)$$

where K = flow coefficient that is a function of head
g = acceleration due to gravity (32.2 ft/s^2)
H = depth of flow above the weir crest
Q = discharge in cubic feet per second
θ = V-notch angle

The V-notch weir at the MBDB and GRB was constructed with a θ value of 120 degrees. French (1985) recommends 0.585 as the K value, resulting in Equation 4.2.

$$Q = 4.33 * H^{\frac{5}{2}} \qquad (4.2)$$

For the trapezoidal weirs at BBFM and BBWR, discharge was generated using Equation 4.3. from French (1985).

$$Q = \frac{2}{3} C_e \sqrt{2g} \left(b + \frac{4}{5} H \tan\left(\frac{1}{2} \theta \right) H^{1.5} \right) \qquad (4.3)$$

where b = length of the weir crest

θ = angle of the trapezoid

C_e = the weir coefficient, which varies according to Equation 4.4.

$$C_e = \beta_1 + \beta_2 \frac{H}{P} \qquad (4.4)$$

where β_1, β_2 = coefficients which are a function of the ratio of the length of the weir crest to the width of the weir

P = distance from the streambed to the crest of the weir.

The accepted accuracy of these structures is ± 1% (Henderson 1966; King and Brater 1963; Vennard and Street 1982). Field measurements were used to verify the accuracy of these equations for determining discharge at our sites.

At the two residential sampling locations, GBBR and GBWD, stage through culverts with known cross sectional shapes was monitored. Free-surface flow through the culverts occurred throughout the study period. The upstream site was inlet controlled (free-fall outlet). In addition, outlet controlled conditions appeared to dominate at the downstream location. Velocity measurements were thus taken over a range of flow conditions in order to develop rating curves for each site. Rating curve (stage-discharge relationship) development was carried out by Vincent Vignally, DCR staff hydrologist. A lookup table was used to convert monitored stage to discharge at each site via Excel.

Offset Measurement

Offset measures were collected periodically throughout the study and were measured from the V-notch or trapezoid base of the weir to the water elevation in the upstream pool. The offset measurement was recorded with a corresponding stage reading off the data recorder. The difference between the two measures yielded a constant offset, which represented the depth between the V-notch or trapezoid base and the pressure sensor of the transducer.

Periodic offsets varied due to horizontal shifting of the PS1200, caused by ice or removal and replacement of the probe for calibration or repair, and the weirs. An average offset for each site was established by taking the arithmetic mean of all offset measures collected over a period between known repositions of the transducer. The average value was used to transform continuous stage (depth) data into depth over the weir, which could then be converted into discharge using the empirical equations above.

Velocity Measurement

Open channel measurements were performed in accordance with USGS methods (Carter and Davidian, 1968) and included reach segmentation, channel cross sectioning, and velocity profiling where necessary. Open channel measurements were necessary at GBBR and GBWD to develop rating curves. At BBFM and BBWR, limited open channel measurements were used to estimate the correct parameter values to use in stage based discharge calculations.

Additional open channel measurements were conducted on MBDB and GRB to calculate hydraulic travel times. The locations of the open channel measurements were selected based on changes in topographical slopes interpreted off a USGS 7.5 minute topographical quadrangle. Reaches of nearly equal horizontal distance and topographical slope were established. Each reach was then divided into three equally spaced measurement stations. The stations were labeled on a USGS topographical map, which was used to aid navigation to the field station location for measurements.

Hydraulic travel times were estimated for several travel routes from various points in each watershed to the weir stations. The hydraulic travel times were compared to lag times (the time from start of precipitation to stream response or rise of the hydrograph) and time-to-peak (time from start of precipitation to hydrograph peak) to provide information on where event stormwater likely originated within each watershed.

CHAPTER 5
OVERVIEW OF WATER QUALITY MONITORING RESULTS

COMPARISON OF EVENT AND MONTHLY DATA

In this section, monthly and event data for nitrate/nitrite, total Kjeldahl nitrogen (TKN), total phosphorus (TP), total suspended solids (TSS), total coliform (TC), and fecal coliform (FC) are presented and discussed. The standard naming abbreviations used for the study sites and sampling locations are repeated in Table 5-1 for convenience. Of particular interest is the relative magnitude of concentrations observed during wet weather events versus monthly baseline sampling for nutrients and standard indicator organisms. These analyses were separated from those focused on *Cryptosporidium* and *Giarda*, Chapter 6, as they are of more interest to regulators in charge of setting and attaining total maximum daily loads (TMDLs) for waters impaired by nutrients or bacteria. These analyses may also be of interest to water supply utilities concerned about total organic carbon (TOC), disinfection byproduct production in their treatment facility, and potential watershed management activities to reduce TOC sources. While occurrence and transport of *Cryptosporidium* and *Giarda* are not the focus of this chapter, these factors are incorporated within the general discussion.

Monthly and event concentrations were compared by overlaying the range of values detected during events and the monthly concentration data, Figures 5-1 through 5-36. The monthly data are represented by a continuous line on each figure while the arithmetic mean, maximum and minimum values observed during each storm event are represented by a high low line. It should be noted that when the high low line is above or below the continuous line, it is extended to connect. Thus the hatches on the high low line rather than its extent indicate minimum and maximum values. Only the wet-weather events per site are included; results from the dry weather sampling events will be presented separately. Non-detect values for each parameter are represented as one half of the detection limit to enable plotting.

These plots provide a qualitative understanding of the water quality response of the studied watersheds to wet weather events. They also provide insight into the success of capturing what would be considered "true" wet weather events. Flow criteria were not used to define events due to a lack of historical data for the watersheds. Instead a forecasted precipitation criterion of 0.25 to 0.5 inches was set. Due to the complexity of preparing and mobilizing for events, it was demoralizing to call off an event once field crews were in place, and particularly after sampling was initiated. Although several events were called off both pre- and after-mobilization, some low accumulation events were sampled. We strove to balance the need to fulfill an event quota at each site over a finite amount of time with laboratory and personnel constraints, the vagaries of the weather, and the desire to capture a wide range of events in terms of rainfall magnitude. The figures presented in this section illustrate the wide range of watershed response which was a product of the underlying variability in rainfall accumulation and runoff response of the sampled events.

Table 5-1
Summary of sampling site locations

Influence	Sampling Location	Description	Code
Wildlife	Middle Branch Dickey Brook	Ranging wildlife	MBDB
	Griswold Brook	Beaver influenced	GRB
Agricultural	Burrow Brook Woods Road	Upstream of cow pasture	BBWR
	Burrow Brook Farm	Downstream of cow pasture	BBFM
Residential	Gates Brook Woodland Street	Upstream of residential and commercial area	GBWD
	Gates Brook Brayton Road	Downstream of residential and commercial area	GBBR

Source: Rees *et al.*, 2005. Transport of *Cryptosporidium* and *Giardia* in Massachusetts Watersheds During Storm Events. In *Managing Watersheds for Human and Natural Impacts: Engineering, Ecological, and Economic Challenges*. Copyright ©2005 ASCE. Reproduced with permission.

Nitrate/Nitrite

Nitrate/Nitrite data are presented on Figures 5-1 through 5-6. Of the 40 wet-weather events represented on the six figures (six to seven events per sampling location), 24 of the nitrate/nitrite concentration high-low lines (which represent the range between minimum and maximum concentration observed during the event) encompass the baseline sample for that month. For eight of the events [three at Burrow Brook (BB) and five at Prescott (PR)] the range of event samples are all greater than the baseline sample for the month during which the event occurred, while for eight of the events (seven at BB and one at PR) the range of event samples are all almost equivalent or less than the baseline sample for that month. Fewer data are available at the PR sites because ammonia was monitored rather than nitrite/nitrate until May 2002.

Dilution, relative to routine sampling, of nitrate concentrations during high flows has often been reported in the literature but, as noted above, was not a consistent feature in our data set. The variability of event data with respect to monthly data (for this study) is likely largely due to variation in the magnitude of precipitation and resulting runoff for each event, as one aim was to monitor transport over a wide range of events identified by forecast precipitation. There does not appear to be any bias with respect to sampling location or season and this relative magnitude.

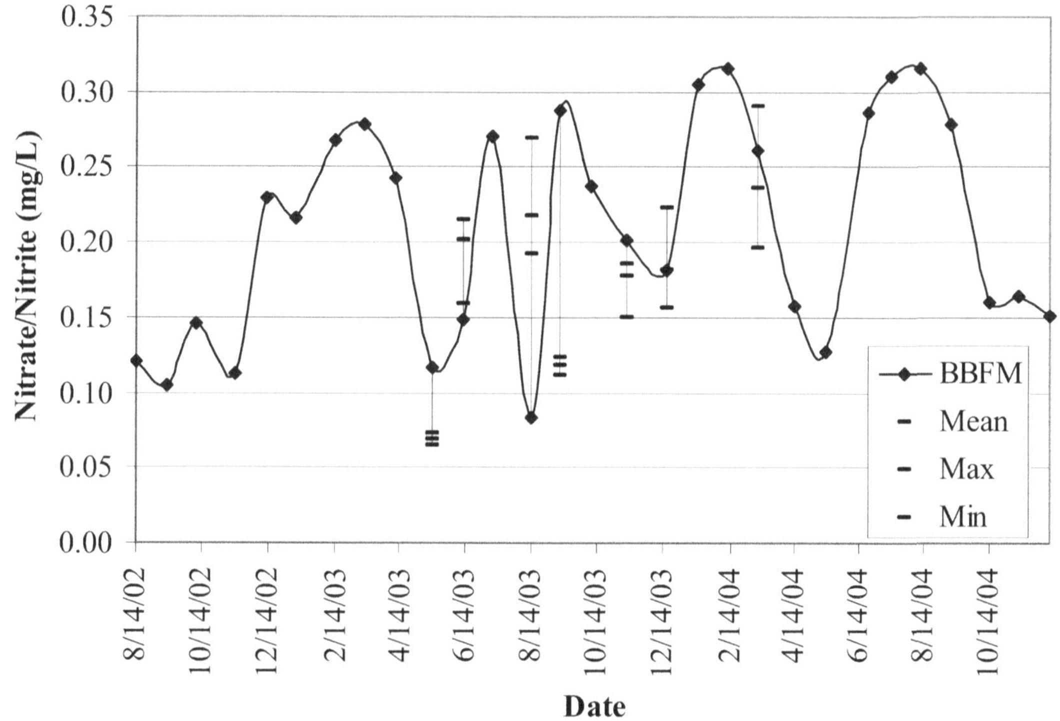

Figure 5-1 Monthly and Event Nitrate/Nitrite Data for BBFM

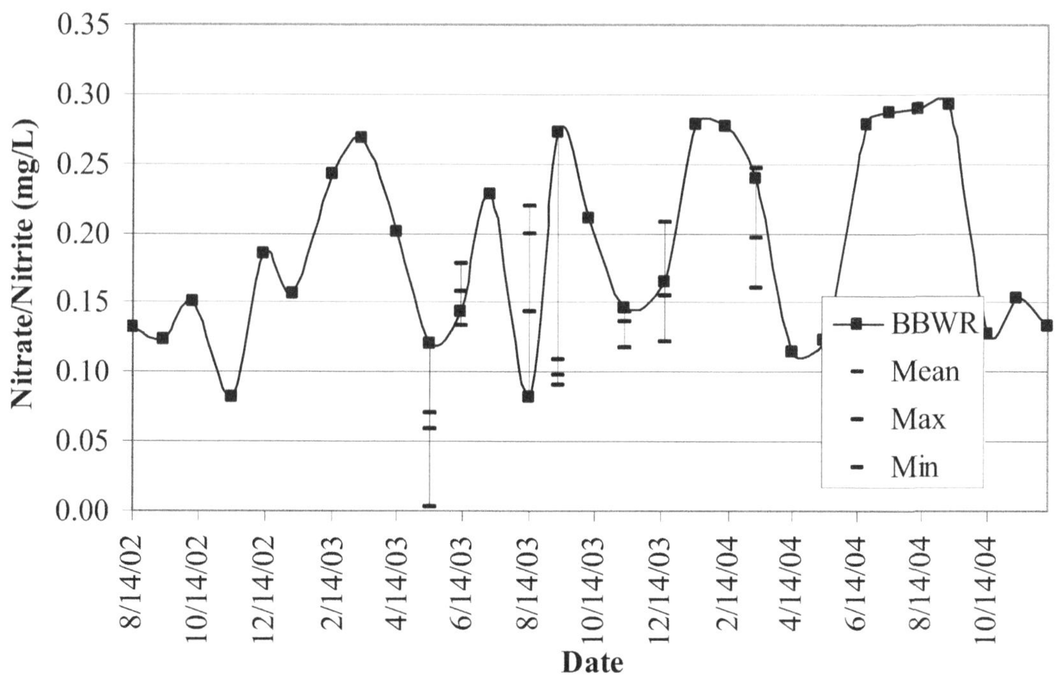

Figure 5-2 Monthly and Event Nitrate/Nitrite Data for BBWR

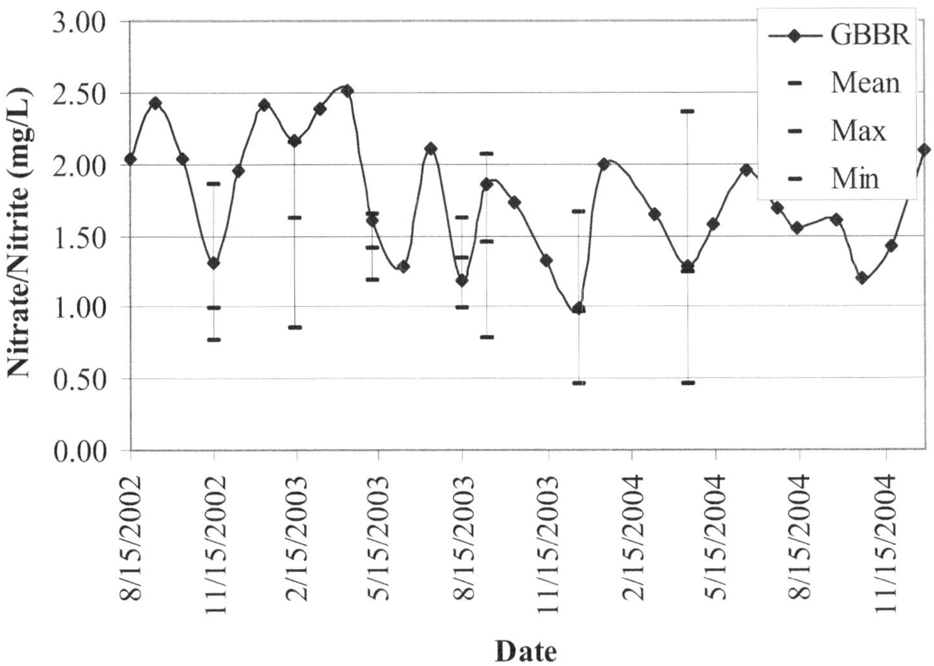

Figure 5-3 Monthly and Event Nitrate/Nitrite Data for GBBR

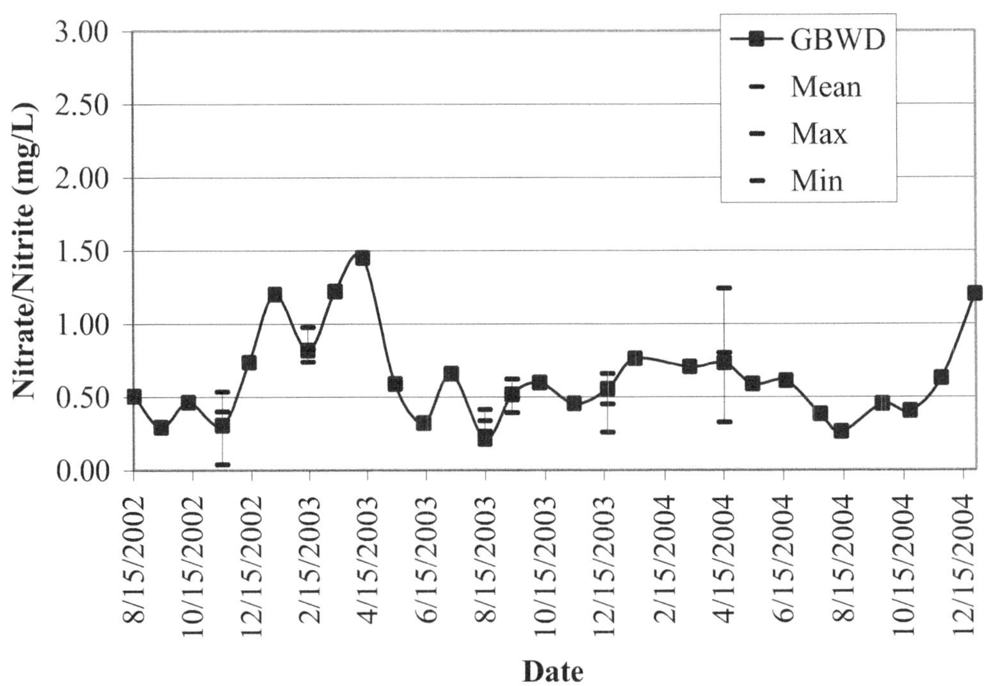

Figure 5-4 Monthly and Event Nitrate/Nitrite Data for GBWD

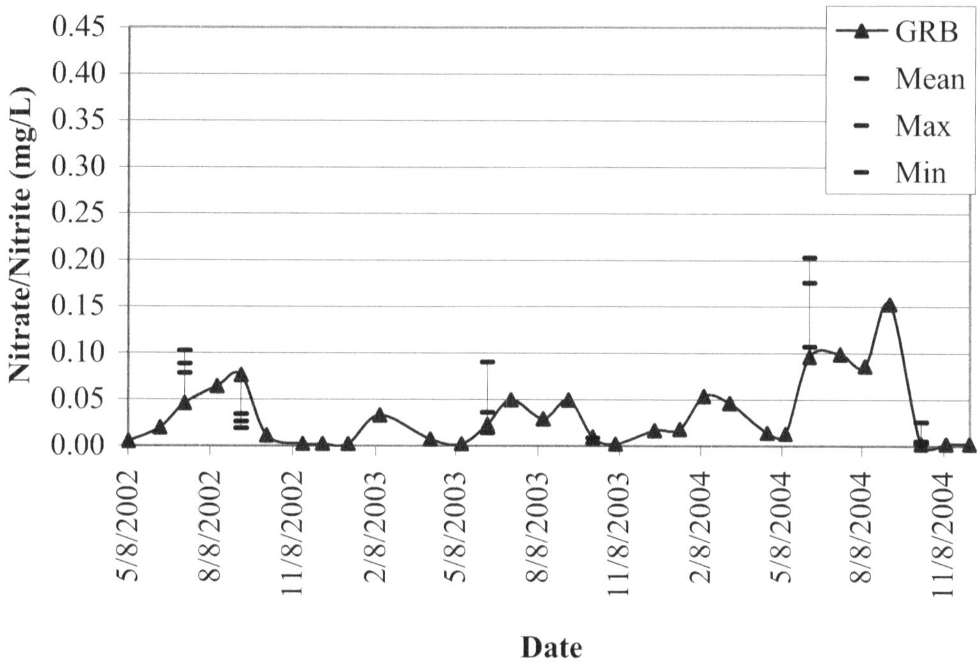

Figure 5-5 Monthly and Event Nitrate/Nitrite Data for GRB

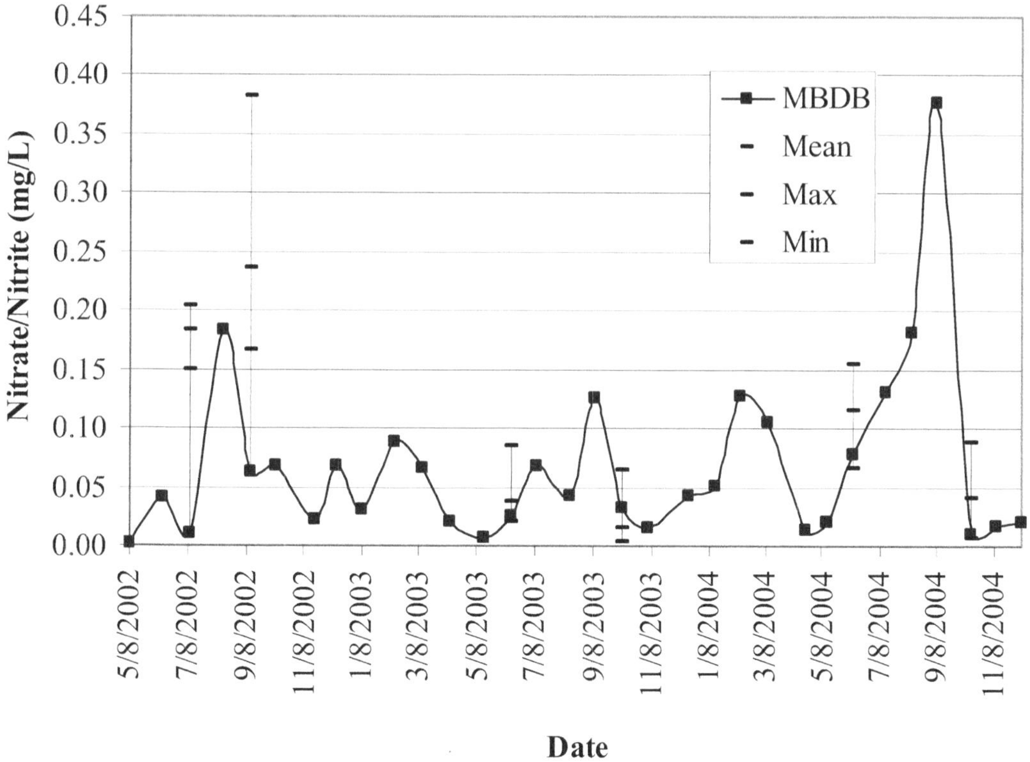

Figure 5-6 Monthly and Event Nitrate/Nitrite Data for MBDB

Total Kjeldahl Nitrogen

Total Kjeldahl Nitrogen data are presented on Figures 5-7 through 5-12. Of the 42 wet-weather events presented on the six figures (seven events per each of the six sampling loctions), 15 of the corresponding high-low lines encompass the baseline sample for that month, although the mean and maximum event values are typically larger. For 26 events (six at BB, eight at Gates Brook, GB,, and twelve at PR) the high low line lies almost entirely above the baseline sample for that month. In August of 2003, the monthly sample for TKN at GBWD (14-August-03) exceeded the highest event concentration for TKN, collected two weeks before on 1-Aug-03. The precipitation accumulation for the 1-August-03 events was fairly low, with approximately 8 mm (0.4 inches) of rain recorded at GBBR and six-tenths of an inch recorded further south at the Worcester Airport NCDC gauge. Two days prior to the monthly sample, however, over 25 mm of rain (more than 1 inch) fell over the entire region, as measured at both GBBR and Worcester. The higher than event TKN concentration associated with this monthly sample were likely due to the preceding rainfall. Increases in TKN concentration during high flows were expected based on literature values. Such increases are typically associated either with overland flow washing organic matter into the stream or with flushing of wetland and riparian areas rich in organics.

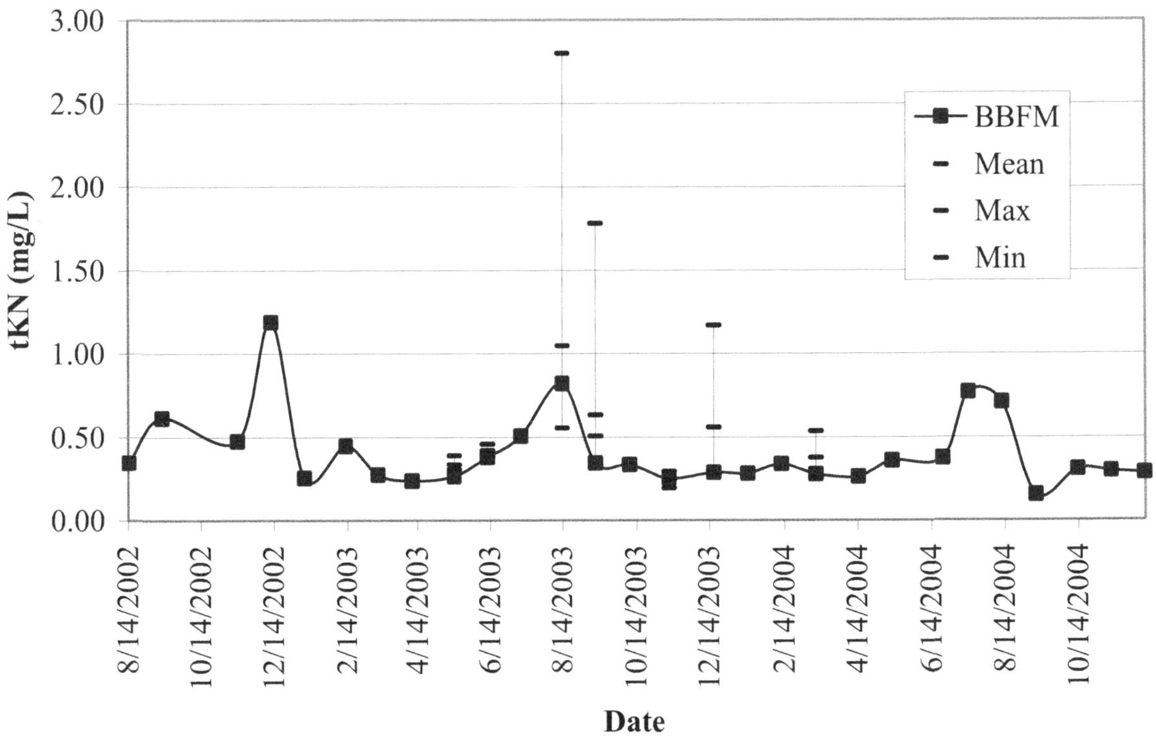

Figure 5-7 Monthly and Event Total Kjeldahl Nitrogen Data for BBFM

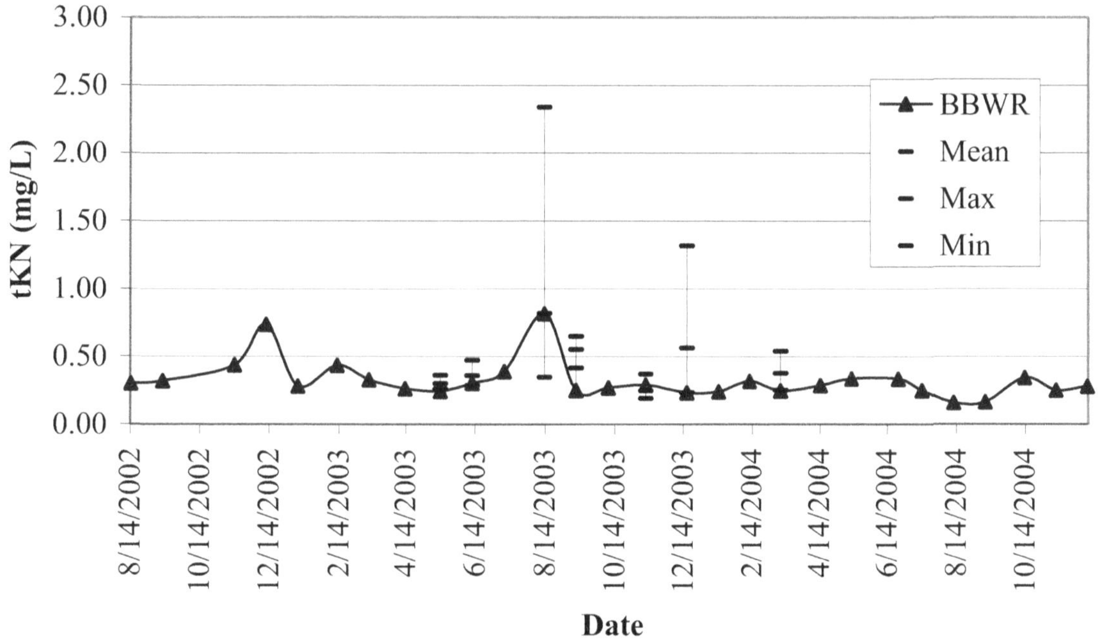

Figure 5-8 Monthly and Event Total Kjeldahl Nitrogen Data for BBWR

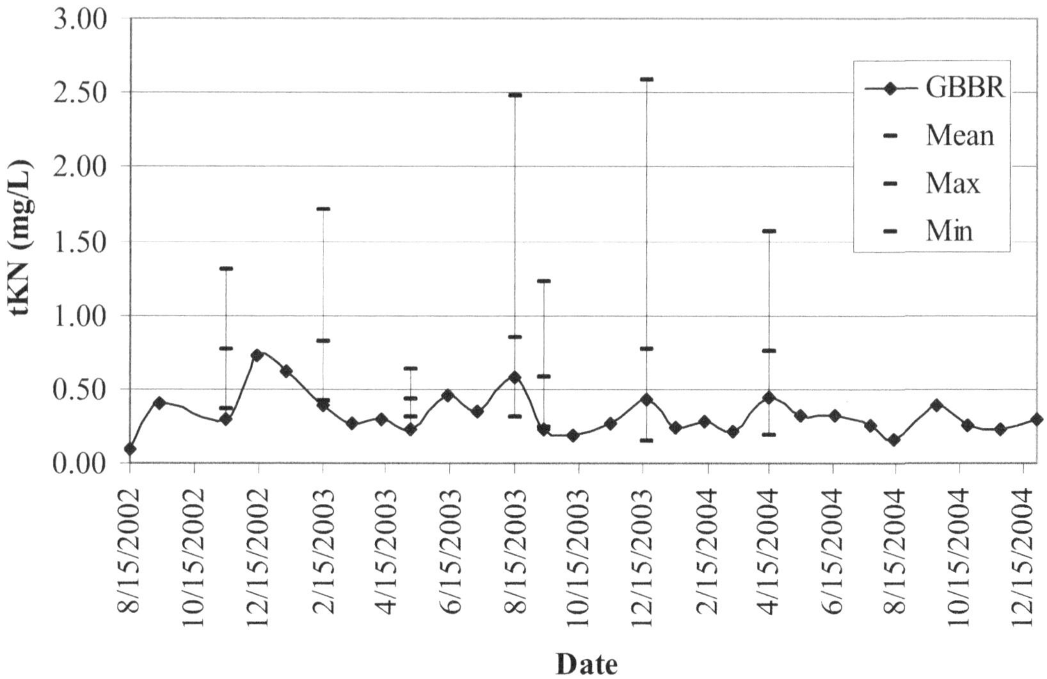

Figure 5-9 Monthly and Event Total Kjeldahl Nitrogen Data for GBBR

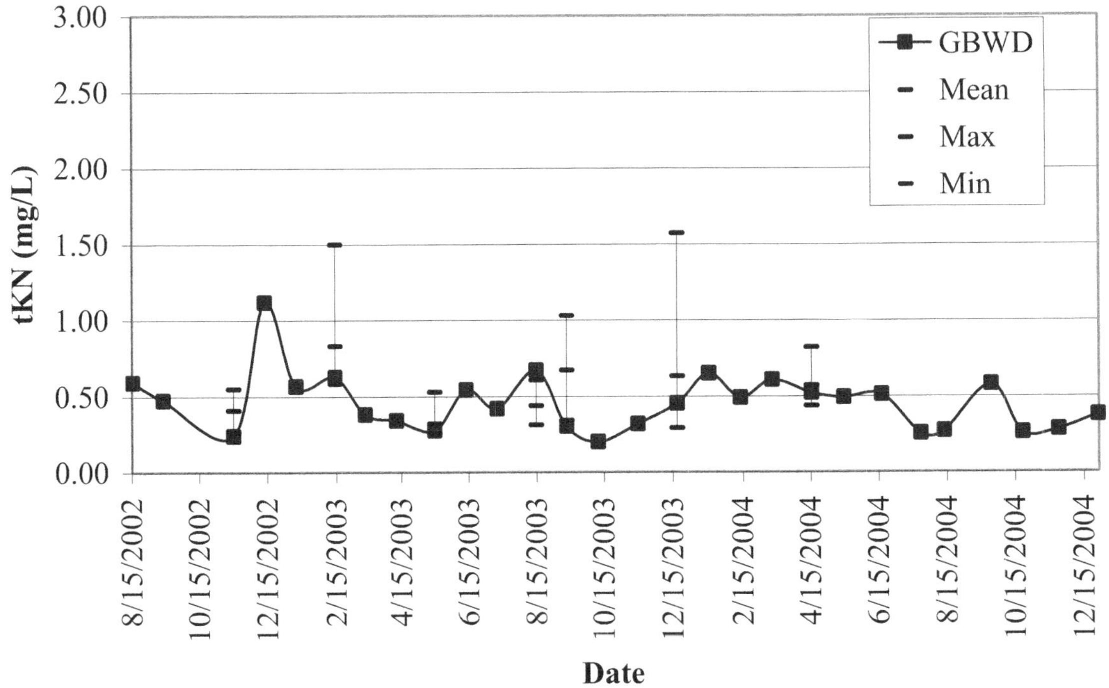

Figure 5-10 Monthly and Event Total Kjeldahl Nitrogen Data for GBWD

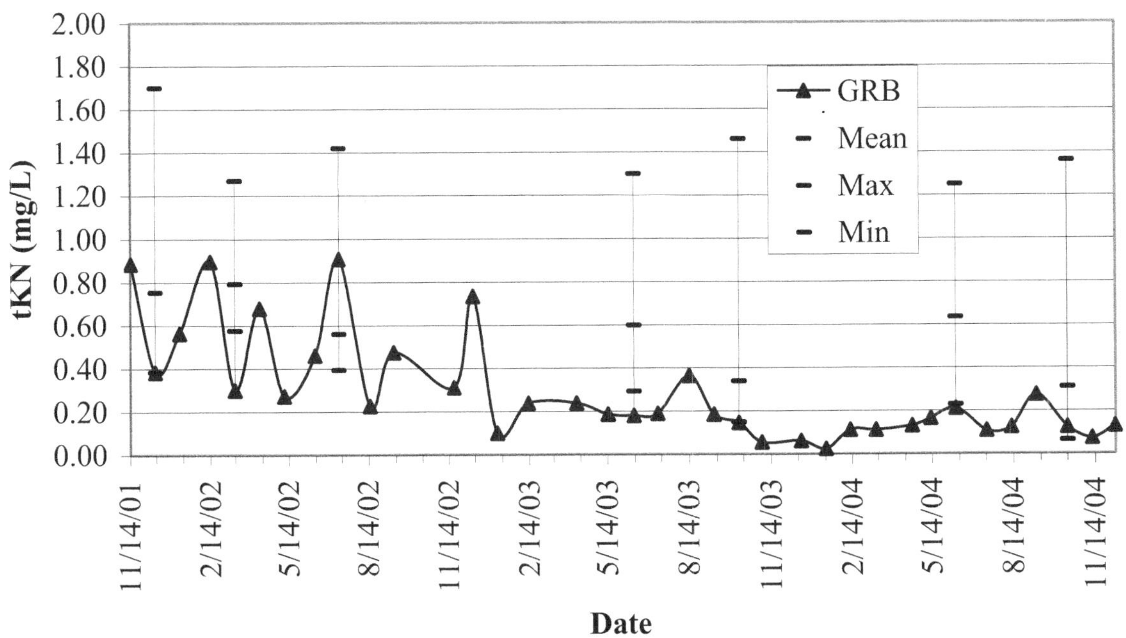

Figure 5-11 Monthly and Event Total Kjeldahl Nitrogen Data for GRB

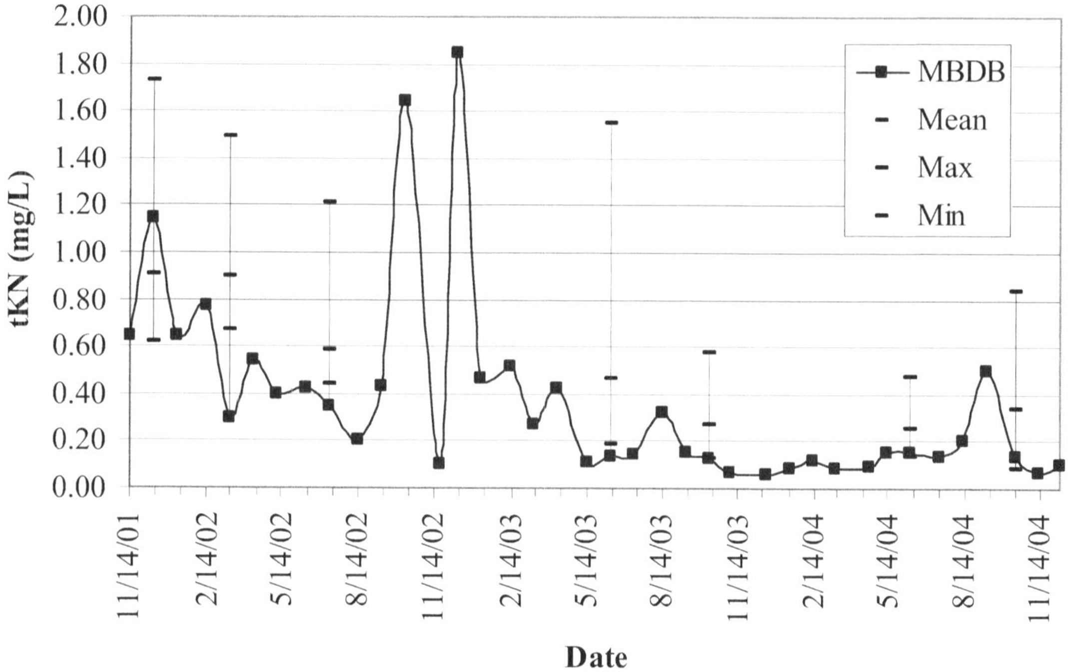

Figure 5-12 Monthly and Event Total Kjeldahl Nitrogen Data for MBDB

Total Phosphorus

Total Phosphorus (TP) data are presented on Figures 5-13 through 5-18. Of the 42 events presented on the six figures (seven events per each of the six sampling locations), 7 of the corresponding high-low lines encompass the baseline sample for that month. For 34 events (ten at BB, thirteen at GB and eleven at PR) the high-low line lies almost entirely above the baseline sample for that month. Higher event concentrations (compared to monthly) were expected based on literature values and the affinity of phosphorus for suspended sediment, which tends to be transported in greater quantities during storm events. The monthly concentration observed at GRB on 10-July-02 was greater than the highest concentration observed during an event occurring approximately two weeks later on 28-July-04. This appears to be due to rainfall, as over 6 mm (.25 inches) of rain fell the day prior to monthly sampling but a total of only 2 mm rainfall occurred during the monitored event. At GB, event concentrations at the downstream site, GBBR, stand out as overall being much higher than at the upstream GBWD site. Differences between the other sites were not as pronounced.

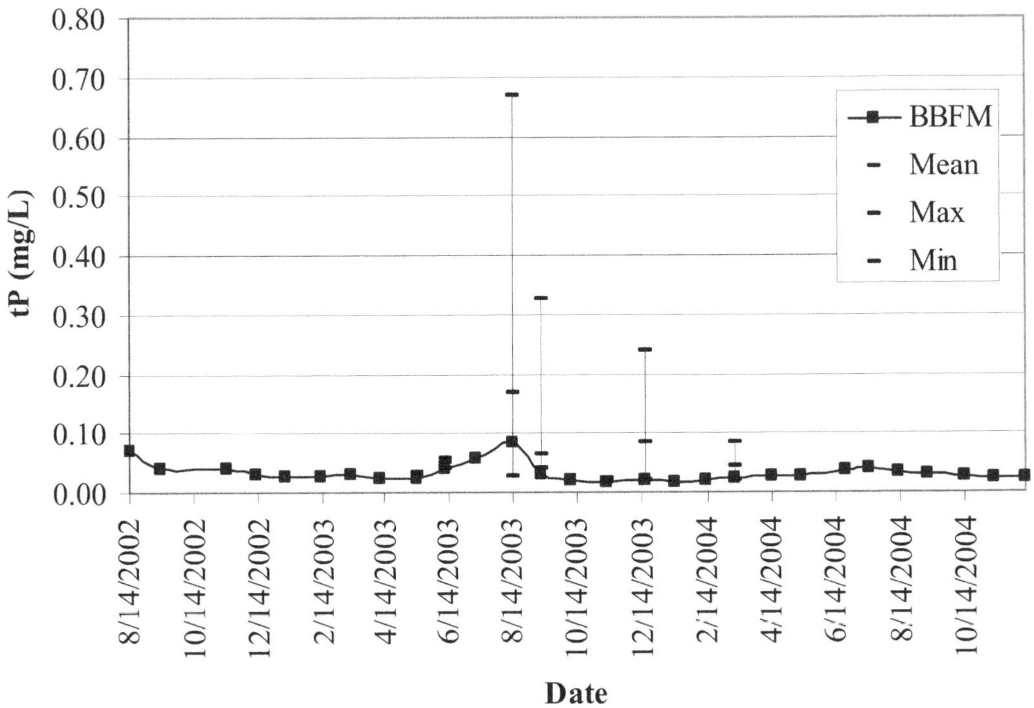

Figure 5-13 Monthly and Event Total Phosphorus Data for BBFM

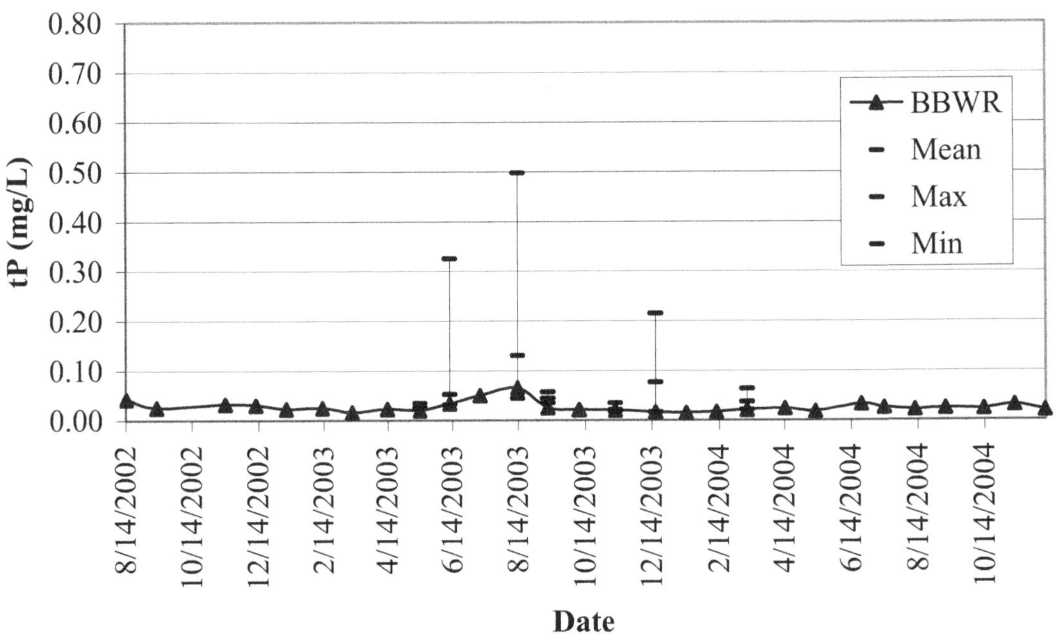

Figure 5-14 Monthly and Event Total Phosphorus Data for BBWR

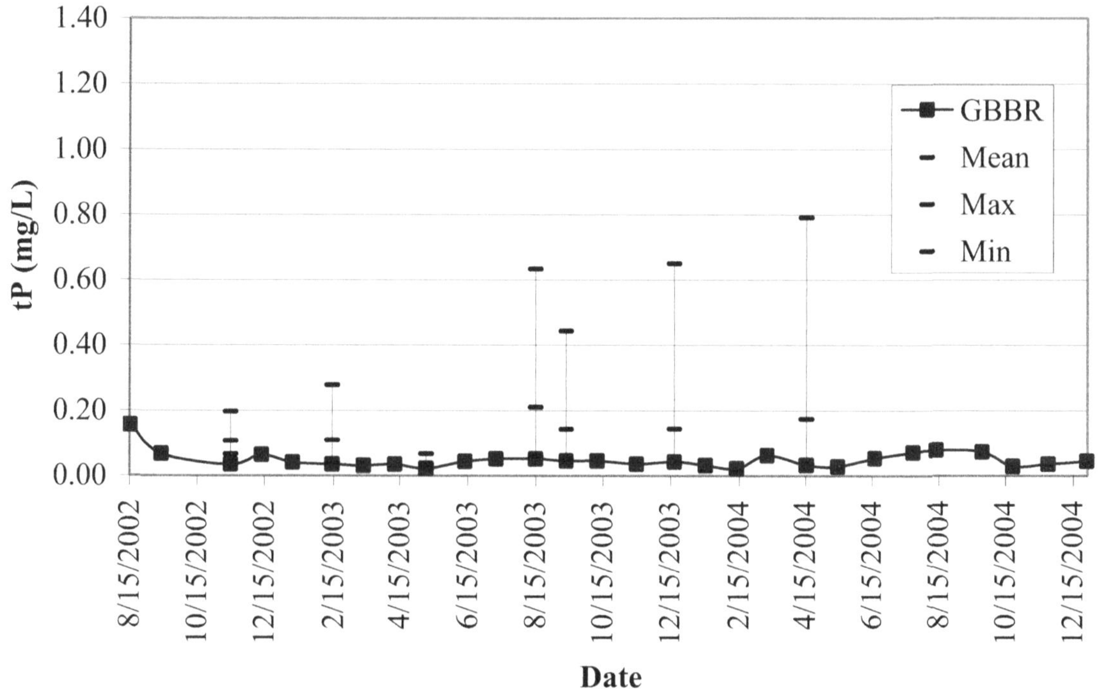

Figure 5-15 Monthly and Event Total Phosphorus Data for GBBR

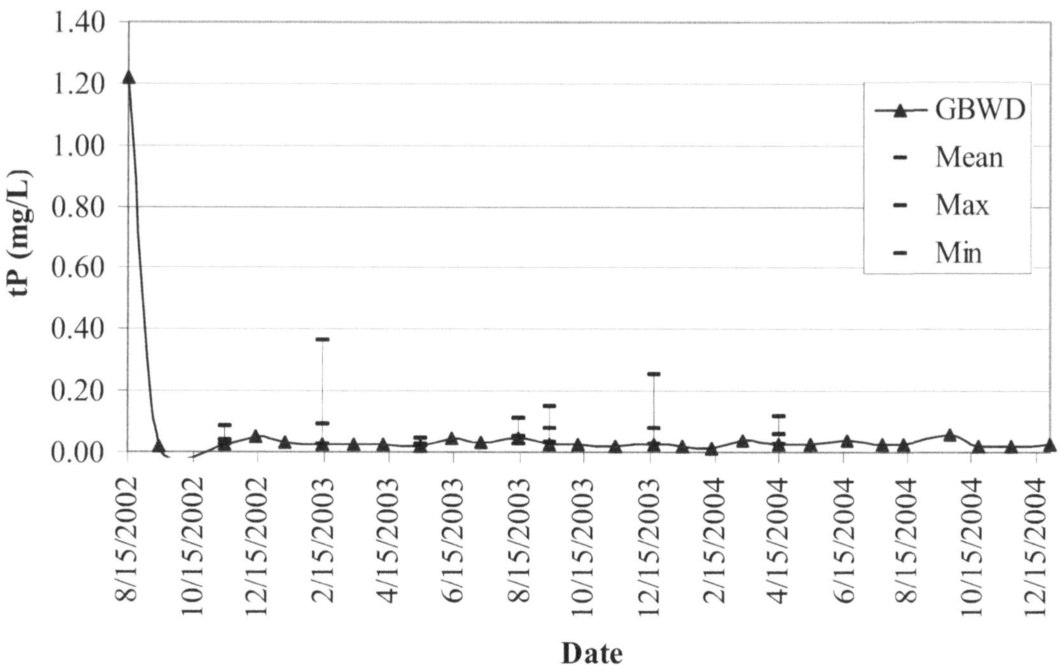

Figure 5-16 Monthly and Event Total Phosphorus Data for GBWD

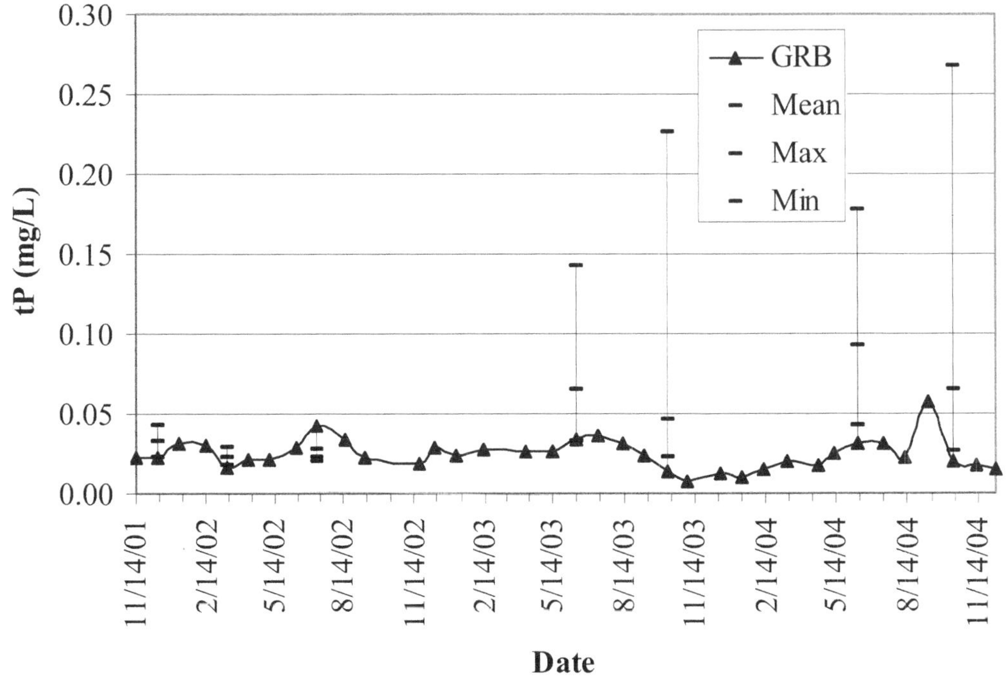

Figure 5-17 Monthly and Event Total Phosphorus Data for GRB

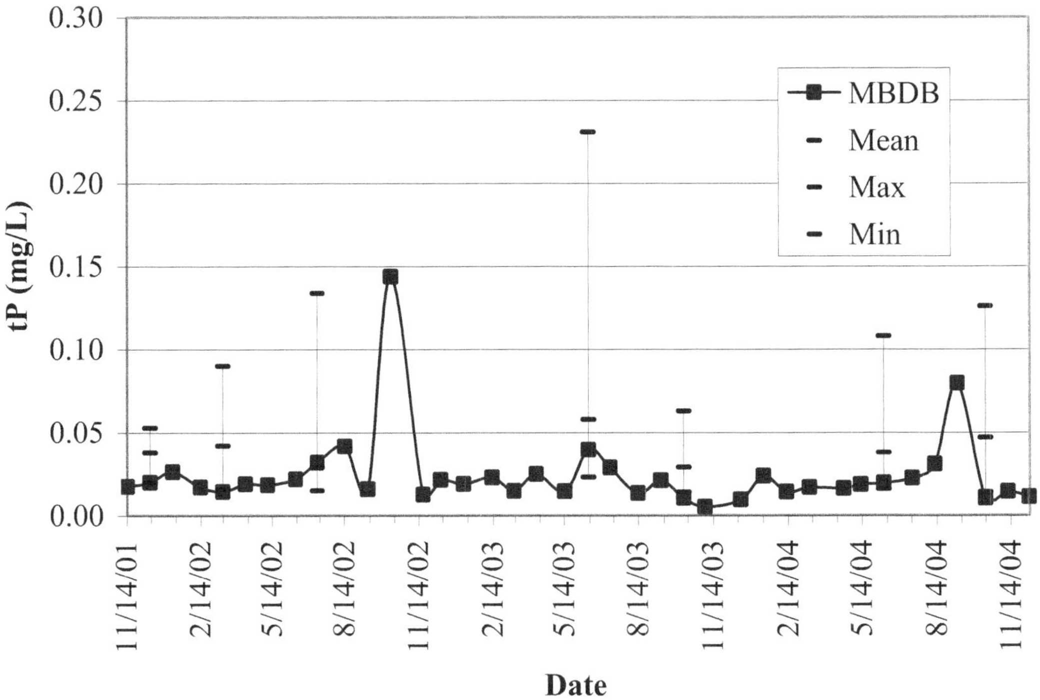

Figure 5-18 Monthly and Event Total Phosphorus Data for MBDB

Total Suspended Solids

Total Suspended Solids data are presented on Figures 5-19 through 5-24. Of the 44 events presented on the six figures (seven per the two sampling sites at both BB and GB and eight per the two sampling sites at PR, due to the practice event in December-01), 22 of the corresponding high-low lines either encompass the baseline sample for that month or have very small ranges near the monthly concentration. For 22 events (five at BB, ten at GB and seven at PR) the range of the high-low line lies almost entirely above the baseline sample for that month. There are no instances where the high low-line range lies completely below the baseline sample. It is important to note that a large majority of the baseline samples were non-detect for TSS and that 30 of the event means were greater than the baseline sample. This trend of increased TSS concentrations during rainfall events was expected and is likely due to elevated streamflow scouring the streambed, resuspending previously settled particles, as well as overland flow washing particulates into the stream.

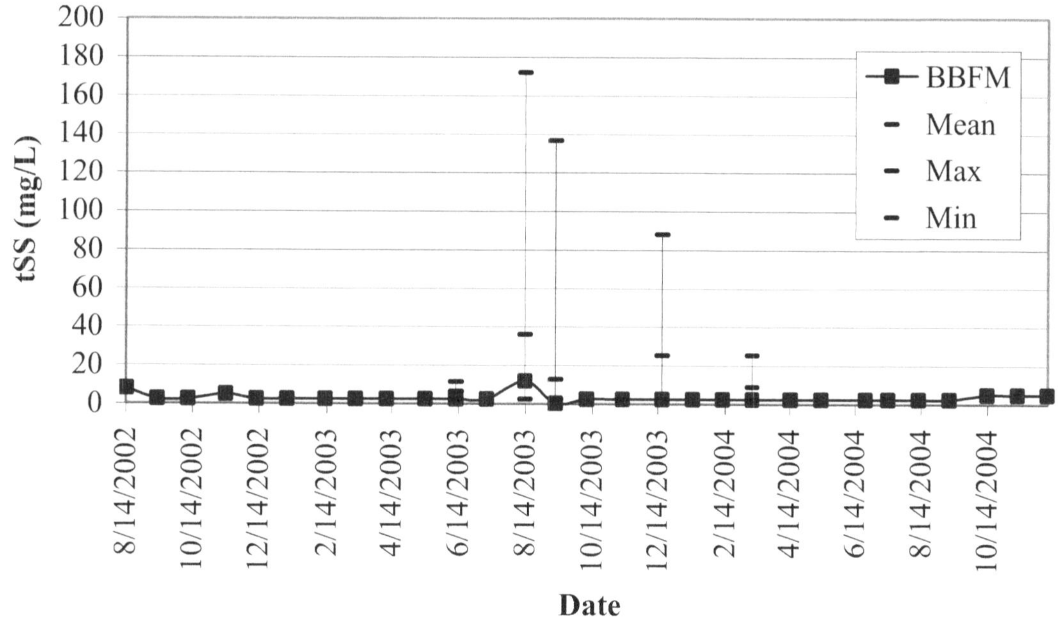

Figure 5-19 Monthly and Event Total Suspended Solids Data for BBFM

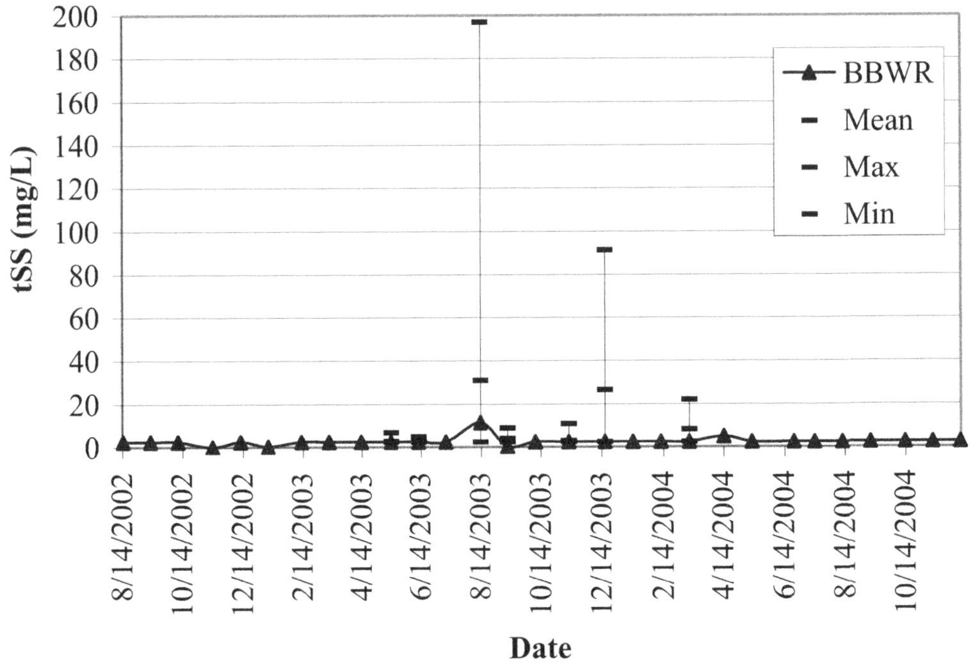

Figure 5-20 Monthly and Event Total Suspended Solids Data for BBWR

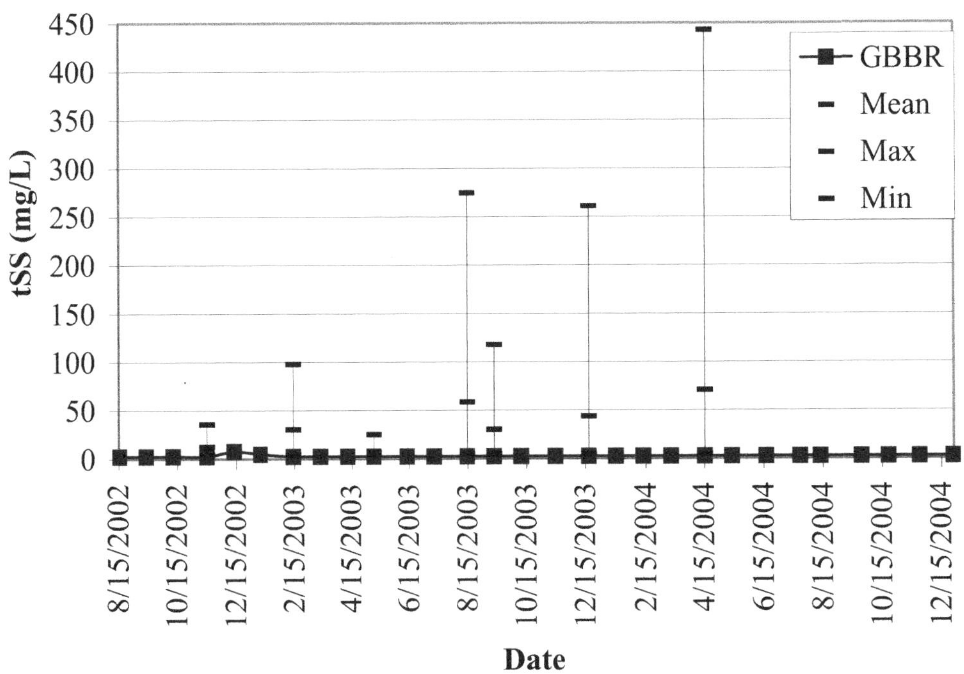

Figure 5-21 Monthly and Event Total Suspended Solids Data for GBBR

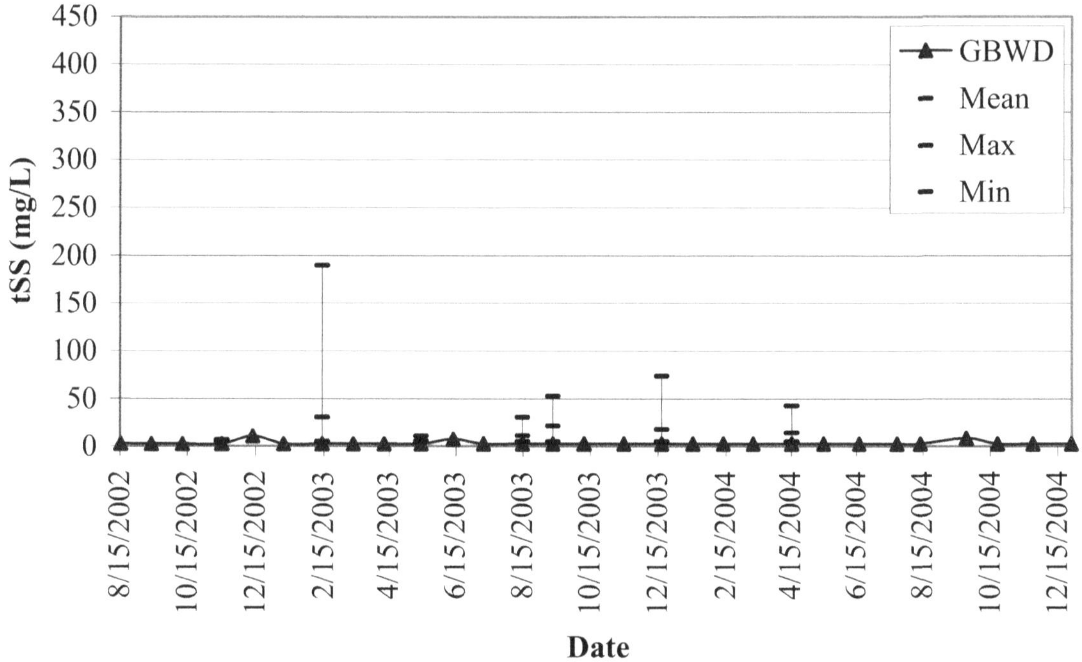

Figure 5-22 Monthly and Event Total Suspended Solids Data for GBWD

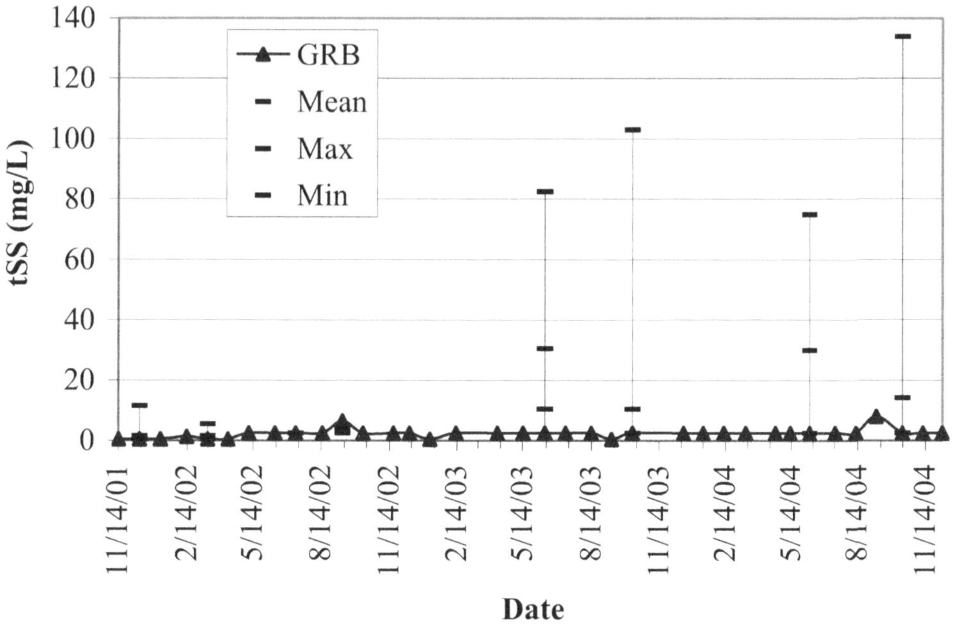

Figure 5-23 Monthly and Event Total Suspended Solids Data for GRB

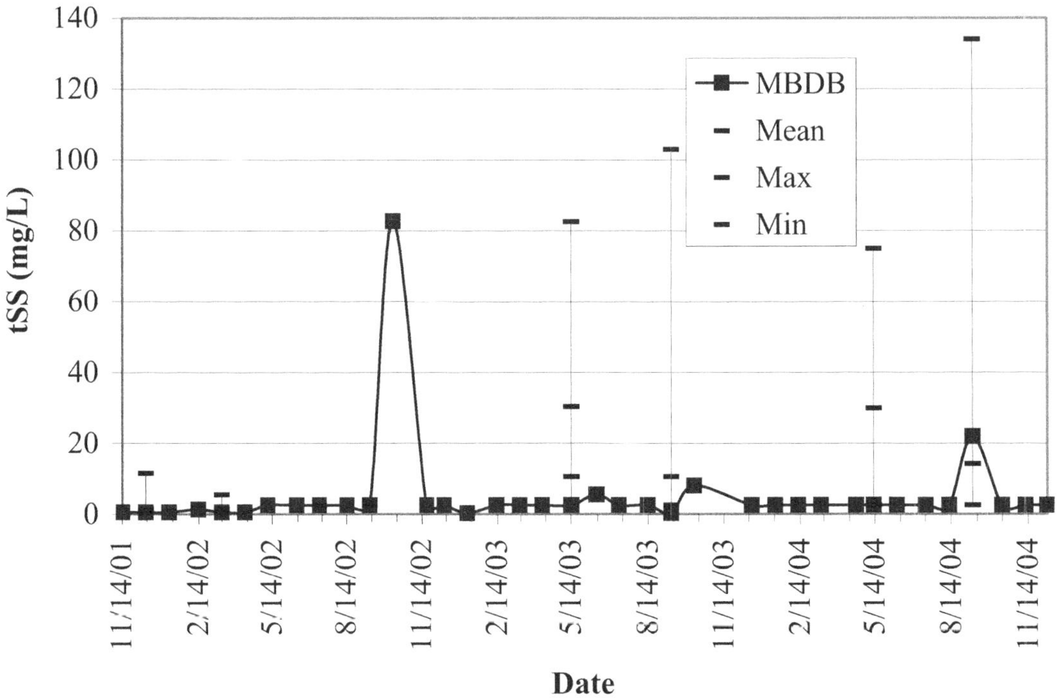

Figure 5-24 Monthly and Event Total Suspended Solids Data for MBDB

Total Coliform

Total Coliform data are presented on Figures 5-25 through 5-30. Of the 44 events presented (seven per sampling location at the BB and GB study sites and eight per sampling location at the PR sampling locations due to the December-01 practice event), 13 of the corresponding high-low lines encompass the baseline sample for that month. Of these thirteen, during six the event mean concentration exceeded the associated baseline monthly sample. For 30 events (nine at BB, eight at GB and thirteen at PR) the range of the high-low line lies almost entirely above the baseline sample for that month. One event, occurring on 11-November-03 had a high-low range lower than the corresponding monthly concentration. Based on literature values, increased TC concentrations were expected during rainfall events both due to transport via overland and due to resuspension from the streambed.

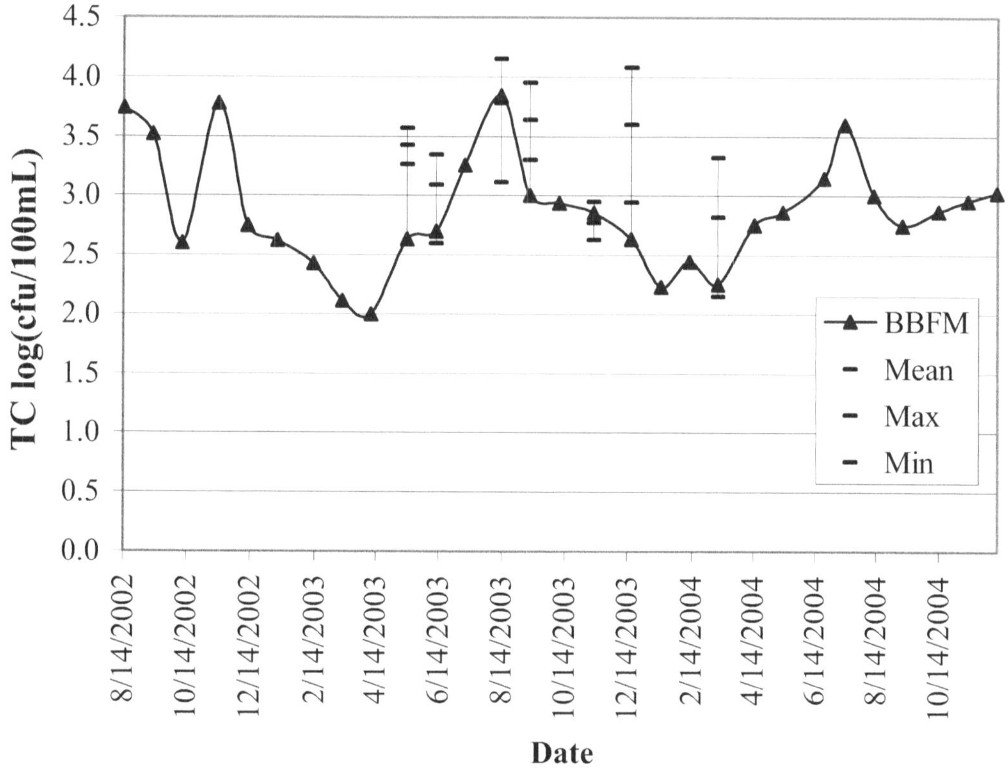

Figure 5-25 Monthly and Event Total Coliform Data for BBFM

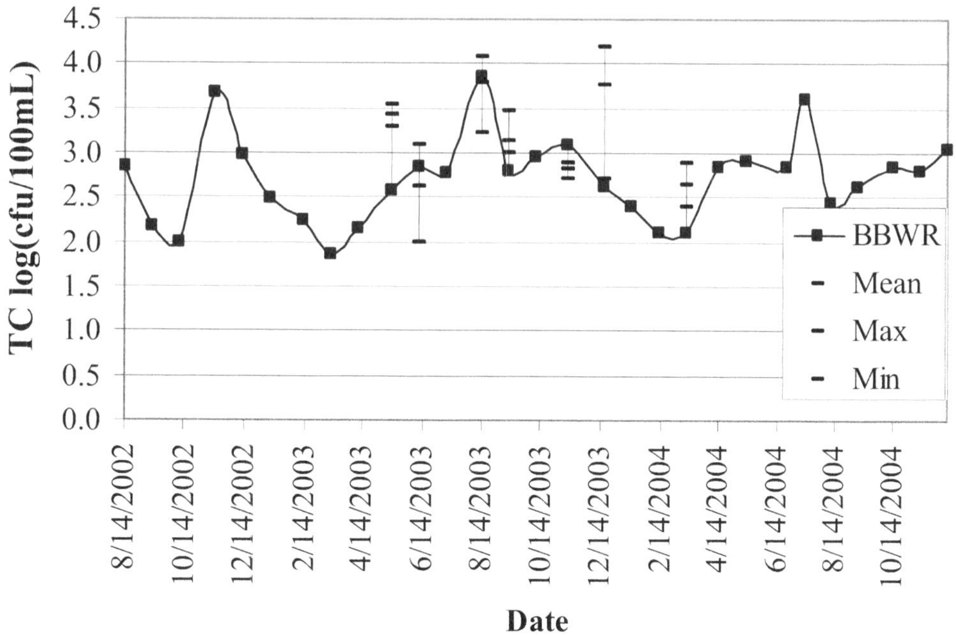

Figure 5-26 Monthly and Event Total Coliform Data for BBWR

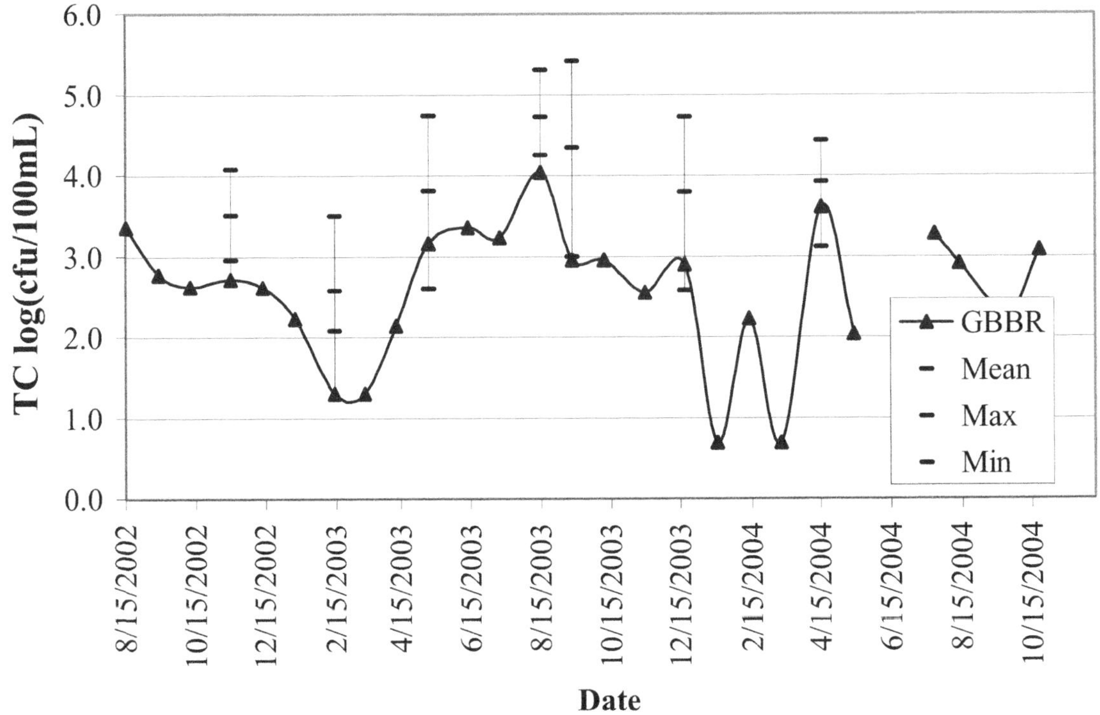

Figure 5-27 Monthly and Event Total Coliform Data for GBBR

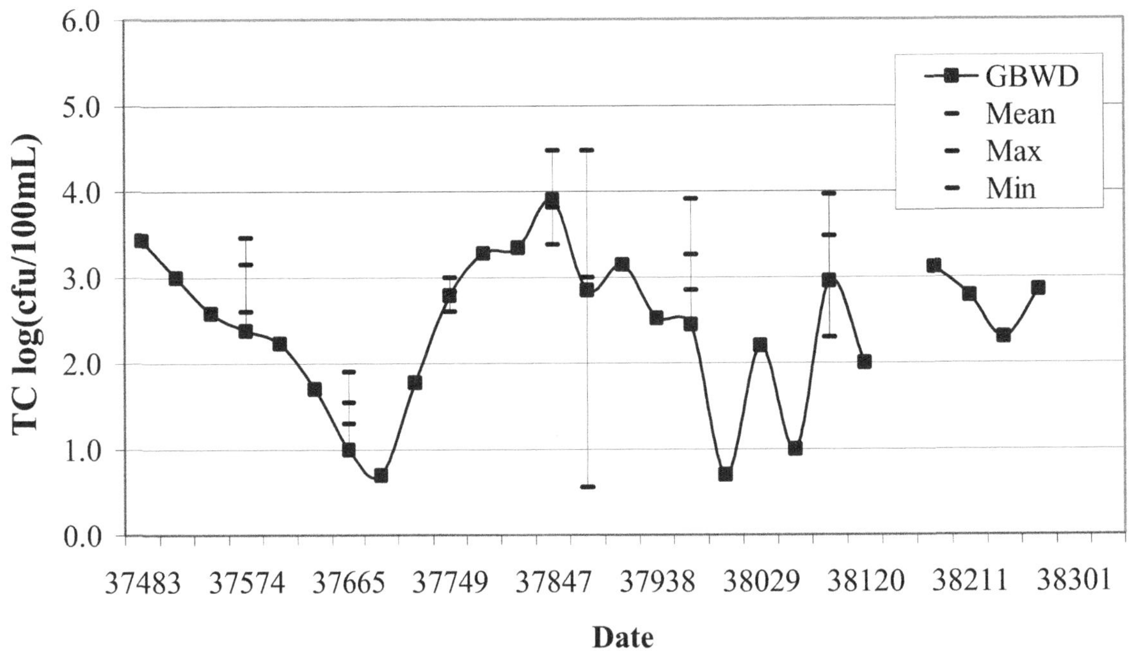

Figure 5-28 Monthly and Event Total Coliform Data for GBWD

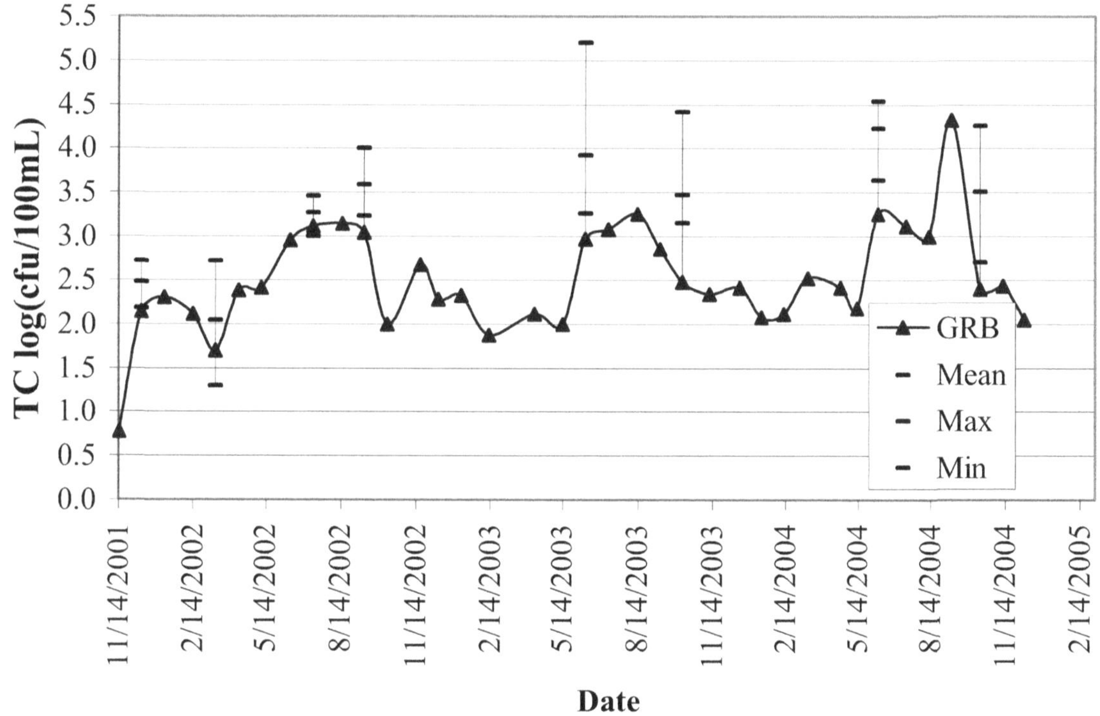

Figure 5-29 Monthly and Event Total Coliform Data for GRB

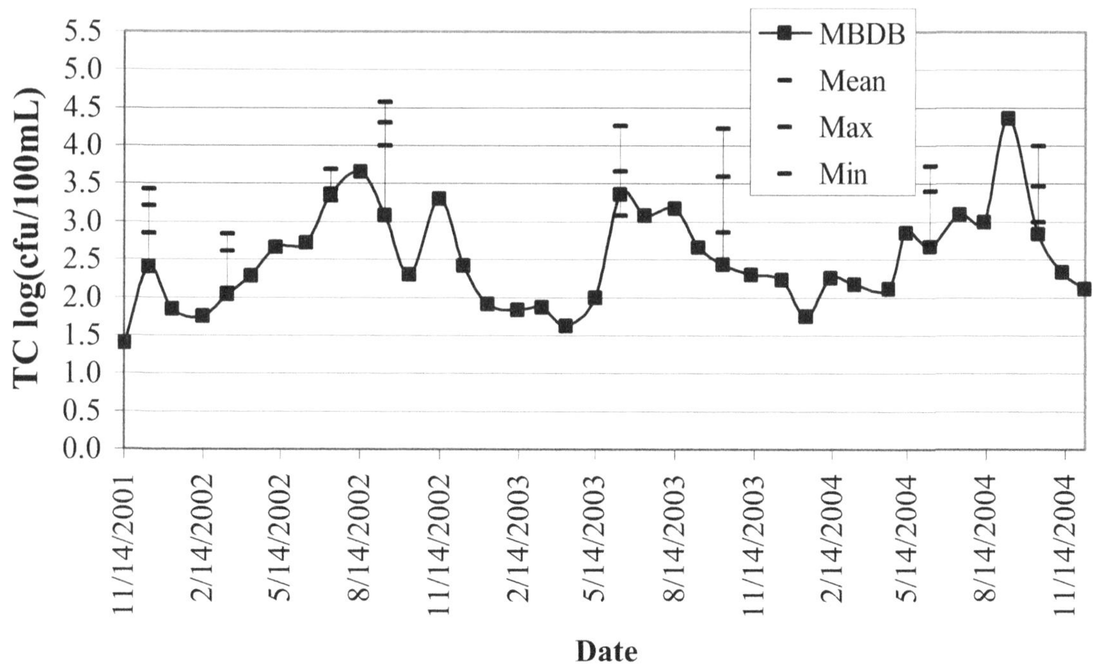

Figure 5-30 Monthly and Event Total Coliform Data for MBDB

Fecal Coliform

Fecal Coliform data are presented on Figures 5-31 through 5-36. Of the 44 events presented on the six figures (seven events at each sampling location for the BB and GB study sites and eight events at the PR sampling locations due to the December-01 practice event), 19 of the corresponding high-low lines encompass the baseline sample for that month. However, for most of these events the event mean concentration was approximately equivalent or greater than the corresponding monthly baseline value. For 23 events (six at BB, eight at GB and nine at PR) the range of the high-low line lies almost entirely above the baseline sample for that month. For one event at GB, occurring on 11-December-03, the high low-line range lies completely below the baseline sample at both GBBR and GBWD. Similar low values of TC were not observed during this event, which had a precipitation accumulation of almost 30 mm. Approximately 13 mm (0.5 inches) of rain had fallen on both December 6th and 7th, preceding the event. Sources of fecal coliform in the watershed may thus have been depleted. One event at GRB could not be compared due to missing monthly data. Increased FC concentrations were expected during rainfall events due to the potential for overland flow to wash particulates with attached fecal matter into the stream and due to resuspension from the streambed.

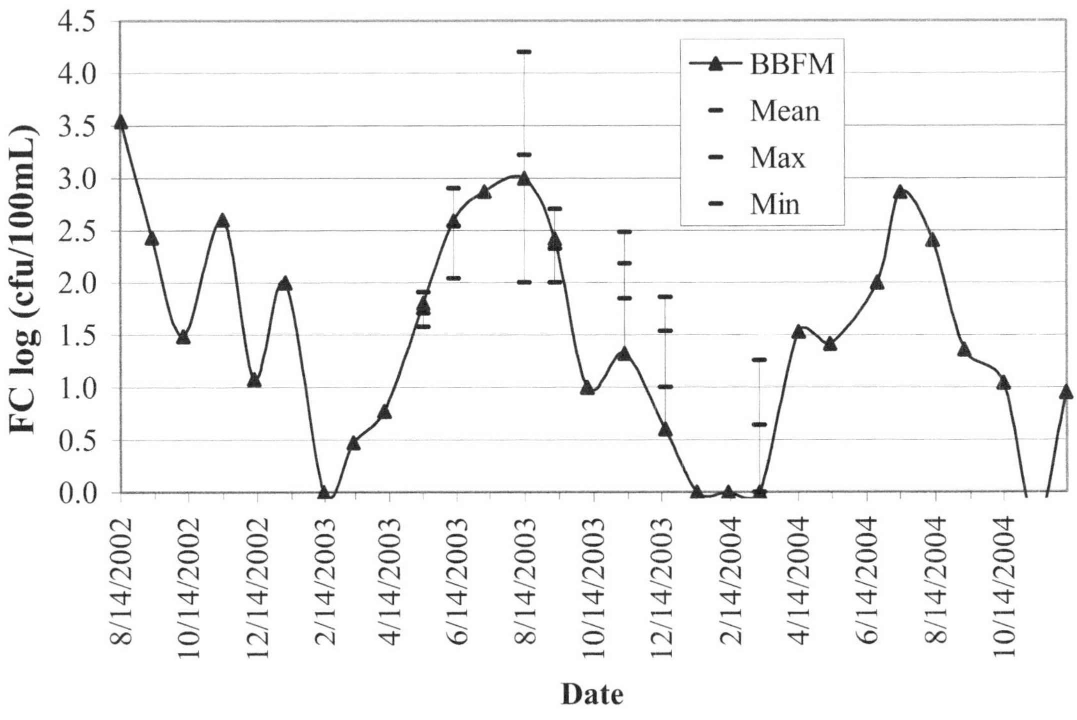

Figure 5-31 Monthly and Event Fecal Coliform Data for BBFM

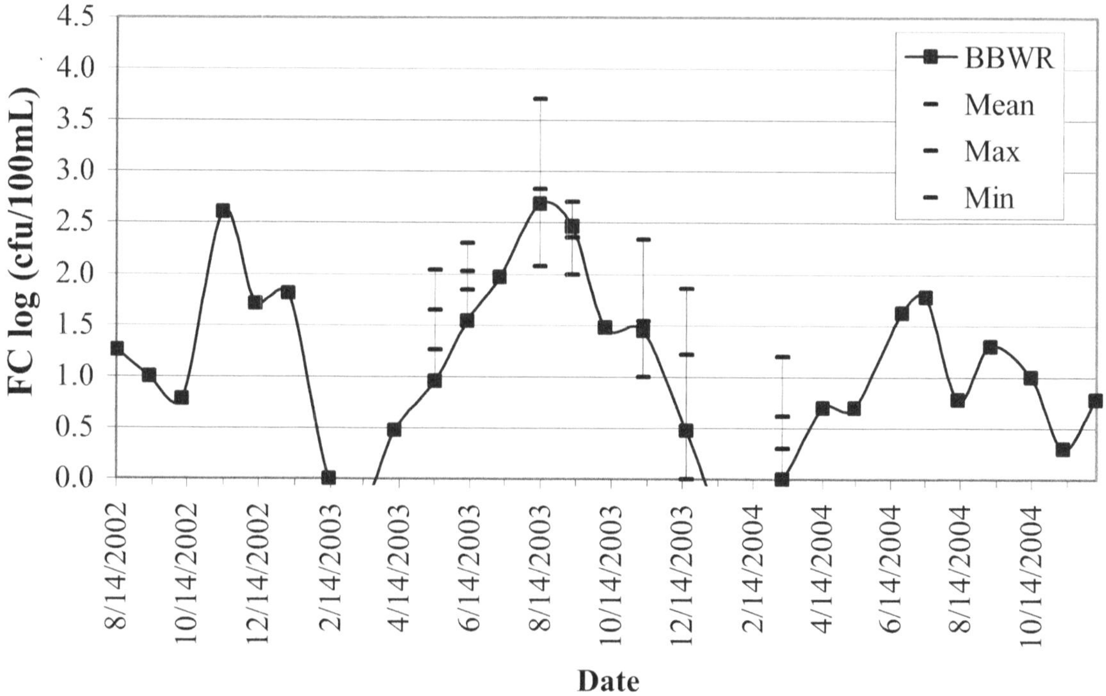

Figure 5-32 Monthly and Event Fecal Coliform Data for BBWR

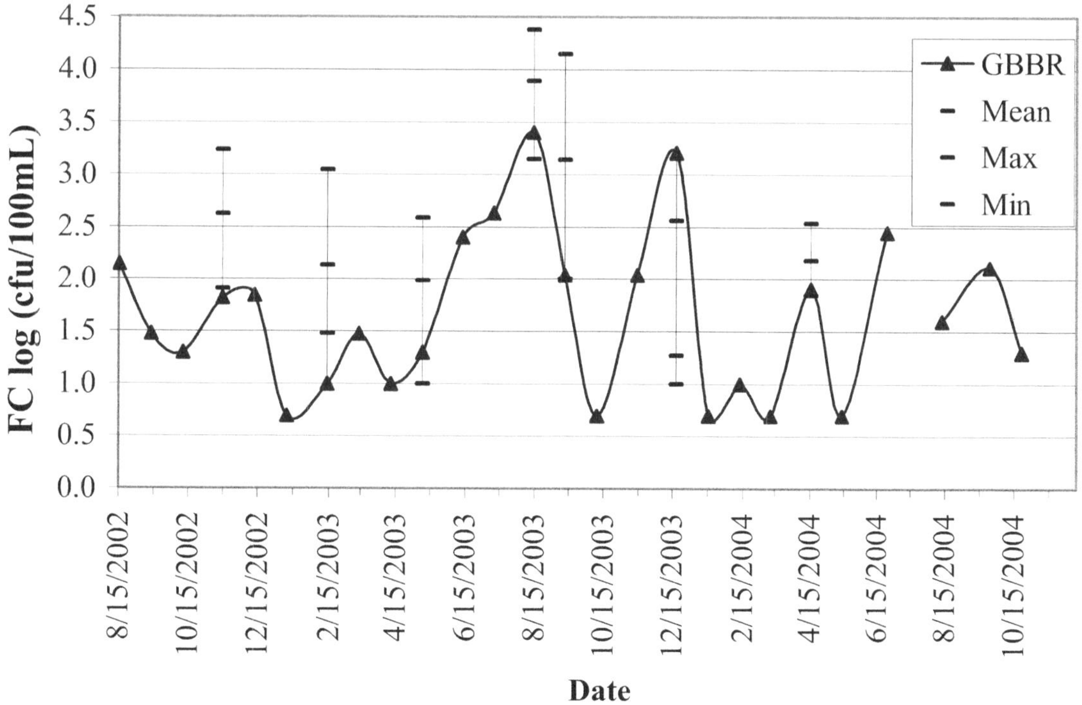

Figure 5-33 Monthly and Event Fecal Coliform Data for GBBR

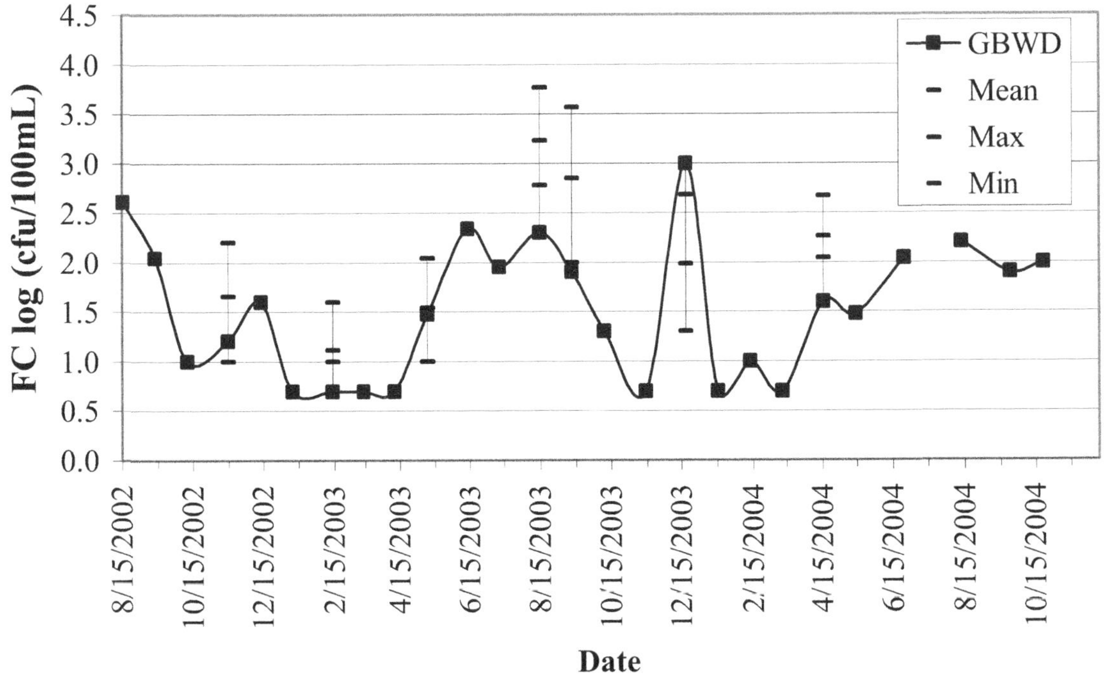

Figure 5-34 Monthly and Event Fecal Coliform Data for GBWD

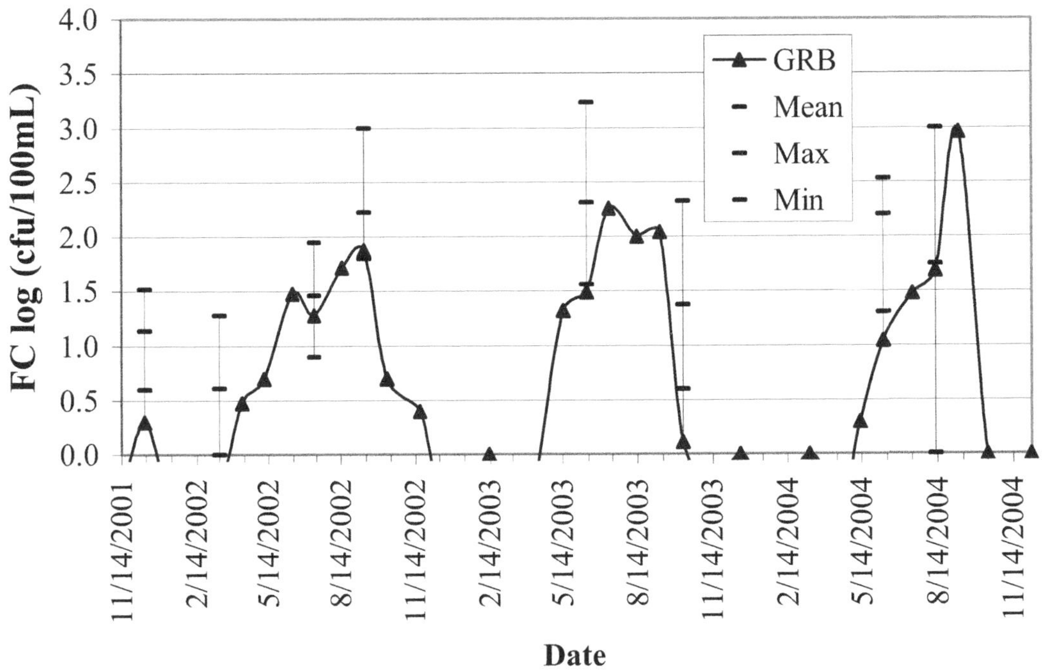

Figure 5-35 Monthly and Event Fecal Coliform Data for GRB

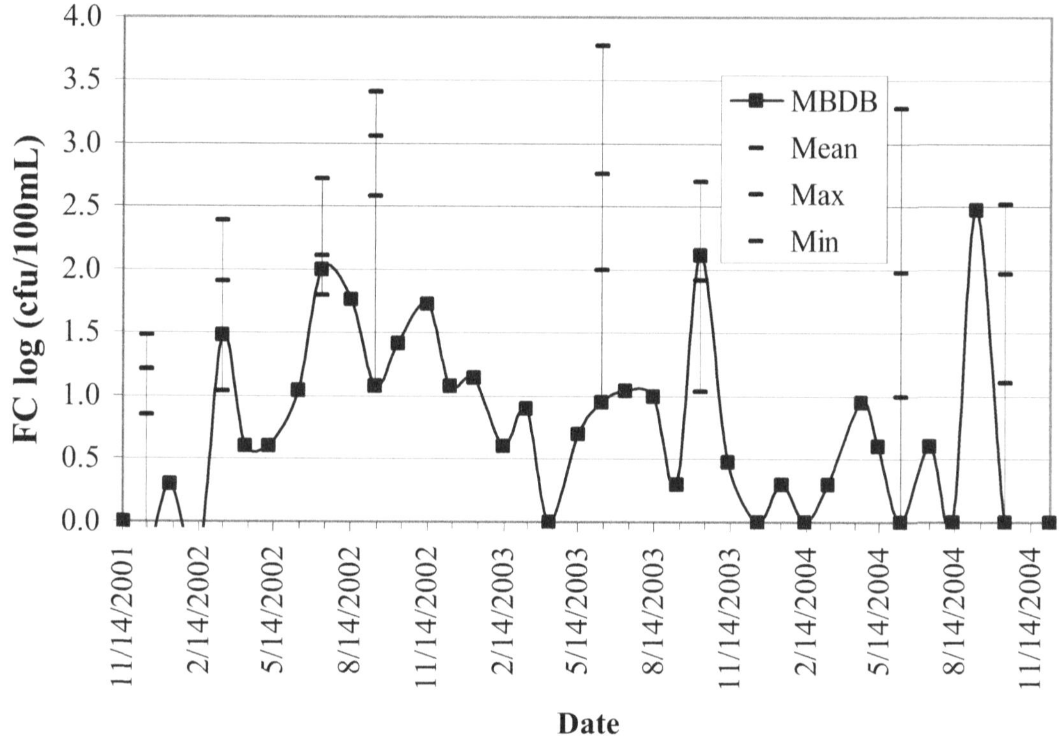

Figure 5-36 Monthly and Event Fecal Coliform Data for MBDB

Antecedent Rainfall

The influence of antecedent precipitation on variations in monthly and event concentration was evaluated in two ways. First, antecedent rainfall was determined for each monthly sampling date and event. Four periods of antecedent rainfall were evaluated:

1. Rainfall on the same day occurring between 7:00 AM and the time the first sample was collected
2. Rainfall in previous 24-hours ending at 7:00 AM
3. Rainfall in the previous 48-hours ending at 7:00 AM
4. Rainfall in the previous 72-hours ending at 7:00 AM

Results for the monthly data are presented in Tables 5-2 through 5-4 while results for the event data are presented in Table 5-5. It should be noted that due to a delay in installation and equipment failures, data for the agricultural study sites are lacking until September 2003. Secondly, rainfall on the same day as sampling but occurring between 7:00 AM and the time of the first sample collection was summed with the 72-hour antecedent rainfall ending at 7:00 AM. These data were used to develop the scatter plots of antecedent rainfall versus concentration. For events, the EMC was used. Results are presented in Appendix A for the monthly data and in Appendix B for the event data.

Review of Tables 5-2 through 5-4 and the figures in Appendix A revealed no strong correlations (as measured by the R^2 statistic) between antecedent rainfall and concentration for any of the parameters. Nitrate/nitrite concentrations typically tended to be slightly lower when antecedent rainfall was high. Mixed results were found for TKN. The wildlife study sites were characterized by several non-detects, which resulted in a fairly flat linear correlation line. Considering greater than detection limit values only, higher antecedent rainfall resulted in some lower TKN values. BBFM also exhibited a slight decrease in TKN with higher antecedent rainfall. TKN concentrations tended to increase slightly with increased antecedent precipitation at the other sites. At most sites, total phosphorus concentration tended to decrease slightly as antecedent rainfall increased. Almost all monthly suspended sediment data were less than the detection limit, thus no correlations were found. At the BB and PR sites, TC showed little variability due to antecedent precipitation. At the GB sites, however, TC results tended to increase with antecedent rainfall. Similar trends were found for FC. Individual dates with high values of bacteria or nutrients could sometimes be attributed to antecedent conditions. For example, in the urban watershed, 5.33mm of rain fell before sampling occurred in September 2004, resulting in higher than average concentrations for both nutrients and coliform.

Review of Table 5-5 and the figures in Appendix B found no strong correlations between the EMC and prior day, 24-, 48- or 72-hour antecedent rainfall conditions for most parameters and sampling locations. The highest correlations (R^2~0.70) were found for nitrate concentration at the agricultural study site, where nitrate concentration was inversely proportional to the expanded 72-hour rainfall total. Although the R^2 values were weak for all other parameters and sites, some general trends may be noted. Nitrate EMC values appeared to increase slightly with antecedent precipitation at the GB sites. The EMC values decreased slightly with antecedent rainfall at MBDB. The TKN EMC values tended to increase slightly with increased antecedent rainfall at GB and PR, but were split for events at BB, resulting in an almost flat trend. Mixed results were also observed for TP, TSS, TC, and FC.

While antecedent rainfall appears to have some influence on monthly and event concentrations, the results suggest that additional influences, such as season and discharge, are key factors in determining stream nutrient and bacteria concentrations.

Table 5-2
Precipitation preceding monthly sampling at the agricultural study sites

Date Collected	Antecedent Rainfall (mm) Prior that Day (07:00 to sample collection)	Previous 24hrs	Previous 48hrs	Previous 72hrs
Burrow Brook				
14-Aug-02	--	--	--	--
11-Sep-02	--	--	--	--
9-Oct-02	--	--	--	--
13-Nov-02	--	--	--	--
11-Dec-02	--	--	--	--
8-Jan-03	--	--	--	--
12-Feb-03	--	--	--	--
12-Mar-03	--	--	--	--
9-Apr-03	--	--	--	--
14-May-03	0	--	--	--
11-Jun-03	NA	0	1.52	1.52
9-Jul-03	NA	NA	NA	NA
13-Aug-03	NA	NA	NA	NA
10-Sep-03	0	NA	NA	NA
8-Oct-03	0	0	0	0
10-Nov-03	0	0	0	0
17-Dec-03	0	0.51	0.51	9.14
14-Jan-04	0	2.54	2.54	2.54
11-Feb-04	0	3.56	3.56	3.56
10-Mar-04	0	2.79	6.10	6.10
14-Apr-04	0	39.62	47.75	47.75
12-May-04	0	2.54	5.59	14.99
23-Jun-04	0	4.57	4.57	5.33
14-Jul-04	0	2.03	2.03	2.29
11-Aug-04	0	2.03	2.03	2.54
8-Sep-04	0	0	0	0
13-Oct-04	0	0	0	0
10-Nov-04	0	0	0	0
8-Dec-04	1.02	16.51	16.51	16.51

Table 5-3
Precipitation preceding monthly sampling at the urban study sites

	Antecedent Rainfall (mm)			
Date Collected	**Prior that Day (07:00 to sample collection)**	**Previous 24hrs**	**Previous 48hrs**	**Previous 72hrs**
Gates Brook				
15-Aug-02	--	--	--	--
12-Sep-02	--	--	--	--
10-Oct-02	NA	NA	NA	NA
14-Nov-02	0	1.524	29.46	32.77
12-Dec-02	1.52	4.064	4.06	4.06
9-Jan-03	NA	NA	NA	NA
13-Feb-03	0	3.05	3.05	3.05
13-Mar-03	NA	NA	NA	NA
10-Apr-03	NA	NA	NA	NA
8-May-03	0	3.56	3.81	3.81
12-Jun-03	0	12.44	12.45	12.70
10-Jul-03	0	6.10	6.10	6.10
14-Aug-03	0	0	25.40	28.19
11-Sep-03	0	0	0	0
9-Oct-03	0	0	0	0
13-Nov-03	0	3.56	8.64	8.64
18-Dec-03	0	44.96	47.75	47.75
15-Jan-04	0	0	4.06	4.06
Invalid Data Removed				
10-Mar-04	0	0.25	3.81	9.14
14-Apr-04	1.27	8.89	70.10	79.50
12-May-04	0	0	0	0
23-Jun-04	0	0	0.25	0.76
14-Jul-04	0	0	0	1.78
11-Aug-04	0	0.51	0.51	0.51
8-Sep-04	0	0	0	0
13-Oct-04	0	0.51	0.51	0.51
10-Nov-04	NA	NA	NA	NA
8-Dec-04	0	0	0	0

Table 5-4
Precipitation preceding monthly sampling at the wildlife study sites

	Antecedent Rainfall (mm)			
Date Collected	**Prior that Day (07:00 to sample collection)**	**Previous 24hrs**	**Previous 48hrs**	**Previous 72hrs**
Prescott				
14-Nov-01	0	0	0	0
12-Dec-01	0.51	6.10	6.86	6.86
9-Jan-02	0	0	5.08	15.24
13-Feb-02	0	0	0	25.40
13-Mar-02	0	0	0	0
10-Apr-02	0	1.78	1.78	1.78
8-May-02	0	0	0	0
12-Jun-02	0	1.52	1.52	1.78
10-Jul-02	0	6.35	6.35	6.35
15-Aug-02	0	0	0	0
11-Sep-02	0	0	0	0
9-Oct-02	0	0	0	0
19-Nov-02	0	7.11	55.88	57.15
11-Dec-02	0	0	0	1.78
8-Jan-03	0	0	0	0
12-Feb-03	0	0	1.02	1.02
Icy Road Conditions				
9-Apr-03	0	5.59	7.11	15.24
14-May-03	0	0.51	3.30	21.08
12-Jun-03	0	11.68	11.68	12.19
9-Jul-03	0	0	0	0
13-Aug-03	0	0	0	0
10-Sep-03	0	0	0	0
8-Oct-03	0	0	0	0.25
3-Nov-03	NA	NA	NA	NA
17-Dec-03	20.32	8.38	8.38	8.38
14-Jan-04	0	3.81	3.81	3.81
11-Feb-04	NA	5.33	7.62	7.62
11-Mar-04	0	0	2.03	5.59
21-Apr-04	0	0	0	0
12-May-04	0	0	0	8.89
9-Jun-04	0	0	0	0.25
14-Jul-04	0	12.70	12.70	12.70
11-Aug-04	0	0	0	0
8-Sep-04	5.33	0	0	0
13-Oct-04	0	0	0	0
10-Nov-04	0	0	0	0
8-Dec-04	0	0	0	0

Table 5-5
Precipitation (mm) preceding sampled events

Event	Prior that Day (07:00 to sample collection)	Previous 24hrs	Previous 48hrs	Previous 72hrs
Burrow Brook				
5/28/2003	2.032	0.762	36.068	36.322
6/18/2003	0.254	1.524	1.524	2.286
8/12/2003	0.508	0	4.572	6.604
9/19/2003	0	0	0.254	21.082
11/12/2003	0	0	0	0
12/17/2003	0	0.508	0.508	9.144
3/31/2004	3.81	0.762	0.762	0.762
7/14/2004	0	2.032	2.032	2.286
Gates Brook				
11/6/2002	2.032	0.762	1.016	1.016
2/4/2003	0.254	5.334	21.844	22.352
5/2/2003	0.254	0.762	NA	NA
8/1/2003	NA	NA	NA	NA
9/2/2003	0	0	0	0
12/11/2003	1.27	0	2.54	2.54
4/12/2004	0.508	0	0	0
8/4/2004	NA	NA	NA	NA
Prescott				
12/14/2001	0	7.366	14.478	20.574
3/26/2002	NA	NA	NA	NA
7/28/2002	0.254	0	0	0
9/26/2002	9.398	0	0	0
6/11/2003	3.048	0	0.508	0.508
10/27/2003	2.032	0.762	0.762	3.048
3/10/2004	NA	NA	NA	NA
6/9/2004	0	0	0	0.254
10/15/2004	0.254	3.81	3.81	3.81

Summary

Depending on the water quality parameter and study site, approximately one- to two-thirds of the observed event concentration ranges appeared to exceed the most temporally correlated monthly baseflow concentration, including 62% of the TKN, 36% of the TP, 50% of the TSS, 68% of the TC, and 52% of the FC event high-low ranges. When the high-low event concentrations spanned the baseflow, the event mean concentration often exceeded the monthly concentration. Higher concentrations during events than during baseflow were anticipated for these parameters. Nitrate/nitrite ranges spanned the monthly baseflow concentration 60% of the time, were lower 20% of the time, and were higher 20% of the time. Lower event concentrations were anticipated, although it was expected that results might vary between watersheds and the literature has reported various effects of wet weather in terms of nitrate. No attempt has been made to compare relative magnitudes during an event across parameters (in other words, were all parameter event concentrations above the monthly concentration, or were mixed results observed).

Based on the small size of the watersheds and the inherent difficulties in identifying a metric for events other than precipitation, it is felt that the captured events adequately represent the study site watershed response to a wide range of hydrologic conditions. The resulting data are thus useful for identifying the magnitude of isolated contributions (of contaminants) from distinct land uses. This study is unique from others not only in terms of monitoring over the rise and the fall of a storm-event hydrograph, but more importantly also in the nature of the watersheds. The focus watersheds for this study are all small (less than 10 km^2). Other studies have focused on much larger (order of magnitude) watersheds with mixed land uses, although some studies have utilized upstream – downstream sampling techniques, as herein, to isolate suspect sources. The probability of presence, particularly of *Cryptosporidium* and *Giardia*, is thus likely lower in watersheds included in this study while the probability of transport to the stream and outlet, when present, is higher. In addition, manure spreading for crop fertilization was not practiced.

SEASONALITY AND SAMPLING LOCATION COMPARISONS

The focus of this chapter is presentation of the wet weather events, with focus on standard water quality parameters and alternative indicator organisms rather than *Cryptosporidium* and *Giardia*. The available monthly data is such that comparisons between study sites and between sampling locations may be conducted. Similarly, comparisons at a single sampling location may be made to assess inter-annual and seasonal variability. Such analyses will be useful for understanding relative contributions by land use. However, due to the volume of data and the peripheral importance in terms of the study, these analyses will be conducted as an element of a graduate student thesis. A few qualitative insights, however, are of interest and are presented in this section.

Site Comparisons

Table 5-6 summarizes the monthly data by parameter and sampling location. Suspended sediment data are excluded because results were typically less than the detection limit. For all parameters monitored on a monthly basis (TC, FC, Nitrate, TKN, and TP) average concentrations were lowest at the pristine wildlife sites, including that influenced by beaver. In general, TC levels were greatest at the agricultural site, including the upstream sampling location. FC levels at the two residential sites were greater than at the farm sites, although differences compared to the downstream agricultural site were smaller. Nitrate concentrations at the residential study site locations were highest. In addition, the downstream residential sampling location average concentration was almost three times the upstream average concentration, particularly at GBBR. In terms of TKN and TP, results at the agricultural and residential sampling locations were mixed. The highest average concentration for both parameters occurred at the upstream residential site (GBWD), located downstream from a swampy area. Total phosphorus at GBBR (downstream residential site) was also higher than at the agricultural sites, but monthly average TKN data, while highest at GBWD, was higher at the farm (BBFM) than at either BBWR or GBBR, which had similar average concentrations. The relative magnitudes described above are summarized qualitatively in Table 5-7. Recall that statistical significance of the differences has yet to be tested.

Table 5-6
Arithmetic averages of the monthly water quality data by parameter and sampling location

Arithmetic Average Monthly Water Quality Parameter (cfu/100 mL or mg/L)					
Site	Total *Coliform	Fecal *Coliform	Nitrate	Total Kjeldahl Nitrogen	Total Phosphorus
BBFM	2.87	1.50	0.209	0.409	0.033
BBWR	2.72	1.09	0.189	0.327	0.027
GBBR	2.60	1.71	1.758	0.332	0.049
GBWD	2.44	1.61	0.611	0.458	0.072
GRB	2.51	0.72	0.034	0.306	0.025
MBDB	2.51	0.73	0.067	0.385	0.024

Note: * Log Base 10 Transformation

Table 5-7
Qualitative comparison of average parameter concentrations at the sampling locations, based on monthly water quality data

Arithmetic Average Monthly Water Quality Parameter (cfu/100 mL or mg/L)					
Site	Total *Coliform	Fecal *Coliform	Nitrate	Total Kjeldahl Nitrogen	Total Phosphorus
BBFM	High	Mid	Mid	High	Mid
BBWR	High	Mid	Mid	Mid	Mid
GBBR	Mid	High	Very High	Mid	High
GBWD	Mid	High	High	High	High
GRB	Low	Low	Low	Low	Low
MBDB	Low	Low	Low	Low	Low

Note: * Log Base 10 Transformation

Seasonal Variability

Time series plots of the monthly data, Figures 5-1 through 5-36, suggest some seasonality in the water quality data. TC and FC were elevated in the summer months at all sites. However, FC monthly data at GBBR, GBWD, and MBDB varied throughout the year, so summer season elevated concentrations were not as prominent. Nitrate data were highly variable, exhibiting five to six month cycles of increases and decreases at many of the sites. Summer concentrations, however, tended to be a bit higher. At the BB locations, except for an unusually high December value in 2002, TKN concentrations were slightly elevated in summer months. TKN concentrations at the GB and PR sampling locations showed no discernable seasonal pattern. However, variability from August-02 through August-03 at GBBR, GBWD, MBDB, and GRB was greater with an overall decrease over the period. At the BB sites, TP and TSS concentrations were elevated during summer months in 2003 and 2002 (incomplete data) but were fairly flat through summer of 2004. At the other sites, SS monthly concentrations were mainly no detect values. Total phosphorus concentrations at GB and PR exhibited no strong seasonal trends. Seasonality of the monthly water quality data is of only peripheral interest and will thus not be explored more fully through statistical analyses herein.

EVENT WATER QUALITY MONITORING RESULTS

The monitoring program for the study was conducted over a three-year period where microbiological, chemical and physical parameters were measured for each watershed as described in Chapter 4. In the previous sections, the relative concentrations between the monthly data and event data were described and the monthly standard water quality data summarized. Tables 5-8 through 5-10 provide a summary of the monitoring program results for the microbiological, physical and chemical parameters per event for PR, BB, and GB respectively. In these tables, data are summarized by the arithmetic mean. The TC and FC data were log transformed prior to calculation of the mean. Alternative indicator organism data are provided as a range rather than a mean. In the next section, event mean concentrations, rather than simple arithmetic averages, are calculated.

Historical Context of Monitoring Program Results

In general, historical data for the sampling sites did not exist. One exception was the Gates Brook watershed, where long-term monitoring has been conducted for standard water quality parameters and alternative indicator organisms. Results for the alternative indicator organism studies are summarized in Long *et al.* (2002) and annual DCR project reports. In general, the historical and new monitoring program results were similar.

Pathogen Detection in Watersheds

The occurrence and detection of pathogens in the watershed was the major focus of this study. These results will be discussed extensively in Chapter 6 and are thus not discussed further here. No site specific *Giardia* or *Cryptosporidium* data were available for comparison to historical records. As noted above, historical alternative indicator organism data exist, but only for Gates Brook.

Table 5-8
Event average (arithmetic) of samples collected during events at PR sites

	14-Dec-01		26-Mar-02		28-Jul-02		26-Sep-02	
Parameter	**MBDB**	**GRB**	**MBDB**	**GRB**	**MBDB**	**GRB**	**MBDB**	**GRB**
TC (cfu/100 mL)*	3.21	2.48	2.61	2.04	3.43	3.25	4.30	3.59
FC (cfu/100 mL)*	1.21	1.14	1.91	0.61	2.11	1.42	3.06	2.22
THPC(cfu/mL)*	2.44	2.08	2.23	2.83	2.33	2.19	3.40	3.09
C. perf(cfu/100 mL)†	4 to 5	<1 to 1	2 to 20	1 to 3	<1 to 1	<1 to 2	1 to 5	<1 to 3
F+ (pfu/100 mL) †	<1 to 3	<1 to 7	<1	<1	<1	<1	14 to 68	<1 to 1
F- (pfu/100 mL) †	2 to 9	<1 to 7	<1 to 2	<1 to 1	<1 to 1	<1	490-855	<1
Rhodo. (cfu/100 mL) †	--	--	< 20	< 20	< 20	< 20	< 5 to 5	< 5 to 5
NH_3 (mg/L)	< 0.005	< 0.005	<0.005	<0.005	--	--	--	--
NO_3 (mg/L)	--	--	--	--	.170	.088	.236	.026
tkN (mg/L)	.903	.754	.902	.792	.582	.542	--	--
tP (mg/L)	.038	.032	.042	.022	.029	.023	--	--
tSS (mg/L)	5.35	1.91	12.3	2.00	7.3§	2.5§	66.9§	15.0§
UV254 (cm^{-1})	0.145	0.170	.094	.141	.123	.102	.158	.133
Crypto (#/100 L)	--	--	<2	<2 to 2	ND	ND	ND	ND
Giardia (#/100 L)	--	--	14 - 32	36 - 256	ND	ND	ND	ND
	11-Jun-03		**27-Oct-03**		**10-Mar-04‡**		**9-Jun-04**	
Parameter	**MBDB**	**GRB**	**MBDB**	**GRB**	**MBDB**	**GRB**	**MBDB**	**GRB**
TC (cfu/100 mL)*	3.66	3.92	3.60	3.47	2.06	2.34	3.40	4.23
FC (cfu/100 mL)*	2.76	2.31	1.92	1.37	0.15	0	1.98	2.20
THPC(cfu/mL)*	5.36	5.33	5.27	5.18	3.86	3.74	4.86	5.29
C. perf(cfu/100mL) †	<1 to 2	<1 to 9	0.5--10	<0.5--19	<0.5 - 1	<0.5-4	<0.5-0.5	<0.5 -- 2
F+ (pfu/100 mL) †	--	--	--	--	--	--	--	--
F- (pfu/100 mL) †	--	--	--	--	--	--	--	--
Rhodo. (cfu/100 mL) †	5 to 20	23 to 67	<20-42	<20-33	--	--	<20-20	20 to 35
NO_3/NO_2 (mg/L)	0.037	0.036	0.016	0.005	0.068	0.062	0.115	0.176
NO_3 (mg/L)	0.037	0.036	0.016	0.005	0.068	0.062	0.115	0.174
tkN (mg/L)	0.465	0.598	0.270	0.337	0.156	0.183	0.256	0.633
tP (mg/L)	0.058	0.065	0.029	0.045	0.022	0.023	0.038	0.092
tSS (mg/L)	22.2	30.4	6.7	12.2	7.4	5.3	7.6	30.1
UV254 (cm^{-1})	0.137	0.159	0.338	0.252	0.032	0.045	0.104	0.117
Crypto (#/100 L)	ND	ND	ND	ND	ND	ND	ND	ND
Giardia (#/100 L)	ND	ND	ND	ND	ND	ND	ND	ND

(continued)

Table 5-8 (Continued)

	15-Oct-04	
Parameter	**MBDB**	**GRB**
TC (cfu/100 mL)*	3.47	3.50
FC (cfu/100 mL)*	1.97	1.74
THPC(cfu/mL)*	5.55	5.74
C. perf(cfu/100mL) †	10 to 40	11 to 50
F+ (pfu/100 mL) †	--	--
F- (pfu/100 mL) †	--	--
Rhodo. (cfu/100 mL) †	<20 to 35	<50 to 120
NO_3/NO_2 (mg/L)	0.040	0.019
NO_3 (mg/L)	0.040	0.018
tkN (mg/L)	0.339	0.309
tP (mg/L)	0.047	0.065
tSS (mg/L)	16.7	31.7
UV254 (cm^{-1})	0.197	0.232
Crypto (#/100 L)	ND	ND
Giardia (#/100 L)	ND	ND

Notes:
* Arithmetic mean of log transform of reported data
† Range of results
‡ Spatial sampling event; arithmetic average of all data collected across watershed
§ Arithmetic mean of 3 data points processed by UMass

Table 5-9
Event average (arithmetic) of samples collected during events at GB sites

	6-Nov-02		4-Feb-03		2-May-03	
Parameter	**GBWD**	**GBBR**	**GBWD**	**GBBR**	**GBWD**	**GBBR**
TC (cfu/100 mL)*	3.15	3.51	1.55	2.58	2.84	3.81
FC (cfu/100 mL)*	1.66	2.62	1.11	2.13	1.54	1.99
THPC(cfu/mL)*	2.81§	2.88§	5.08§	5.25§	5.16§	5.57§
C. perf (cfu/100 mL)†	2 to 7	42 to 59	5.5 to 10	10 to 295	0.5 to 5	5 to 42
F+ (pfu/100 mL)†	<1	2 to 6	<1	<1 to 1	<1 to 2.5	<1
F- (pfu/100 mL)†	<1	<1 to 2	<1	1 to 13	<1	<1 to 3.6
Bifido. (cfu/100 mL)†	<10 to <50	<50 to <500	<4	<4 to 3	<4	<4 to 4
NO_3 (mg/L)	0.422	.988	.806	1.603	.644	1.413
tkN (mg/L)	0.408	.766	.830	.822	.320	.434
tP (mg/L)	0.042	.108	.092	.110	.027	.036
tSS (mg/L)	5.29	15.0	51.4	42.4	8.0	12.7
UV254 (cm^{-1})	0.258	.207	0.0875	0.0790	.143	.131
Crypto (#/100 L)	ND	ND	ND	ND	ND	ND
Giardia (#/100 L)	ND	ND	<5 to 19	64 to 208	ND	ND
	1-Aug-03		**2-Sept-03**		**11-Dec-03**	
Parameter	**GBWD**	**GBBR**	**GBWD**	**GBBR**	**GBWD**	**GBBR**
TC (cfu/100 mL)*	3.82	4.73	3.81	4.35	3.05	2.77
FC (cfu/100 mL)*	3.23	3.89	2.85	3.14	1.99	1.27
THPC(cfu/mL)*	5.39§	5.91§	5.78§	5.80§	3.27	NA
C. perf (cfu/100 mL)†	0.5 to 7	2 to 60	0 to 6	0.5 to 7	2 to 17	NA
F+ (pfu/100 mL)†	--	--	--	--	--	--
F- (pfu/100 mL)†	--	--	--	--	--	--
Bifido. (cfu/100 mL)†	<2	<2 to 33	<2	<2	<2	<2
NO_3 (mg/L)	0.334	1.097	0.491	1.46	0.444	0.961
tkN (mg/L)	0.439	0.850	0.671	0.581	0.631	0.761
tP (mg/L)	0.053	0.210	0.077	0.144	0.080	0.145
tSS (mg/L)	12.9	76.4	21.8	30.4	18.4	43.9
UV254 (cm^{-1})	0.1427	0.1367	0.358	0.541	0.171	0.147
Crypto (#/100 L)	ND	ND	ND	ND	ND	ND
Giardia (#/100 L)	ND	ND	ND	ND	ND	ND

(continued)

Table 5-9 (Continued)

	12-Apr-04		4-Aug-04‡
Parameter	**GBWD**	**GBBR**	**GB**
TC (cfu/100 mL)*	3.48	3.92	3.83
FC (cfu/100 mL)*	2.26	2.18	2.02
THPC(cfu/mL)*	5.19	5.29	5.18
C. perf (cfu/100 mL) †	0.5 to 15	9 to 76	<0.5 to 2
F+ (pfu/100 mL) †	--	--	--
F- (pfu/100 mL) †	--	--	--
Bifido. (cfu/100 mL) †	<2	<2 to 75	<2
NO_3 (mg/L)	0.795	1.239	1.36
tkN (mg/L)	0.571	0.751	0.779
tP (mg/L)	0.057	0.175	0.129
tSS (mg/L)	14.2	70.9	147.9
UV254 (cm^{-1})	0.154	0.147	0.204
Crypto (#/100 L)	ND	ND	ND
Giardia (#/100 L)	ND	ND	ND

Notes: * Arithmetic mean of log transform of reported data
† Range of results
‡ Spatial sampling event; arithmetic average of all data collected across watershed
§ Arithmetic mean of 3 data points processed by UMass

Table 5-10

Event average (arithmetic) of samples collected during events at BB sites

	28-May-03		18-Jun-03		12-Aug-03	
Parameter	**BBWR**	**BBFM**	**BBWR**	**BBFM**	**BBWR**	**BBFM**
TC (cfu/100 mL)*	3.43	3.42	2.63	3.09	3.79	3.78
FC (cfu/100 mL) *	1.65	1.71	2.03	2.54	2.83	3.22
THPC(cfu/mL)*	5.07§	5.30§	2.26§	2.57§	6.01§	6.06§
C. perf(cfu/100 mL)†	1 to 2.5	<1 to 3.5	NA	NA	1.5	0 to 1
F+ (pfu/100 mL)†	10 to 26	6 to 23	NA	NA	NA	NA
F- (pfu/100 mL)†	10 to 17	8 to 17	NA	NA	NA	NA
Rhodo. (cfu/100 mL)†	<5 to 200	<5	NA	NA	<20 to 33	<20 to 73
NO_3 (mg/L)	.062	.069	.158	.774	0.193	0.209
tkN (mg/L)	.295	.334	.364	.422	0.819	1.032
tP (mg/L)	.027	.030	.035	.050	0.127	0.134
tSS (mg/L)	<5	5.0	5	7.1	30.3	35.1
UV254 (cm^{-1})	0.2722	0.2779	.3059	.2934	0.4619	0.4753
Crypto	ND	<10 to 100	ND	ND	ND	ND
Giardia	ND	<10 to 698	ND	ND	ND	ND
	19-Sept-03		**12-Nov-03**		**17-Dec-03**	
Parameter	**BBWR**	**BBFM**	**BBWR**	**BBFM**	**BBWR**	**BBFM**
TC (cfu/100 mL) *	3.13	3.64	2.81	2.77	3.76	3.60
FC (cfu/100 mL) *	2.36	2.32	1.54	2.18	1.22	1.54
THPC(cfu/mL) *	5.33	5.42	4.70	4.92	NA	NA
C. perf(cfu/100 mL)†	0 to 4.5	0 to 4	0 to 0.5	0 or <0.5	NA	NA
F+ (pfu/100 mL)†	NA	NA	NA	NA	NA	NA
F- (pfu/100 mL)†	NA	NA	NA	NA	NA	NA
Rhodo. (cfu/100 mL)†	503 to 920	255 to 763	80 to 93	27-153	NA	NA
NO_3 (mg/L)	0.098	0.117	0.136	0.179	0.155	0.181
tkN (mg/L)	0.552	0.633	0.250	0.243	0.564	0.559
tP (mg/L)	0.045	0.064	0.021	0.019	0.077	0.085
tSS (mg/L)	5.6	13.7	5.4	5	27.0	25.9
UV254 (cm^{-1})	1.0083	1.0385	0.297	0.462	0.259	0.271
Crypto	<5 to 10	ND	ND	ND	ND	ND
Giardia	<5 to 5	<5 to 70	ND	ND	ND	ND

(continued)

Table 5-10 (Continued)

	31-Mar-04		14-Jul-04‡
Parameter	**BBWR**	**BBFM**	**BB**
TC (cfu/100 mL)*	2.65	2.82	3.34
FC (cfu/100 mL)*	0.62	0.73	1.66
THPC(cfu/mL)*	4.29	4.55	4.75
C. perf(cfu/100 mL)†	0.5 to 1	0.5 to 4	1 to 6
F+ (pfu/100 mL)†	--	--	--
F- (pfu/100 mL)†	--	--	--
Rhodo. (cfu/100 mL)†	57 to 93	20 to 67	<20 to 14
NO_3 (mg/L)	0.197	0.236	0.213
tkN (mg/L)	0.379	0.377	0.541
tP (mg/L)	0.038	0.043	0.046
tSS (mg/L)	9.2	9.8	6.1
UV254 (cm^{-1})	0.2446	0.2126	0.395
Crypto	ND	ND	ND
Giardia	ND	ND	ND

Notes: * Arithmetic mean of log transform of reported data
† Range of results
‡ Spatial sampling event; arithmetic average of all data collected across watershed
§ Arithmetic mean of 3 data points processed by UMass

EVENT MEAN CONCENTRATIONS

Flow weighted mean concentrations (FWMC) of nutrients and microorganisms were calculated for all of the events sampled during this study, while time weighted mean concentrations (TWMC) were calculated for each of the continuous physical parameters. Both of these concentrations are interchangeably referred to as event mean concentration (EMC). Flow weighted mean concentrations were calculated using Equation 5.1

$$EMC = \frac{\sum_{i=t_1}^{t_n} Q_i * t_i * C_i}{\sum_{i=t_1}^{t_n} Q_i * t_i} \tag{5.1}$$

where:
- t = time interval over which the sample is representative
- I = incremental event sample
- n = last event sample
- Q = stream flow
- C = concentration

Time weighted concentrations were calculated as above except that flow was not included. The results of these calculations are presented in Tables 5-7 through 5-12. For each site the events are broken down by season. As noted in Chapter 6, due to numerous non-detects, analyses for phages were discontinued in the summer of 2003. During a few events, alternative indicator organism analyses had to be dropped due to timing issues (refer to Table 6-10).

Event mean concentrations provide a mechanism for summarizing a storm event by normalizing via flow, thus facilitating comparison between events and across season, site location, and research study. Such comparison can improve understanding of the magnitude of runoff quality issues related to various land use types. Additional analyses, such as comparison of loads, may provide further insight and are addressed in a later section.

Gates Brook Residential Study Site

The EMC values for events at GBBR are presented in Table 5-11. As presented in Table 5-11, total and fecal coliform EMCs are greater in the summer than the other seasons, as previously noted from the monthly data. Higher coliform counts in summer have typically been linked to the higher temperatures and microbial growth-death dynamics (see, for instance, Hunter *et al.*, 1999). Similarly, pH values are greater in the summer than during the other seasons. Conversely, the lowest EMCs for TC were recorded in the winter. Event mean concentrations of TP and TKN were lowest during the spring. *Clostridium perfringens* were detected at the highest levels during the winter event. Water temperatures followed the expected seasonal variations. Other parameters tended to exhibit variability across the seasons in terms of their EMC values. It should be noted that low discharge associated with the winter 2003 event is likely erroneous, caused by reading errors associated with icing. The only detection of *Giardia* occurred during the winter event.

Event mean concentrations for events at GBWD are presented in Table 5-12. Similar to GBBR, TC, FC, and pH were greater during summer events while TC and FC were lower in winter events. Both TP and TKN were lower during spring events than the other seasons. A high EMC value for *Clostridium perfringen*, however was not detected during the winter event but *Giardia* was detected during this event. In general, patterns between the two sampling locations were similar.

The EMC values are also useful for highlighting differences in concentrations between the up- and downstream sampling locations. In general, EMC values for all parameters (standard water chemistry, standard microbial, alternative indicator organisms, and physical data) were higher (or for a few very similar) at the downstream GBBR sampling location than upstream at GBWD. A few exceptions are noteworthy. Conductivity values at GBWD were higher than at GBBR during the summer events. During the 11-December-03 event, FC, TC, *Bifidobacteria*, and *Clostridium perfingen* EMC values were higher at GBWD. Lastly, during the 2-September-2003 event, TSS and TKN values at GBWD were slightly higher. The results suggest that overall, the heavily industrialized area between GBWD and GBBR had a negative impact on water quality during wet weather events, and this negative impact was greatest in the summer months, based on concentration data alone. This is explored further in the following section.

Major investigations into urban runoff quality were undertaken by the U.S. EPA through the Nationwide Urban Runoff Program (NURP) and by the U.S. Geological Survey (USGS; Urban Stormwater-Quality Investigations) in addition to other studies reported in the literature. Adams and Papa (2000) summarize EMC values collected from 16 representative catchment outfalls in Ontario, nine of which were separated storm sewer catchments such as found in Gates Brook. Values reported here for SS, nitrate, TKN, and TP were on the low end of the range of values reported for this study. Values for FC were very low in comparison.

Table 5-11
Flow Weighted and Time Weighted Mean Concentration for Events at GBBR

GBBR	FALL		WINTER	SPRING		SUMMER	
Parameter	11-6-02	12-11-03	2-4-03	5-2-03	4-13-04	8-1-03	9-2-03
Event Mean Concentration – Flow weighted							
Total suspended solids (mg/L)	11.78	40.94	26.39	4.43	16.15	47.55	17.30
Nitrate (mg/L)	1.02	0.83	1.59	1.42	1.74	1.13	1.45
Total Phosphorus (mg/L)	0.11	0.14	0.10	0.03	0.05	0.19	0.10
Total Kjeldahl Nitrogen (mgL)	0.76	0.78	0.76	0.37	0.30	0.78	0.49
Fecal Coliform (count/100 mL)	525	70.1	196	89.4	213	11,800	2,320
Total Coliform (count/100 mL)	4,324	956	494	9,360	15,200	92,000	42,500
F+ Phages	3.66	NA	0.85	0.50	NA	NA	NA
F- Phages	1.02	NA	8.96	2.06	NA	NA	NA
Rhodococcus (cfu/100 mL)	NA	NA	NA	NA	NA	NA	NA
Bifidobacteria (cfu/100 mL)	ND	1.00	2.33	3.37	44.60	10.50	1.00
C. perfringens	48.33	8.47	151.88	27.04	56.05	37.24	5.30
Cryptosporidium oocyst (per 100L)	ND	ND	ND	ND	ND	ND	ND
Giardia cysts (per 100L)	ND	ND	119.06	ND	ND	ND	ND
Event Mean Concentration – Time weighted							
Discharge ($ft^3 s^{-1}$)	46.32	48.86	0.11	40.61	53.26	36.24	36.11
pH	7.31	7.31	7.37	NA	7.26	7.48	7.64
Conductivity	539.20	391.20	524.84	NA	503.31	512.28	516.87
Temperature (°C)	6.89	2.89	2.04	NA	7.65	17.25	15.60

Note: ND replaced with half detection limit in calculations for all but *Cryptosporidium* and *Giardia*

Table 5-12
Flow Weighted and Time Weighted Mean Concentration for Events at GBWD

GBWD	FALL		WINTER	SPRING		SUMMER	
Parameter	11-6-02	12-11-03	2-4-03	5-2-03	4-13-04	8-1-03	9-2-03
Event Mean Concentration – Flow weighted							
Total suspended solids (mg/L)	3.19	11.57	20.06	3.20	3.77	10.14	20.65
Nitrate (mg/L)	0.41	0.32	0.82	0.59	0.90	0.33	0.46
Total Phosphorus (mg/L)	0.04	0.06	0.07	0.02	0.02	0.05	0.07
Total Kjeldahl Nitrogen (mgL)	0.42	0.58	0.79	0.32	0.34	0.43	0.68
Fecal Coliform (count/100 mL)	67.1	130	15.0	36.0	216	3,280	979
Total Coliform (count/100 mL)	1,550	1,450	53.2	633	6,080	12,700	6,670
F+ Phages	0.50	NA	0.50	1.46	NA	NA	NA
F- Phages	0.50	NA	0.50	0.50	NA	NA	NA
Rhodococcus (cfu/100 mL)	NA	NA	NA	NA	NA	NA	NA
Bifidobacteria (cfu/100 mL)	NA	56.48	2.00	2.00	1.57	0.38	1.00
C. perfringens	3.74	10.41	8.52	2.78	14.57	5.21	4.16
Cryptosporidium oocyst (per 100L)	ND	ND	ND	ND	ND	ND	ND
Giardia cysts (per 100L)	ND	ND	11.61	ND	ND	ND	ND
Event Mean Concentration – Time weighted							
Discharge ($ft^3 s^{-1}$)	0.95	4.54	1.56	1.48	8.63	0.50	1.30
pH	6.85	6.52	5.97	7.24	6.56	7.30	7.26
Conductivity	188.93	309.99	NA	779.04	372.06	637.95	608.88
Temperature (°C)	5.70	1.66	1.42	14.68	7.51	16.21	14.80

Burrow Brook Agricultural Site

Event mean concentrations for BBFM are presented in Table 5-13. In general, EMC values were variable across the seasons, suggesting few seasonal trends. Water temperature followed the expected seasonal variations. Conductivity values appeared to be greatest in the summer. High TC and FC values were also observed in the summer. However, values on the same order were also observed at least once in a different season. *Cryptosporidium* oocysts were observed once each in the spring and summer and *Giardia* cysts were observed during the same spring event.

Event mean concentrations for BBWR are presented in Table 5-14. As observed for BBFM, EMC values were variable across the seasons, suggesting few seasonal trends other than for water temperature. Fecal coliforms were found in higher concentrations during summer events while the lowest concentrations were observed during the winter event. A fall event exhibited the largest discharge and one of the highest total suspended solids concentrations. However the highest EMC total suspended solids concentration was associated with one of the lowest discharges and occurred in the summer. *Cryptosporidium* and *Giarda* were found during a later summer event, characterized by the lowest discharge of the seven events.

In general, EMC values for standard water chemistry and standard microbial indicator organisms were higher (or for a few very similar) at the downstream BBFM sampling location than upstream at BBWR. During the 17-December-03 and 31-March-04 events, FC levels were slightly elevated at BBWR. During the spring and summer events, sometimes *Rhodococcus* EMC values at BBWR were higher than at BBFM. Physical water quality TWMC values were variable. Conductivity values at GBWD tended to be higher, in terms of their EMC, during spring months. Values for pH were sometimes higher at BBWR and sometimes higher at BBFM, with no clear patterns. However, the results suggest that overall, the dairy pasture area between GBWD and GBBR had a negative impact on water quality during wet weather events. The negative impact in terms of nutrients was less than previously observed in the residential study site. At BB, the negative impact was greatest for the microbial parameters, particularly standard indicator organisms. This is explored further in the following section.

Table 5-13
Flow Weighted and Time Weighted Mean Concentration for Events at BBFM

BBFM	FALL		WINTER	SPRING		SUMMER	
Parameter	11/12/03	12/17/03	3/31/04	5/28/03	6/18/03	8/12/03	9/19/03
Event Mean Concentration – Flow weighted							
Total suspended solids (mg/L)	2.5	24.90	5.95	2.60	5.76	54.30	8.87
Nitrate (mg/L)	0.16	0.17	0.21	0.07	0.18	0.21	0.12
Total Phosphorus (mg/L)	0.02	0.09	0.03	0.03	0.05	0.26	0.06
Total Kjeldahl Nitrogen (mgL)	0.25	0.58	0.34	0.33	0.42	1.28	0.59
Fecal Coliform (count/100 mL)	143	45.9	6.95	55.9	440	8,040	242
Total Coliform (count/100 mL)	599	6,030	951	2,740	1,240	8,130	4,250
F+ Phages	NA	NA	NA	15.40	NA	NA	NA
F- Phages	NA	NA	NA	10.94	NA	NA	NA
Rhodococcus (cfu/100 mL)	78.1	NA	70.4	2.5	NA	11.8	492
Bifidobacteria (cfu/100 mL)	NA	NA	NA	NA	NA	NA	NA
C. perfringens	0.25	NA	2.62	1.66	NA	6.30	0.94
Cryptosporidium oocyst (per 100L)	ND	ND	ND	41.38	ND	ND	31.2
Giardia cysts (per 100L)	ND	ND	ND	236.56	ND	ND	ND
Event Mean Concentration – Time weighted							
Discharge (ft^3 s^{-1})	4.56	32.76	20.58	13.35	5.73	4.88	2.50
pH	5.77	4.99	6.51	NA	5.95	6.22	6.24
Conductivity	42.01	19.08	38.08	55.00	50.77	88.03	77.06
Temperature (°C)	3.64	0.30	3.45	13.47	13.38	22.08	17.19

Table 5-14.
Flow Weighted and Time Weighted Mean Concentration for events at BBWR

BBWR	FALL		WINTER	SPRING		SUMMER	
Parameter	11/12/03	12/17/03	3/31/04	5/28/03	6/18/03	8/12/03	9/19/03
Event Mean Concentration – Flow weighted							
Total suspended solids (mg/L)	2.66	22.65	5.74	2.69	2.56	47.99	3.75
Nitrate (mg/L)	0.13	0.13	0.18	0.06	0.15	0.18	0.10
Total Phosphorus (mg/L)	0.02	0.07	0.03	0.03	0.03	0.18	0.04
Total Kjeldahl Nitrogen (mgL)	0.27	0.52	0.32	0.30	0.34	1.00	0.54
Fecal Coliform (count/100 mL)	49.4	28.2	7.07	48.4	97.3	1,900	289
Total Coliform (count/100 mL)	633	9,430	550	2,680	390	9,510	1,550
F+ Phages	NA	NA	NA	18.80	NA	NA	NA
F- Phages	NA	NA	NA	12.53	NA	NA	NA
Rhodococcus (cfu/100 mL)	86.1	NA	42.1	87.2	NA	10.3	667
Bifidobacteria (cfu/100 mL)	NA	NA	NA	NA	NA	NA	NA
C. perfringens	0.36	NA	0.89	1.64	NA	1.5	2.34
Cryptosporidium oocyst (per 100L)	ND	ND	ND	ND	ND	ND	3.35
Giardia cysts (per 100L)	ND	ND	ND	ND	ND	ND	3.76
Event Mean Concentration – Time weighted							
Discharge ($ft^3 s^{-1}$)	2.81	23.75	16.18	11.14	4.22	2.78	1.17
pH	5.88	4.88	5.84	5.77	6.22	5.94	6.31
Conductivity	55.00	37.63	26.61	64.91	72.84	81.80	76.13
Temperature (°C)	4.09	0.67	3.37	13.47	15.40	21.20	16.84

Prescott Peninsula Wildlife Study Site

Event mean concentrations for MBDB are presented in Table 5-15. Results from the 4-December-2001 event, a practice event where samples were not collected for *Cryptosporidium* and *Giardia* analyses, are also provided. In general, nitrate levels were highest during the summer events while *Rhodococcus* EMC values were higher in the fall. *Clostridium perfringen* EMC values were higher in both the fall and winter than during summer and spring events. Water temperature followed the expected seasonal trends. The highest average discharge value occurred during the 26-September-2002 event, when the highest EMC values for TSS, FC, TC, and phages (although not analyzed for during all events) also occurred. However, no *Giardia* or *Cryptosporidium* were observed during this event and *Rhodococcus* levels were lower than in many of the other events. The event mean concentrations of other parameters tended to vary across seasons.

Event mean concentration values for events at GRB are presented in Table 5-16. Spring events demonstrated the highest TC and TSS EMC values. *Rhodococcus* and *C. perfringens* were found in greatest concentration during one of the fall events, but *Cryptosporidium* and *Giardia* were detected only during the winter event. Overall, EMC values were variable across season, with no other clear seasonal patterns suggested by the limited event data.

In comparing MBDB and GRB, differences occur across the events. Event mean concentrations were greater for all available parameters, including flow, at MBDB than GRB during the one winter event. However, *Cryptosporidium* and *Giardia* were detected at GRB during this event and not at MBDB. Most EMC values were also higher at MBDB than GRB during summer events, although pH values were higher at GRB. During the spring events, EMC values were typically higher at GRB than MBDB, although conductivity values were slightly higher at MBDB and FC values were split. During the fall events, results were typically variable, although nitrate and conductivity values at MBDB tended to be larger and TP and pH values tended to be lower. The effect of beaver on water quality thus appears to be variable by season and largely event driven. This is discussed further below and is the focus of a graduate student master's thesis.

Table 5-15.
Flow Weighted and Time Weighted Mean Concentration for events at MBDB

MBDB								
	FALL			WINTER	SPRING		SUMMER	
Parameter	12/4/01	10/27/03	10/15/04	3/26/02	6/11/03	6/9/04	7/28/02	9/26/02
Event Mean Concentration – Flow weighted								
Total suspended solids (mg/L)	4.69	4.50	11.26	10.25	18.72	5.07	6.07	45.73
Nitrate (mg/L)	NA	0.02	0.02	NA	0.04	0.10	0.18	0.18
Total Phosphorus (mg/L)	0.04	0.03	0.04	0.09	0.05	0.04	0.03	NA
Total Kjeldahl Nitrogen (mgL)	0.88	0.29	0.34	0.97	0.44	0.22	0.55	NA
Fecal Coliform (count/100 mL)	12.53	153	152	132	949	189	138	1,294
Total Coliform (count/100 mL)	1,295	6,130	3,087	526	4,394	2,716	2,585	23,628
F+ Phages	2.61	NA	NA	1.00	NA	NA	0.50	32.95
F- Phages	6.45	NA	NA	1.17	NA	NA	0.80	648.74
Rhodococcus (cfu/100 mL)	NA	26.87	24.04	NA	12.10	11.51	NA	4.20
Bifidobacteria (cfu/100 mL)	NA	NA	NA	NA	NA	NA	NA	NA
C. perfringens	4.61	4.32	26.26	9.09	1.12	0.28	0.77	2.08
Cryptosporidium oocyst (per 100L)	NA	ND	ND	ND	ND	ND	ND	ND
Giardia cysts (per 100L)	NA	ND	ND	22.25	ND	ND	ND	ND
Event Mean Concentration – Time weighted								
Discharge ($ft^3 s^{-1}$)	0.41	1.66	0.95	1.62	0.29	0.33	0.18	1.78
pH	6.61	6.13	5.81	6.31	6.10	5.76	6.94	6.83
Conductivity	73.85	39.34	51.88	73.91	23.98	40.35	39.31	28.39
Temperature (°C)	5.49	11.23	11.91	2.28	13.77	15.98	16.78	9.51

Table 5-16
Flow Weighted and Time Weighted Mean Concentration for events at GRB

GRB		FALL		WINTER	SPRING		SUMMER	
Parameter	12/4/01	10/27/03	10/15/04	3/26/02	6/11/03	6/9/04	7/28/02	9/26/02
Event Mean Concentration – Flow weighted								
Total suspended solids (mg/L)	1.99	17.20	8.28	1.42	27.52	20.34	2.5	3.77
Nitrate (mg/L)	NA	0.005	0.005	NA	0.03	0.15	0.09	0.01
Total Phosphorus (mg/L)	0.03	0.06	0.06	0.02	0.06	0.07	0.02	NA
Total Kjeldahl Nitrogen (mgL)	0.76	0.45	0.32	0.76	0.56	0.49	0.55	NA
Fecal Coliform (count/100 mL)	14.87	70.81	451	10.3	306	186	40.06	668
Total Coliform (count/100 mL)	294	5,482	6,189	141	20,702	15,006	2,133	7,044
F+ Phages	6.52	NA	NA	0.50	NA	NA	0.50	0.71
F- Phages	2.80	NA	NA	0.70	NA	NA	0.50	0.79
Rhodococcus (cfu/100 mL)	NA	17.08	61.99	NA	58.30	25.90	NA	3.96
Bifidobacteria (cfu/100 mL)	NA	NA	NA	NA	NA	NA	NA	NA
C. perfringens	0.67	12.08	38.74	2.27	2.08	1.31	0.87	1.24
Cryptosporidium oocyst (per 100L)	NA	ND	ND	1.54	ND	ND	ND	ND
Giardia cysts (per 100L)	NA	ND	ND	120.63	ND	ND	ND	ND
Event Mean Concentration – Time weighted								
Discharge (ft^3 s^{-1})	0.13	0.79	0.32	0.67	1.06	0.23	0.03	0.17
pH	6.94	6.33	5.65	NA	6.57	6.38	6.98	7.07
Conductivity	71.10	30.12	50.53	NA	13.58	39.75	38.06	11.48
Temperature (°C)	5.81	9.79	11.98	NA	10.78	15.95	16.14	3.26

Event EMC Comparison between Sampling Locations at a Study Site

Tables 5-17, 5-18 and 5-19 provide a comparison over all events between the two sampling locations at GB, BB and PR respectively. Each is discussed individually below. Differences between sampling locations have not yet been evaluated in terms of statistical significance (work in progress as part of a PhD thesis). It should be noted that with the exception of nitrate and discharge at Gates Brook, *C. perfringens* at Burrow Brook, and F+ and F- Phages at Prescott, overlap between the EMC mean + or – the EMC standard deviation at the upstream and downstream sampling sites overlap to some extent, suggesting that statistically significant differences may not exist.

For events at GB, the upstream site (GBWD) was characterized by lower average EMC values and ranges than the downstream site (GBBR) for all evaluated parameters with the exception of *Bifidobacteria* and temperature, which were similar. While the standard deviation for the data was for the most part greater at GBBR than GBWD, the coefficient of variation (CV), which is the standard deviation normalized by the mean and thus provides a dimensionless measure of variability, was sometimes greater at GBWD. These data further support the earlier noted decrease in water quality attributable to the heavily residential area between the two sites. Variation across the events tended to be high (CV values ranging from 0.25 to 2.0 for standard water quality parameters), which emphasizes the complexity inherent in event based water quality data.

For events at Burrow Brook, the upstream site (BBWR) was characterized by lower average EMC values and ranges than the downstream site (BBFM) for all standard water quality parameters (excluding the physical parameters) other than TC, which was similar. Average pH, conductivity, and temperature values were slightly lower at BBFM. Alternative indicator organisms, *Cryptosporidium*, and *Giardia* means and ranges are difficult to compare due to non-detects and reduced sample numbers. In general, the available data suggest that phages were present in slightly higher concentrations upstream at BBWR, as were *Rhodococcus*. *Clostridium perfringens* appeared to be present in higher numbers downstream, as did *Cryptosporidium* and *Giardia*. Both the standard deviation and CV of the EMC data from BBFM tended to be greater. Total coliform was a notable exception, with higher variation observed at BBWR. The across-event summary of the EMC data further suggest that the dairy farm had an overall slight negative impact on water quality during wet weather events.

For events at Prescott, MBDB was characterized by higher EMC values for all parameters except TKN, TC, *Rhodococcu*s and *C. perfringens*. The standard deviation and CV were also typically greater at MBDB. The CV was greater at GRB than MBDB for nitrate, *C. perfingens*, *Rhodococcu*s, and conductivity. *Cryptosporidium* and *Giardia*, however, were observed at higher levels at GRB. Earlier it was noted that comparative water quality between the two sites appeared influenced by season. Thus the effect of beaver on water quality appears mixed. Beaver may contribute to larger microbial concentrations, particularly TC, but appear to slightly improve concentrations in terms of nutrients and suspended sediment. Sudden impoundment breaches, however, could have severe negative impacts on downstream water quality. The effect of beaver on water quality was the focus of a master's thesis stemming from this research and a journal paper on the topic is in preparation.

Table 5-17
Comparison of FWMC and TWMC for events at GB

	GBWD				GBBR			
Parameter	Range	Mean	Std Dev	CV	Range	Mean	Std Dev	CV
Event Mean Concentration – Flow weighted								
Total suspended solids (mg/L)	3.19-20.65	10.37	7.61	0.73	4.43-47.55	23.51	15.73	0.67
Nitrate (mg/L)	0.32-0.9	0.55	0.23	0.43	0.83-1.74	1.31	0.33	0.25
Total Phosphorus (mg/L)	0.02-0.07	0.05	0.02	0.45	0.03-0.19	0.10	0.05	0.52
Total Kjeldahl Nitrogen (mgL)	0.32-0.79	0.51	0.18	0.35	0.3-0.78	0.61	0.21	0.35
Fecal Coliform (count/100 mL)	15-3,280	675	1,197	1.77	70.1-11,800	2,173	4,319	1.99
Total Coliform (count/100 mL)	53.2-12,700	4,162	4,590	1.10	494-92,000	23,548	33,485	1.42
F+ Phages	0.5-1.46	0.82	0.55	0.68	0.5-3.66	1.67	1.73	1.04
F- Phages	0.5-0.5	0.5	0.00	0.00	1.02-8.96	4.01	4.32	1.08
Rhodococcus (cfu/100 mL)	NA				NA			
Bifidobacteria (cfu/100 mL)	0.38-56.48	10.57	22.50	2.13	1-44.6	10.47	17.09	1.63
C. perfringens	2.78-14.57	7.06	4.30	0.61	5.3-151.88	47.76	49.66	1.04
Cryptosporidium oocyst (per 100L)	NA				NA			
Giardia cysts (per 100L)	11.61				119.06			
Event Mean Concentration – Time weighted								
Discharge ($ft^3 s^{-1}$)	0.5-8.65	2.71	2.92	1.08	0.11-53.26	37.36	17.64	0.47
pH	6.52-7.30	6.96	0.33	0.05	7.26-7.64	7.40	0.14	0.02
Conductivity	188.93-779.04	483	226	0.47	391.2-539.2	498	53.69	0.11
Temperature (°C)	1.42-16.21	8.85	6.35	0.72	2.04-17.25	8.72	6.37	0.73

Table 5-18
Comparison of FWMC and TWMC for events at BB

Parameter	BBWR Range	BBWR Mean	BBWR Std Dev	BBWR CV	BBFM Range	BBFM Mean	BBFM Std Dev	BBFM CV
Event Mean Concentration – Flow weighted								
Total suspended solids (mg/L)	2.56-47.99	12.58	17.21	1.37	2.5-54.3	14.98	18.96	1.27
Nitrate (mg/L)	0.06-0.18	0.13	0.04	0.32	0.07-0.21	0.16	0.05	0.31
Total Phosphorus (mg/L)	0.02-0.18	0.06	0.06	0.99	0.02-0.26	0.08	0.08	1.09
Total Kjeldahl Nitrogen (mgL)	0.27-0.54	0.47	0.26	0.55	0.25-1.28	0.54	0.35	0.65
Fecal Coliform (count/100 mL)	7.07-1,900	346	692	2.00	6.95-8,040	1,282	2,984	2.33
Total Coliform (count/100 mL)	390-9,510	3,535	4,130	1.17	599-8,130	3,420	2,856	0.84
F+ Phages	18.8				15.4			
F- Phages	12.53				10.94			
Rhodococcus	10.3-667	179	275	1.54	2.5-492	131	205	1.56
Bifidobacteria	NA				NA			
C. perfringens	0.36-2.34	1.35	0.75	0.56	0.25-630	128	281.13	2.21
Cryptosporidium oocyst (per 100L)	3.35				31.2-41.38	36.29	7.20	0.20
Giardia cysts (per 100L)	3.76				236.56			
Event Mean Concentration – Time weighted								
Discharge ($ft^3 s^{-1}$)	1.17-23.75	8.86	8.51	0.96	2.5-32.76	12.05	11.12	0.92
pH	4.88-6.31	5.83	0.47	0.08	4.99-6.24	5.95	0.53	0.09
Conductivity	26.61-81.80	59.27	20.66	0.35	19.08-88.03	52.86	23.48	0.44
Temperature (°C)	0.67-21.2	10.72	7.91	0.74	0.3-22.08	10.50	8.13	0.77

Table 5-19
Comparison of FWMC and TWMC for events at PR

	MBDB				GRB			
Parameter	Range	Mean	Std Dev	CV	Range	Mean	Std Dev	CV
Event Mean Concentration – Flow weighted								
Total suspended solids (mg/L)	4.5-45.73	14.51	14.62	1.01	1.42-27.52	11.58	10.17	0.88
Nitrate (mg/L)	0.02-0.18	0.09	0.08	0.84	0.005-0.15	0.05	0.06	1.23
Total Phosphorus (mg/L)	0.03-0.09	0.05	0.02	0.48	0.02-0.07	0.05	0.02	0.46
Total Kjeldahl Nitrogen (mgL)	0.22-0.97	0.47	0.27	0.58	0.32-0.76	0.52	0.15	0.28
Fecal Coliform (count/100 mL)	12.53-1,294	430	483	1.13	10.3-668	247	244	0.98
Total Coliform (count/100 mL)	526-23,628	6,152	7,896	1.28	141-20,702	8,100	7,270	0.90
F+ Phages	0.5-32.95	11.48	18.59	1.62	0.5-6.52	0.57	0.12	0.21
F- Phages	0.8-648.74	217	374	1.72	0.5-2.8	0.66	0.15	0.22
Rhodococcus	4.2-26.87	15.74	9.45	0.60	3.96-61.99	33.45	25.63	0.77
Bifidobacteria	NA				NA			
C. perfringens	0.28-26.26	6.27	9.32	1.49	0.67-38.74	8.37	13.96	1.67
Cryptosporidium oocyst (per 100L)	NA				1.54			
Giardia cysts (per 100L)	22.25				120.63			
Event Mean Concentration – Time weighted								
Discharge (ft^3 s^{-1})	0.18-1.78	0.97	0.71	0.73	0.03-1.06	0.47	0.38	0.81
pH	5.76-6.94	6.27	0.46	0.07	5.65-7.07	6.50	0.51	0.08
Conductivity	28.39-73.91	42.45	16.54	0.39	11.48-71.1	30.59	15.44	0.50
Temperature (°C)	2.28-16.78	11.64	4.87	0.42	3.26-16.14	11.32	4.75	0.42

Event EMC Comparison Across Study Sites

Tables 5-20 through 5-22 provide a qualitative comparison across study sites for the EMC data summarized in Tables 5-17 through 5-19 of the previous section. As noted previously for the between sampling location comparisons, statistical significance of the observed differences has not yet been tested. This assessment is also underway as part of a Ph.D. thesis stemming from the research. Comparisons for some parameters were not included due to non-detects or missing data. *Cryptosporidium* and *Giardia* are excluded as they are discussed in the next chapter.

Table 5-20 summarizes differences observed between the more impacted sampling location at each of the three study sites (GBBR – downstream residential site, BBFM – downstream agriculture site, and GRB – beaver influenced ranging wildlife site). The EMC values at GBBR are the greatest across the study sites for each parameter studied. The BBFM site has the second highest EMC values for parameters other than total coliform, which are greater at GRB. Elevated total coliform levels at GRB are likely the result of both the large population of free ranging wildlife on the Prescott Peninsula and the presence of beaver in the watershed. These findings indicate that GBBR is the most impacted sampling location in this study in terms of water quality, with degradation related to urbanization. Water quality degradation decreases with a reduction in anthropogenic activity.

The trends are not as straight forward for the less impacted sites, summarized in Table 5-21 (GBWD – upstream residential site, BBWR – upstream agriculture site, MBDB – ranging wildlife site). This was not unexpected as some similarities in land use exist, particularly between BBWD and GBWD. Based on the EMC data collected for this study, GBWD is the most impacted in terms of fecal coliform, nitrate, and TKN, while MBDB is the most impacted in terms of total coliform and total suspended solids. BBWR is the most impacted in terms of total phosphorus. Of the three sites, GBWD is the most human influenced (43% urban). The elevated total coliform concentrations at MBDB are likely due to the large population of free ranging wildlife on the Prescott Peninsula while the elevated TSS concentration at MBDB are likely due in part to its grade, which is the steepest of the three compared sites.

Table 5-20
Ranking of EMC values across study sites for the most impacted sampling location

Parameter	Comparison								
Fecal Coliform	GBBR	>	BBFM	GBBR	>	GRB	BBFM	>	GRB
Total Coliform	GBBR	>	BBFM	GBBR	>	GRB	BBFM	<	GRB
Nitrate	GBBR	>	BBFM	GBBR	>	GRB	BBFM	>	GRB
TKN	GBBR	>	BBFM	GBBR	>	GRB	BBFM	>	GRB
Total Phosphorus	GBBR	>	BBFM	GBBR	>	GRB	BBFM	>	GRB
Total Suspended Solids	GBBR	>	BBFM	GBBR	>	GRB	BBFM	>	GRB
Rhodococcus	GBBR	NA	BBFM	GBBR	NA	GRB	BBFM	>	GRB
F+ Phages	GBBR	NA	BBFM	GBBR	>	GRB	BBFM	NA	GRB
F- Phages	GBBR	NA	BBFM	GBBR	>	GRB	BBFM	NA	GRB

The two PR sites were also compared in this manner, Table 5-22, although the comparison was discussed in the preceding section. As noted, MBDB was characterized by higher EMC values except for TKN, TC, *Rhodococcus* and *C. perfringens*. Earlier it was noted that comparative water quality between the two sites varied by season. Thus the effect of beaver on water quality appears mixed. Beaver may contribute to larger microbial concentrations, particularly TC, but appear to slightly improve concentrations in terms of nutrients and suspended sediment during storm events. Sudden impoundment breaches, however, could have severe negative impacts on downstream water quality. The effect of beaver on water quality was the focus of a master's thesis stemming from this research and a journal paper on the topic is in preparation.

Table 5-21
Ranking of EMC values across study sites for the less impacted sampling location

Parameter	Comparison								
Fecal Coliform	GBWD	>	BBWR	GBWD	>	MBDB	BBWR	<	MBDB
Total Coliform	GBWD	>	BBWR	GBWD	<	MBDB	BBWR	<	MBDB
Nitrate	GBWD	>	BBWR	GBWD	>	MBDB	BBWR	>	MBDB
TKN	GBWD	>	BBWR	GBWD	>	MBDB	BBWR	=	MBDB
Total Phosphorus	GBWD	<	BBWR	GBWD	=	MBDB	BBWR	>	MBDB
Total Suspended Solids	GBWD	<	BBWR	GBWD	<	MBDB	BBWR	<	MBDB
Rhodococcus	GBWD	NA	BBWR	GBWD	NA	MBDB	BBWR	>	MBDB
F+ Phages	GBWD	NA	BBWR	GBWD	<	MBDB	BBWR	NA	MBDB
F- Phages	GBWD	NA	BBWR	GBWD	<	MBDB	BBWR	NA	MBDB

Table 5-22
Ranking of EMC values across study sites for the wildlife locations

Parameter	Comparison		
Fecal Coliform	MBDB	>	GRB
Total Coliform	MBDB	<	GRB
Nitrate	MBDB	>	GRB
TKN	MBDB	<	GRB
Total Phosphorus	MBDB	=	GRB
Total Suspended Solids	MBDB	>	GRB
Rhodococcus	MBDB	<	GRB
F+ Phages	MBDB	>	GRB
F- Phages	MBDB	>	GRB

EVENT LOADS

Also of interest is the mass of contaminant, rather than just the concentration, transported during wet weather events. Event loads were calculated based on the paired concentration and discharge data and are presented in Tables 5-23 through 5-28. Data across all events for a single parameter are provided in each table. The tables are also organized by season and study site. Event loads reflect the volume of runoff occurring during each event and the event duration. Because each event differs in duration and volume, direct comparisons are difficult. Future work will include transformation of the event loads to a common time unit and normalization based on drainage area to facilitate comparison against each other and against loads estimated based on the monthly data. The data are offered here to provide an overall understanding of differences between sites and events, but no attempt is made to draw conclusions based on the above stated deficiencies. Loads calcuated for GBBR should be considered preliminary pending review of the rating curve, as previously noted, as discharge estimates were higher than expected.

Table 5-23
Event Loading of Nitrate (kg)

	FALL			WINTER			SPRING			SUMMER		
Gates	**Date**	**GBWD**	**GBBR**	**Date**	**GBWD**	**GBBR**	**Date**	**GBWD**	**GBBR**	**Date**	**GBWD**	**GBBR**
	11-6-02	33.8	89.0	2-4-03	1.72	92.3	5-2-03	1.27	84.2	8-1-03	0.120	54.7
	12-11-03	4.36	113				4-13-04	156	2030	9-2-03	1.58	80.7
Burrow	**Date**	**BBWR**	**BBFM**	**Date**	**BBWR**	**BBFM**	**Date**	**BBWR**	**BBFM**	**Date**	**BBWR**	**BBFM**
	11/12/03	0.958	1.94	3/31/04	48.3	67.1	5/28/03	0.750	1.16	8/12/03	1.07	1.73
	12/17/03	19.7	26.8				6/18/03	1.30	2.12	9/19/03	0.241	0.584
Prescott	**Date**	**MBDB**	**GRB**	**Date**	**MBDB**	**GRB**	**Date**	**MBDB**	**GRB**	**Date**	**MBDB**	**GRB**
	12/14/01	NA	NA	3/26/02	NA*	NA*	6/11/03	0.010	0.028	7/28/02	0.011	0.002
	10/27/03	0.082	0.013				6/9/04	0.046	0.044	9/26/02	0.467	0.007
	10/15/04	0.047	0.003									

* Ammonia rather than nitrate was monitored for until May 2002

Table 5-24
Event Loading of TKN (kg)

	FALL			WINTER			SPRING			SUMMER		
Gates	**Date**	**GBWD**	**GBBR**	**Date**	**GBWD**	**GBBR**	**Date**	**GBWD**	**GBBR**	**Date**	**GBWD**	**GBBR**
	11-6-02	35.102	66.557	2-4-03	1.669	44.341	5-2-03	0.692	22.013	8-1-03	0.156	37.554
	12-11-03	7.951	106.775				4-13-04	59.352	353.965	9-2-03	2.351	26.974
Burrow	**Date**	**BBWR**	**BBFM**	**Date**	**BBWR**	**BBFM**	**Date**	**BBWR**	**BBFM**	**Date**	**BBWR**	**BBFM**
	11/12/03	2.047	2.963	3/31/04	88.535	106.960	5/28/03	3.810	5.730	8/12/03	5.878	10.729
	12/17/03	77.292	91.564				6/18/03	2.888	4.839	9/19/03	1.312	2.941
Prescott	**Date**	**MBDB**	**GRB**	**Date**	**MBDB**	**GRB**	**Date**	**MBDB**	**GRB**	**Date**	**MBDB**	**GRB**
	12/14/01	0.684	0.193	3/26/02	2.1745	0.637	6/11/03	0.119	0.490	7/28/02	0.033	0.010
	10/27/03	1.558	1.077				6/9/04	0.105	0.145	9/26/02	2.751	0.0275
	10/15/04	0.640	0.204									

Table 5-25
Event Loading of TP (kg)

	FALL			WINTER			SPRING			SUMMER		
Gates	**Date**	**GBWD**	**GBBR**	**Date**	**GBWD**	**GBBR**	**Date**	**GBWD**	**GBBR**	**Date**	**GBWD**	**GBBR**
	11-6-02	3.705	9.197	2-4-03	0.156	5.884	5-2-03	0.051	1.900	8-1-03	0.018	9.197
	12-11-03	0.848	18.857				4-13-04	4.286	60.819	9-2-03	0.258	5.710
Burrow	**Date**	**BBWR**	**BBFM**	**Date**	**BBWR**	**BBFM**	**Date**	**BBWR**	**BBFM**	**Date**	**BBWR**	**BBFM**
	11/12/03	0.145	0.219	3/31/04	8.525	10.593	5/28/03	0.340	0.520	8/12/03	1.066	2.159
	12/17/03	10.507	13.617				6/18/03	0.294	0.549	9/19/03	0.104	0.275
Prescott	**Date**	**MBDB**	**GRB**	**Date**	**MBDB**	**GRB**	**Date**	**MBDB**	**GRB**	**Date**	**MBDB**	**GRB**
	12/14/01	0.028	0.008	3/26/02	0.0910	0.018	6/11/03	0.015	0.053	7/28/02	0.002	0.000
	10/27/03	0.163	0.142				6/9/04	0.017	0.021	9/26/02	0.272	0.062
	10/15/04	0.082	0.036									

Table 5-26
Event Loading of TSS (kg)

	FALL			WINTER			SPRING			SUMMER		
Gates	**Date**	**GBWD**	**GBBR**	**Date**	**GBWD**	**GBBR**	**Date**	**GBWD**	**GBBR**	**Date**	**GBWD**	**GBBR**
	11-6-02	265.231	1029.777	2-4-03	42.133	1530.111	5-2-03	6.930	262.196	8-1-03	3.677	2302.283
	12-11-03	157.398	5570.794				4-13-04	657.304	18900	9-2-03	71.184	959.712
Burrow	**Date**	**BBWR**	**BBFM**	**Date**	**BBWR**	**BBFM**	**Date**	**BBWR**	**BBFM**	**Date**	**BBWR**	**BBFM**
	11/12/03	19.889	29.510	3/31/04	1571.866	1882.974	5/28/03	33.850	44.520	8/12/03	280.768	453.393
	12/17/03	3361.626	3964.985				6/18/03	21.656	66.030	9/19/03	9.189	44.061
Prescott	**Date**	**MBDB**	**GRB**	**Date**	**MBDB**	**GRB**	**Date**	**MBDB**	**GRB**	**Date**	**MBDB**	**GRB**
	12/14/01	3.626	0.503	3/26/02	23.090	1.187	6/11/03	5.066	24.180	7/28/02	0.364	0.046
	10/27/03	24.166	41.024				6/9/04	2.371	6.044	9/26/02	118.164	1.837
	10/15/04	21.004	5.229									

Table 5-27
Event Loading of Total Coliform (cfu)

	FALL			WINTER			SPRING			SUMMER		
Gates	**Date**	**GBWD**	**GBBR**	**Date**	**GBWD**	**GBBR**	**Date**	**GBWD**	**GBBR**	**Date**	**GBWD**	**GBBR**
	11-6-02	3.10E10	4.22E12	2-4-03	1.26E9	3.12E11	5-2-03	1.36E10	5.13E12	8-1-03	9.79E10	4.92E13
	12-11-03	1.93E10	3.23E11				4-13-04	1.57E12	2.10E13	9-2-03	4.37E11	7.63E13
Burrow	**Date**	**BBWR**	**BBFM**	**Date**	**BBWR**	**BBFM**	**Date**	**BBWR**	**BBFM**	**Date**	**BBWR**	**BBFM**
	11/12/03	4.77E10	7.17E10	3/31/04	5.79E11	1.18E12	5/28/03	3.52E11	5.04E11	8/12/03	7.46E11	7.39E11
	12/17/03	8.62E12	5.25E12				6/18/03	5.52E10	2.36E11	9/19/03	3.53E10	1.65E11
Prescott	**Date**	**MBDB**	**GRB**	**Date**	**MBDB**	**GRB**	**Date**	**MBDB**	**GRB**	**Date**	**MBDB**	**GRB**
	12/14/01	1.06E10	7.66E8	3/26/02	1.18E10	8.51E8	6/11/03	1.21E10	1.66E11	7/28/02	1.18E9	4.34E8
	10/27/03	9.62E12	1.27E11				6/9/04	1.63E10	6.33E10	9/26/02	4.77E11	3.06E10
	10/15/04	5.52E10	3.86E10									

Table 5-28
Event Loading of Fecal Coliform (cfu)

	FALL			WINTER			SPRING			SUMMER		
Gates	**Date**	**GBWD**	**GBBR**	**Date**	**GBWD**	**GBBR**	**Date**	**GBWD**	**GBBR**	**Date**	**GBWD**	**GBBR**
	11-6-02	1.34E9	5.13E11	2-4-03	3.57E8	1.24E11	5-2-03	7.71E8	4.90E10	8-1-03	2.53E10	6.32E12
	12-11-03	1.73E9	2.37E10				4-13-04	5.58E10	2.96E11	9-2-03	6.42E10	4.17E12
Burrow	**Date**	**BBWR**	**BBFM**	**Date**	**BBWR**	**BBFM**	**Date**	**BBWR**	**BBFM**	**Date**	**BBWR**	**BBFM**
	11/12/03	3.72E9	1.71E10	3/31/04	7.43E9	8.66E9	5/28/03	6.35E9	1.03E10	8/12/03	1.49E11	7.31E11
	12/17/03	2.57E10	3.99E10				6/18/03	1.38E10	8.40E10	9/19/03	6.56E9	9.40E9
Prescott	**Date**	**MBDB**	**GRB**	**Date**	**MBDB**	**GRB**	**Date**	**MBDB**	**GRB**	**Date**	**MBDB**	**GRB**
	12/14/01	1.03E8	3.87E7	3/26/02	2.95E9	6.24E7	6/11/03	2.61E9	2.46E9	7/28/02	6.29E7	8.16E6
	10/27/03	2.14E8	1.64E9				6/9/04	1.13E9	7.86E8	9/26/02	2.61E10	2.90E9
	10/15/04	2.71E9	2.81E9									

SPATIAL EVENTS

At each study site, one event was conducted that was spatial in nature. For these events, additional sampling locations within the watershed were identified based on potential upstream sources of contamination. Rather than collecting samples (16 nutrient, 8 standard indicator, 3 alternative indicator, and 3 *Cryptosporidium* and *Giardia*) at two locations over the rise and fall of the hydrograph during wet weather, the identified sampling sites were sampled during a period of no rainfall. Due to access and time constraints, the final number of samples analyzed for each parameter in the watersheds varied. The results of these analyses are summarized by parameter in Figures 5-37 through 5-43. For clarification, Burrow Brook drains from the bottom to top, Gates Brook drains from left to right, Griswold Brook drains from top to bottom and then to the right, and Middle Branch Dickey Brook drains from right to left. Typical sampling location GBWD, GBBR, BBBR, BBFM, MBDB, and GRB are shown on the figures. Spatial sampling occurred at GB on 4-August-2004, at BB on 14-July-2004, and at PR on 10-March-2004. Overall, spatial sampling results corresponded well with event and monthly data that indicated overall water quality parameter levels were higher in GB and BB than in GRB and MBDB.

One goal of the sampling was to determine if *Cryptosporidium* and *Giardia* were present within the stream during non-event conditions. If detected during dry weather, their presence would be anticipated during wet weather. Thus the lack of more detections would suggest either sampling protocols or environmental conditions were causing false negatives. No *Cryptosporidium* and *Giardia* were detected during these spatial sampling events. While this does not rule out the potential for false negatives, the spatial data combined with the analyses in Chapter 6 suggest a lack of presence as the dominant factor for lack of detection.

Nitrate concentrations (Figure 5-37) increased along the main stem from headwaters to outlet in all watersheds except for Griswold Brook, where concentrations were higher in the headwaters above the main beaver impoundment. In Gates Brook, the increase in nitrate was dramatic below the GBWD sampling location. In Burrow Brook, some upstream sampling locations had slightly higher concentrations. Total Kjeldahl nitrogen concentrations (Figure 5-38) tended to be high throughout Burrow Brook and increased between BBWR and BBFM. Gates Brook was characterized by low TKN values except at two locations near swampy areas. In Griswold Brook, TKN concentrations remained fairly constant at mid-range values along the main stem. Concentrations along MBDB decreased from the headwaters to the outlet, with the exception of the sampling location immediately upstream from MBDB. Spatial variability of TP within the study watersheds is shown on Figure 5-39. The results were almost identical to TKN in terms of spatial variability. Along MBDB, however, concentrations remained elevated rather than decreasing. Total suspended sediment concentrations were less than the detection limit at most locations, but isolated elevated values were detected. Values for UV254 (Figure 5-41) tended to be fairly high throughout BB, with some higher values in the headwaters. Similarly, UV254 values along GB were slightly higher in the headwaters than near the outlet. Levels in both GRB and MBDB tended to be very low, although some slightly elevated values were detected near the pond located in the headwaters of MBDB. The levels of TC were higher in BB and GB than at GRB or MBDB, Figure 5-42. Levels of TC tended to increase overall along BB. In GRB and MBDB, values tended to alternate along the stream between non-detect and low levels. Similarly, FC levels were higher in BB and GB than GRB or MBDB. Values tended to increase in a downstream direction within BB. At GB, values rose and fell several times along the mainstem. Values within GRB and MBDB were typically very low or non-detects.

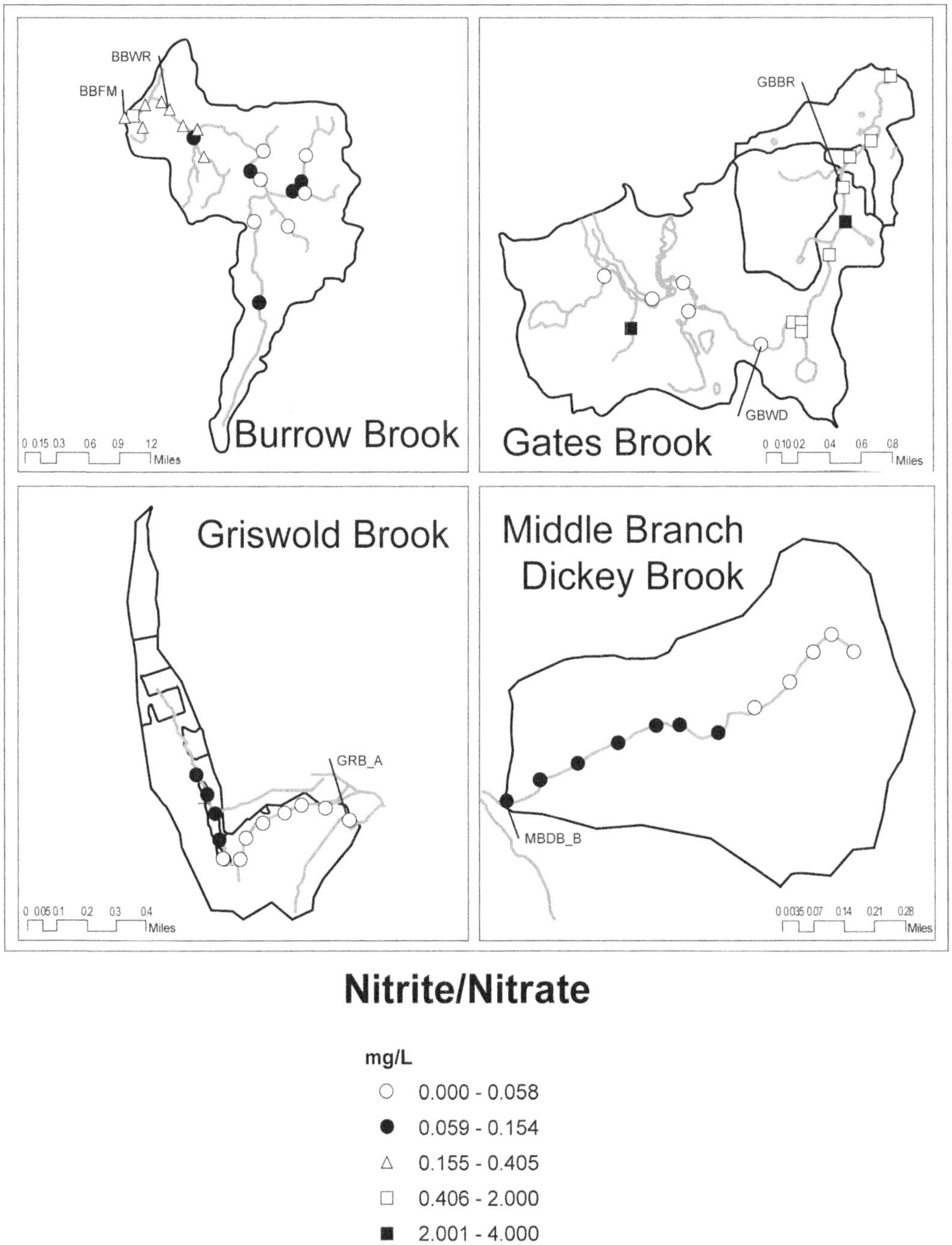

Figure 5-37 Nitrate/Nitrite Spatial Event Concentrations

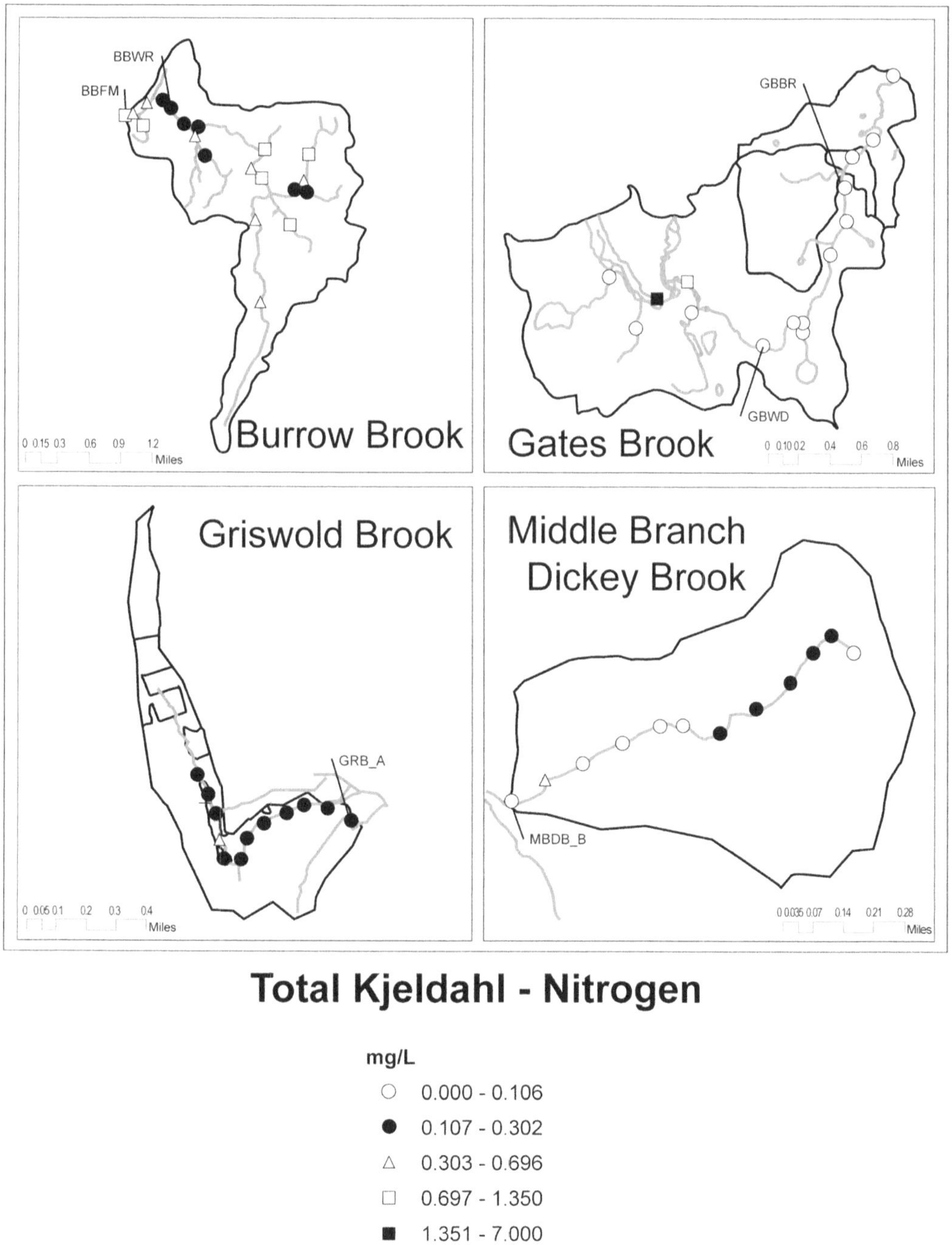

Figure 5-38 Total Kjeldahl Nitrogen Spatial Event Concentrations

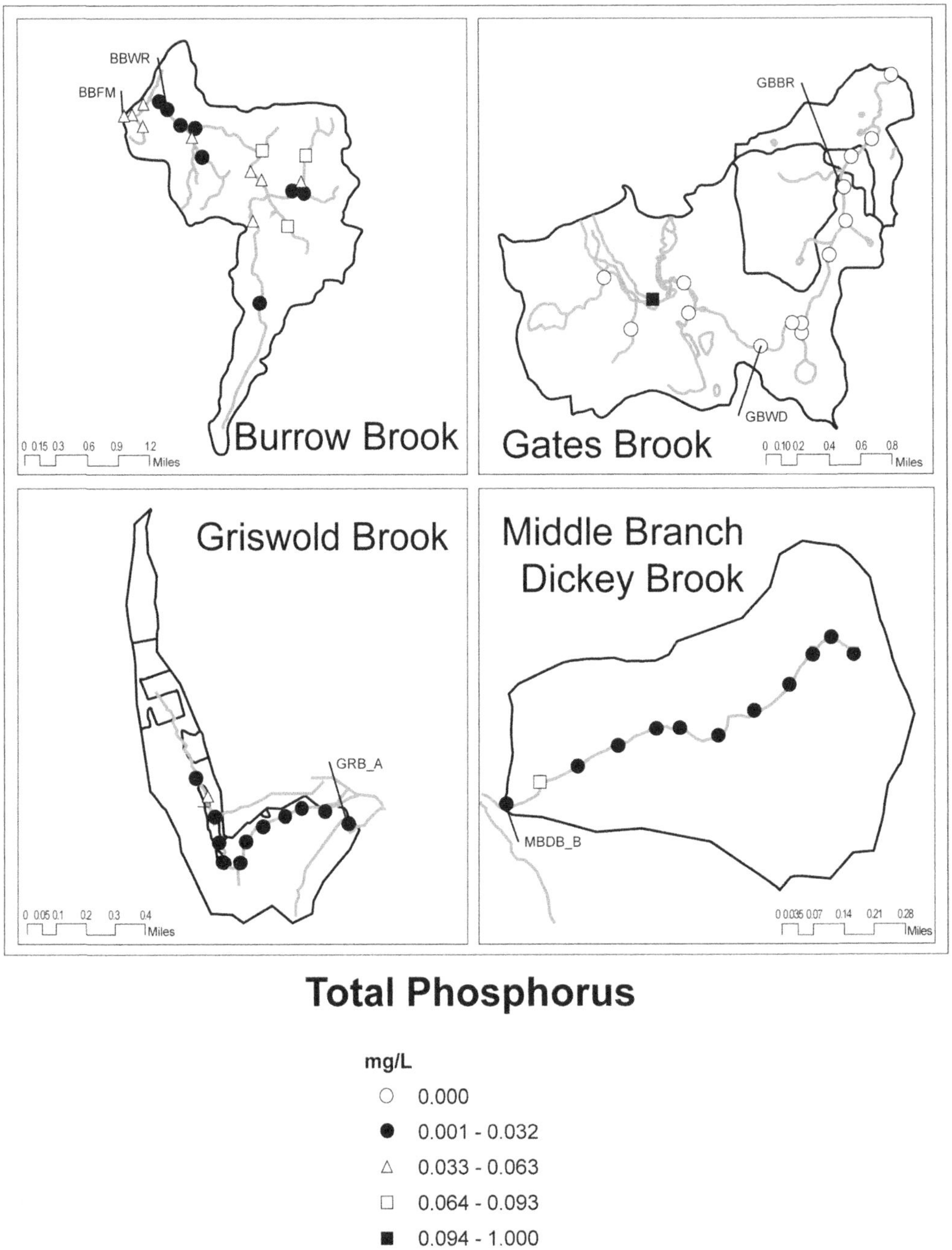

Figure 5-39 Total Phosphorus Spatial Event Concentrations

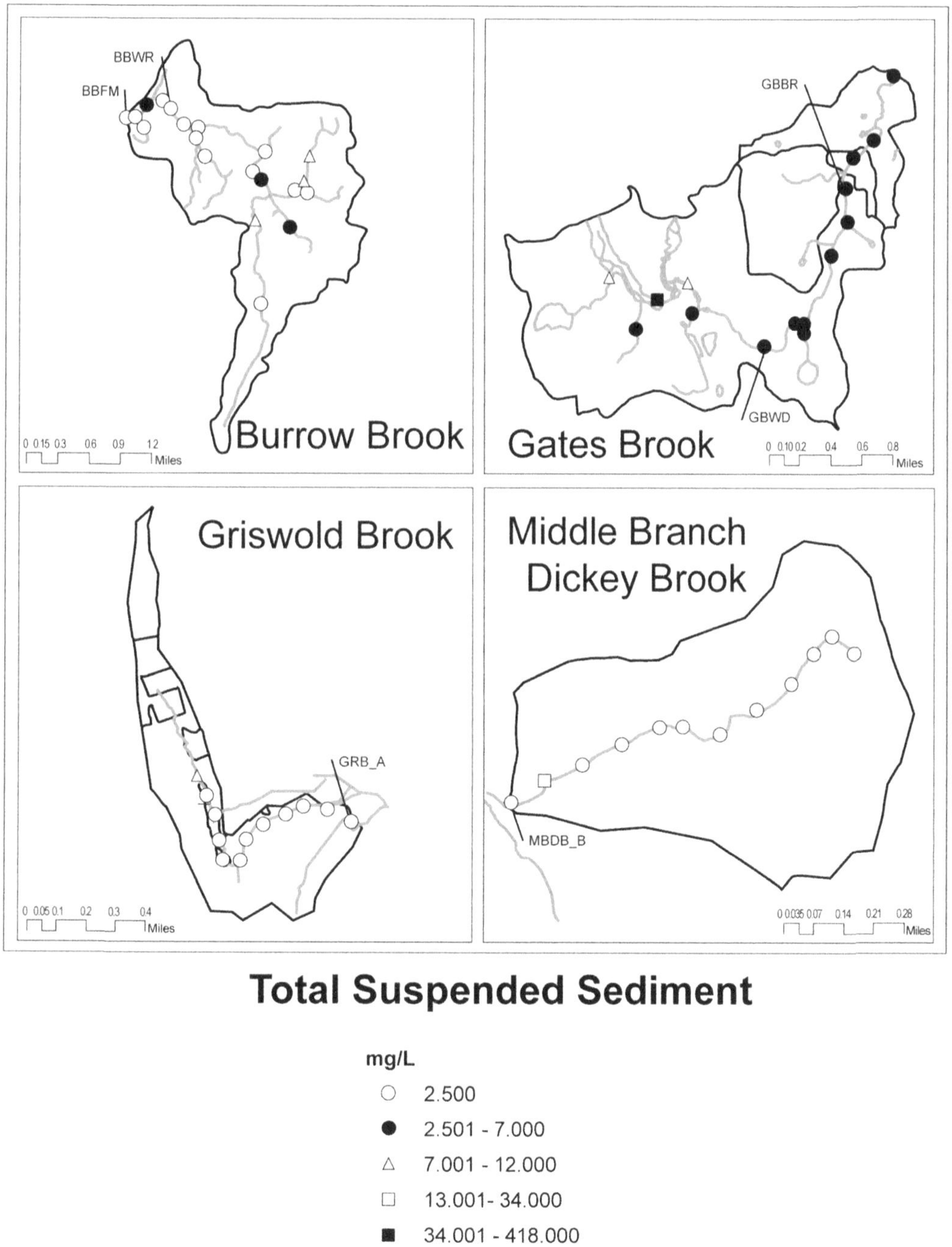

Figure 5-40 Total Suspended Sediment Spatial Event Concentrations

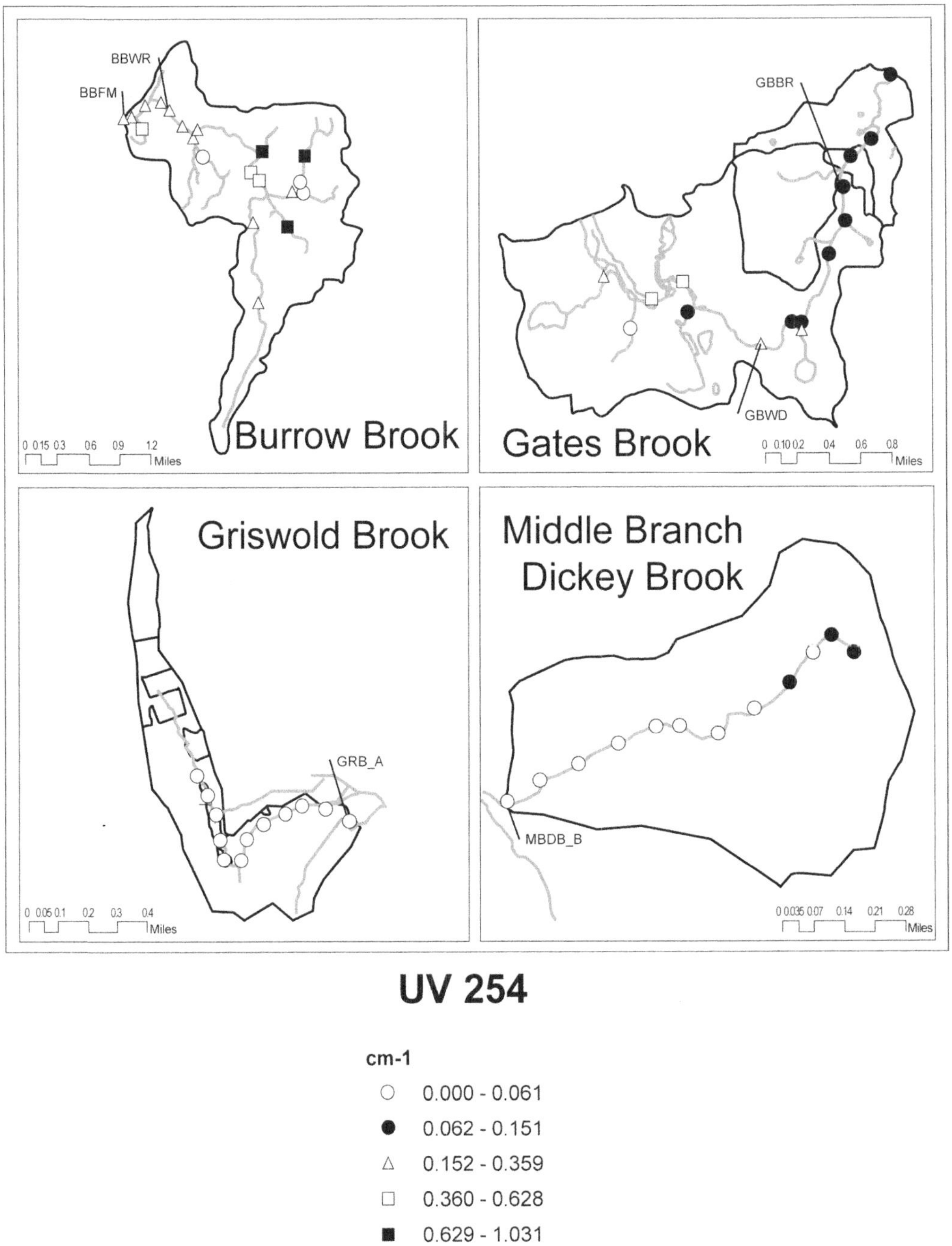

Figure 5-41 UV 254 Spatial Event Results

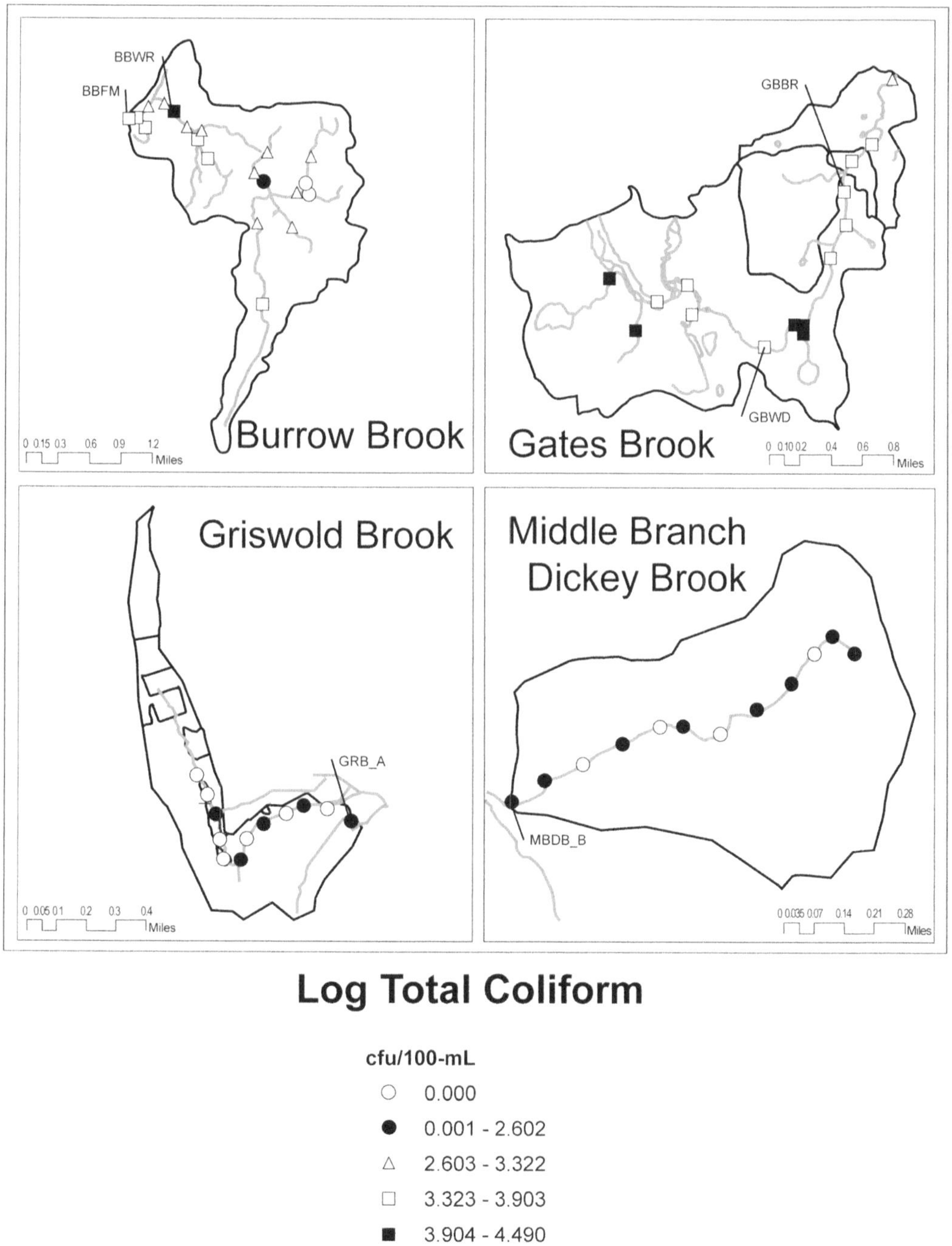

Figure 5-42 Log Total Coliform Spatial Event Concentrations

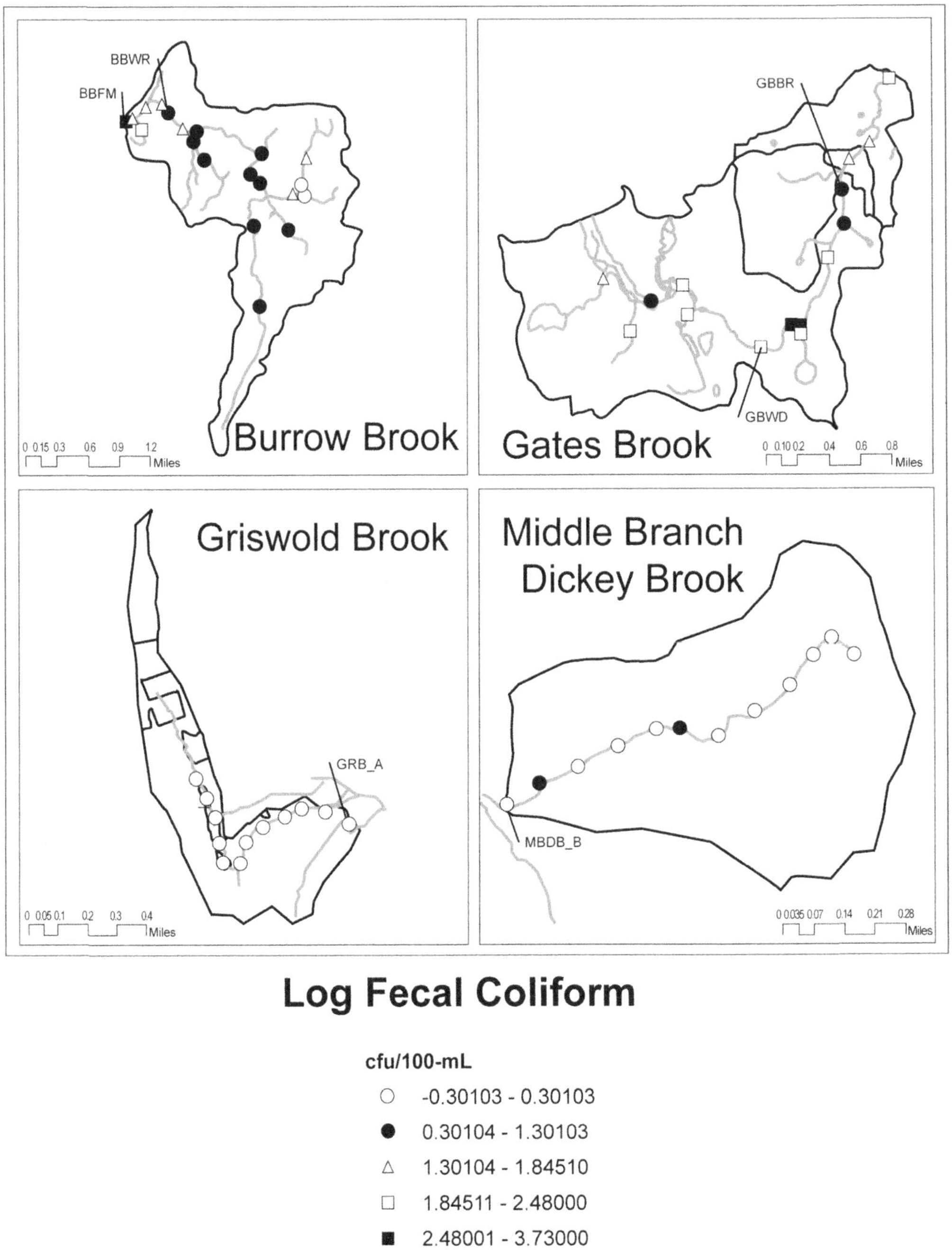

Figure 5-43 Log Fecal Coliform Spatial Event Concentrations

CORRELATIONS

Product-Moment

Product-moment correlation coefficients were estimated using S-Plus for the event parameters at each of the three sampling sites. Equation 5.2 was used to calculate the correlations:

$$r = \frac{\sum_{i=1}^{n}(x_i - \bar{x})(y_i - \bar{y})}{\sqrt{\sum_{i=1}^{n}(x_i - \bar{x})^2 \sum_{i=1}^{n}(y_i - \bar{y})^2}} \qquad (5\text{-}2)$$

where:
- r = sample correlation
- i = first sample
- n = last sample
- x = random variable
- y = random variable

Only the rainfall event data was utilized for this analysis. Flow weighted and time weighted mean concentration/values were used as the random variables. Thus for each event (dry or wet) one concentration value was available per sampling location in the watershed. For each study (GB, BB, and PR), the EMC values were lumped together across events and sampling locations, resulting in a maximum of sixteen values for use in the analysis. Data from all the watersheds were then combined to examine correlations for the entire study data set. This correlation analysis incorporated a maximum of 44 data points. All parameters meeting the following criterion were used in the correlation analysis. The parameter must have values for greater than 50% of the samples being evaluated, none of the parameter values can be equal to zero and the values for one parameter cannot differ by more than two orders of magnitude. Therefore, some of the parameters (*ie.*, *Cryptosporidium*, *Giardia* and F+ and F- phages) had to be removed from the data. Separate correlation analyses were conducted for *Cryptosporidium* and *Giardia* incorporating only the events corresponding to detected concentrations for the microorganism being evaluated. These results are presented in Chapter 6.

Correlation analyses were conducted to determine the degree of dependence among the pollutants. Correlations close to one indicate strong relationships while correlations close to zero indicate no dependence between the parameters. Correlations are useful for identifying hydrologic characteristics contributing to high pollutant levels and for assisting in developing monitoring strategies. Correlations greater than 0.75 are highlighted in the tables. Select weak correlations (in the range of +/- 0.4 to 0.75) are noted below.

The correlations specific to GB are provided in Table 5-29. As may be expected, strong positive correlations were observed between TSS and TP, FC and TC, and FC and temperature. The other correlations were not expected and included negative correlations between TC and TKN, Rain and TKN, and FC and TSS. A positive correlation between conductivity and FC was also observed. Strong correlations with alternative indicator organisms were not observed. *Clostridium perfingens* were weakly correlated with nitrate (r=0.619). *Bifidobacteria* were weakly negatively correlated with TKN (r=-.402). Rainfall was weakly correlated with TP (r=-.656), TSS (r=-.499), and TC (r=0.605).

The correlations specific to BB are provided in Table 5-30. A larger number of strong (greater than 0.75) correlations were observed in the data for this watershed. Rainfall accumulation was found to only be correlated with nitrate. Total Kjeldahl nitrogen was found to be positively correlated with TP, TSS, TC, FC, conductivity, and temperature (perhaps indicating a seasonal variation). Total phosphorus was positively correlated with TSS, TC, FC, and temperature. In addition to TKN and TP, TSS was positively correlated with TC and FC. Total coliform was positively correlated with FC, conductivity and temperature in addition to the previously noted correlations with TKN, TP, and TSS. FC was also correlated with conductivity and temperature. Lastly, temperature and conductivity appeared to be positively correlated. Strong correlations with alternative indicator organisms were also not observed for BB. *Clostridium perfingens* were weakly correlated with pH (r=0.567). *Bifidobacteria* were weakly correlated with nitrate (r=-0.529), rain (r=-.453), and discharge (r=-.429). Additional weak correlations existed between TC and discharge (r=-.402) and FC and discharge (r=-.764).

The correlations specific to PR are provided in Table 5-31. A positive correlation between rainfall accumulation and FC and negative correlations between TP, TC, FC, and pH were found. In addition, FC was positively correlated with both TP and TC. Positive correlations between TSS and TKN and between TC and TP were also found. As for the other study watersheds, strong correlations with alternative indicator organisms were also not observed for PR. *Clostridium perfingens* were weakly negatively correlated with nitrate (r=-.442), *Bifidobacteria* (r=0.551) and conductivity (r=0.662). *Bifidobacteria* were weakly correlated with TC (r=-.487) and FC (r=-.47). Rainfall was weakly correlated with TKN (r=0.544), TP (r=0.741), TC (r=0.642), FC (r=0.820), pH (r=-0.639), and temperature (r=-0.584).

When all of the EMC data from the study were combined for correlation analyses, only a few correlations greater than 0.8 were observed, Table 5-32. These were between TP and TKN, TSS and TKN, TSS and TP, and conductivity and nitrate. The first three were not unexpected based on the literature. Even with the combined data set, strong correlations with alternative indicator organisms were not observed. However, the only indicator organism which had enough data spanning all the sites for inclusion was *Clostridium perfingens*, which had slight positive correlations with conductivity (r=0.53) and nitrate (r=0.64).

While the results of the correlation analysis are of interest, usefulness in terms of defining monitoring strategies is limited. Of most interest were correlations between rain, discharge and the monitored water quality parameters, in addition to correlations with the alternative indicator organisms. Few strong correlations with these parameters (rain, discharge, and the alternative indicator organisms) occurred and, when present, were not consistent across the study watersheds or in the lumped analysis. The lack of correlations and observed level of inconsistency are most likely related both to the relatively small number of samples included in the analysis and to the underlying variability of the hydrologic data (wide range of rainfall and discharge values). Non-parametric correlation analyses, such as the Spearman's rank correlation coefficients were also evaluated and are discussed in the next section.

Table 5-29
Correlation between different pollutants in an urban watershed

	NO_3^-	TKN	TP	TSS	TC	FC	*C. perf*	Bifido	Cond.	Temp.	Rain	Q
NO_3^-	1.000											
TKN	-0.341	1.000										
TP	0.109	0.680	1.000									
TSS	0.079	0.711	0.913	1.000								
TC	0.242	-0.823	-0.324	-0.420	1.000							
FC	0.249	-0.587	-0.281	-0.762	0.839	1.000						
C. perf.	0.619	0.132	0.123	0.179	-0.190	0.038	1.000					
Bifido	-0.073	-0.402	-0.370	-0.340	0.250	0.019	-0.008	1.000				
Cond.	0.343	-0.271	0.144	0.096	0.690	0.790	0.310	-0.066	1.000			
Temp.	0.119	-0.406	-0.080	-0.232	0.731	0.836	-0.333	-0.267	0.645	1.000		
Rain	0.235	-0.859	-0.656	-0.499	0.605	0.229	-0.090	0.355	0.116	0.087	1.000	
Q	0.482	-0.368	0.322	0.346	0.451	0.034	-0.148	0.167	0.244	0.186	0.409	1.000

Note: Correlations greater than 0.75 have been shaded

Table 5-30
Correlation between different pollutants in an agricultural watershed

	NO_3^-	TKN	TP	TSS	TC	FC	*C. perf*	*Bifido*	Cond.	Temp.	Rain	Q
NO_3^-	1.000											
TKN	0.405	1.000										
TP	0.489	0.980	1.000									
TSS	0.527	0.957	0.986	1.000								
TC	0.027	0.849	0.801	0.799	1.000							
FC	0.050	0.839	0.791	0.759	0.793	1.000						
C. perf.	-0.162	-0.206	-0.279	-0.259	-0.019	-0.342	1.000					
Bifido	-0.529	-0.088	-0.277	-0.351	0.015	0.135	0.368	1.000				
Cond.	-0.210	0.768	0.675	0.631	0.876	0.884	-0.052	0.333	1.000			
Temp.	-0.113	0.841	0.751	0.717	0.939	0.859	0.021	0.281	0.948	1.000		
Rain	0.780	0.273	0.338	0.378	0.022	-0.256	0.223	-0.453	-0.319	-0.103	1.000	
Q	0.331	-0.378	-0.285	-0.267	-0.402	-0.764	0.390	-0.429	-0.702	-0.552	0.681	1.000

Note: Correlations greater than 0.75 have been shaded

Table 5-31
Correlation between different pollutants in a forested watershed

	NO_3^-	TKN	TP	TSS	TC	FC	*C. perf*	*Bifido*	Cond.	Temp.	Rain	Q
NO_3^-	1.000											
TKN	0.590	1.000										
TP	0.269	0.729	1.000									
TSS	0.642	0.866	0.345	1.000								
TC	0.276	0.597	0.922	0.180	1.000							
FC	0.191	0.650	0.899	0.287	0.825	1.000						
C. perf.	-0.442	-0.321	-0.243	-0.261	-0.384	-0.148	1.000					
Bifido	-0.370	-0.305	-0.366	-0.072	-0.487	-0.470	0.551	1.000				
Cond.	0.036	-0.471	-0.595	-0.290	-0.549	-0.487	0.662	0.230	1.000			
Temp.	0.347	-0.406	-0.722	0.013	-0.701	-0.665	0.020	0.200	0.570	1.000		
Rain	0.204	0.544	0.741	0.192	0.642	0.820	0.247	-0.226	-0.062	-0.584	1.000	
Q	0.190	0.444	-0.092	0.490	-0.056	-0.119	-0.096	-0.019	0.016	-0.160	0.044	1.000

Note: Correlations greater than 0.75 have been shaded

Table 5-32
Correlation between different pollutants across all watersheds

	NO_3^-	TKN	TP	TSS	TC	FC	*C. perf*	Cond.	Temp.	Rain	Q
NO_3^-	1.000										
TKN	0.131	1.000									
TP	0.173	0.850	1.000								
TSS	0.163	0.843	0.833	1.000							
TC	-0.137	0.128	0.308	0.183	1.000						
FC	0.043	0.493	0.605	0.438	0.737	1.000					
C. perf.	0.639	0.083	0.091	0.092	-0.197	0.038	1.000				
Cond.	0.837	0.155	0.190	0.139	-0.025	0.168	0.529	1.000			
Temp.	-0.302	0.236	0.260	0.252	0.441	0.469	-0.309	-0.189	1.000		
Rain	0.261	0.055	0.210	0.180	0.276	0.080	0.048	0.172	-0.197	1.000	
Q	0.663	0.000	0.158	0.115	0.013	-0.096	0.184	0.602	-0.242	0.379	1.000

Note: Correlations greater than 0.75 have been shaded

Spearman's Rho

Spearman's correlation coefficients were estimated using MINITAB for the event parameters at each of the three sampling sites. The equation used to calculate the correlations is presented below as Equation 5-3.

$$rho = \frac{\sum_{i=1}^{n} \left(R_{xi} - R_{yi}\right) - n * \left(\frac{n+1}{2}\right)^2}{n\left(n^2 - 1\right) / 12} \tag{5-3}$$

where: rho = sample correlation
i = first sample
n = last sample
x = random variable
y = random variable
R_{xi} = ranks of x
R_{yi} = ranks of y

Ties in x or y were assigned average ranks.

Only the event data were utilized for this analysis. The spatial sampling events were included. Flow weighted and time weighted mean concentration/values were used as the random variables. Thus for each event, one concentration value was available per sampling location in the watershed. For each study site (GB, BB, and PR), the EMC values were lumped together across events and sampling locations, resulting in a maximum of sixteen values for use in the analysis. Data from all the watersheds were then combined to examine correlations for the entire study data set. For inclusion in the correlation analysis, a complete set of positive (non-detects were not acceptable) parameter values was necessary. Therefore, some of the parameters for which non-detects frequently occurred (*ie.*, *Cryptosporidium*, *Giardia* and F+ and F- phages) had to be removed from the analysis. This removal criterion was more limiting than the criteria for the product moment correlations, thereby restricting the observations/conclusions that could be made. Separate correlation analyses were conducted for *Cryptosporidium* and *Giardia* incorporating only the events corresponding to detected concentrations for the microorganism being evaluated. These results are presented in Chapter 6.

Similar to the product moment correlation analyses, these analyses were conducted to determine the degree of dependence among the pollutants. Rho values close to one indicate strong relationships while rho values close to zero indicate no dependence between the parameters. Correlations are useful for identifying hydrologic characteristics contributing to high pollutant levels and for assisting in developing monitoring strategies. Parameters with correlations greater than 0.75 are highlighted in the results tables. Select weak correlations (in the range of +/- 0.4 to 0.75) are noted below.

The correlations specific to GB are provided in Table 5-29. As may be expected, a strong positive correlation was observed between TSS and both TKN and TP. Also as might be expected, TP and TKN showed a relatively strong correlation as did TC and FC. Strong correlations with alternative indicator organisms were not observed. *Clostridium perfingens* were weakly correlated with nitrate (r=0.701) as well as with discharge (r=0.424). No other strong or otherwise noteworthy correlations were evidenced by this table.

The correlations specific to BB are provided in Table 5-30. A larger number of strong (greater than 0.75) correlations were observed in the data for this watershed. Total Kjeldahl nitrogen was found to be positively correlated with TP, TSS, TC, FC, conductivity, rain, and temperature (perhaps indicating a seasonal variation), with TP and TSS being the only strong correlations. Total phosphorus was positively correlated with TSS, and weakly correlated with TC (r=0.742) and FC (r=0.469). In addition TSS was positively correlated with TC. Fecal coliform was strongly correlated with conductivity and temperature and weakly negatively correlated with discharge (r=-0.644). Temperature and conductivity appeared to be positively correlated while both temperature and conductivity were negatively correlated with discharge. Nitrate showed weak correlations with TP (r=0.410), TSS (r=0.503) and rain (r=0.549). Rainfall accumulation was also found to be weakly positively correlated with TP and TSS.

The correlations specific to PR are provided in Table 5-31. Strong correlations for this site existed between TP and both TC and FC and between TC and FC. As for the other study watersheds, strong correlations with alternative indicator organisms were not observed. The only such correlation was a week positive correlation between *Clostridium perfingens* and discharge (r=0.591). Rainfall was weakly correlated with TP (r=0.666), TC (r=0.420) and FC (r=0.586). Similar to the other sites although weaker, TP and TSS (r=0.696) were correlated. TSS was also weakly correlated with TC (r=0.612), FC (r=0.676) and discharge (r=0.459).

When all of the EMC data from the study were combined for correlation analyses, only one correlation greater than 0.8 was observed, Table 5-32. This was between TP and TSS. Several weak correlations were observed, including TKN with both TP (r=0.628) and TSS (r=0.538), as well as TP with both TC (r=0.514) and FC (r=0.545). Similarly, TSS also showed weak correlations with TC (r=0.507) and FC (r=0.443). In addition a weak correlation was observed between TC and FC (r=0.668). Due to the restrictive criterion for evaluating Spearman's rho, indicator organisms were not included in this evaluation.

While the results of the correlation analysis are of interest, similar to those of the product moment correlation analyses, usefulness in terms of defining monitoring strategies is limited. Of most interest were correlations between rain, discharge and the monitored water quality parameters, in addition to correlations with the alternative indicator organisms. Few strong correlations with these parameters (rain, discharge, and the alternative indicator organisms) occurred and, when present, were not consistent across the study watersheds or in the lumped analysis. The lack of correlations and observed level of inconsistency are most likely related both to the relatively small number of samples included in the analysis and to the underlying variability of the hydrologic data (wide range of rainfall and discharge values) as well as lack of data for key water quality parameters.

Table 5-33
Spearman's correlation between different pollutants in an urban watershed

Gates Brook Watershed (GBBR and GBWD)

	NO_3^-	TKN	TP	TSS	TC	FC	*C. perf*	Q
NO_3^-	1.000							
TKN	-0.055	1.000						
TP	0.284	0.798	1.000					
TSS	0.367	0.763	0.873	1.000				
TC	0.270	-0.310	0.160	0.095	1.000			
FC	0.138	-0.051	0.314	0.204	0.833	1.000		
C. perf.	0.701	0.160	0.341	0.385	0.187	0.158	1.000	
Q	0.486	-0.042	0.297	0.209	0.376	-0.046	0.424	1.000

Note: Correlations greater than 0.75 have been shaded

Table 5-34
Spearman's correlation between different pollutants in an agricultural watershed

Burrow Brook Watershed (BBWR and BBFM)

	NO_3^-	TKN	TP	TSS	TC	FC	Cond.	Temp.	Rain	Q
NO_3^-	1.000									
TKN	0.301	1.000								
TP	0.410	0.967	1.000							
TSS	0.503	0.824	0.872	1.000						
TC	-0.002	0.719	0.742	0.767	1.000					
FC	0.086	0.552	0.469	0.209	0.336	1.000				
Cond.	-0.154	0.458	0.295	0.128	0.282	0.766	1.000			
Temp.	-0.110	0.473	0.310	0.121	0.255	0.805	0.986	1.000		
Rain	0.549	0.505	0.573	0.709	0.292	-0.292	-0.297	-0.266	1.000	
Q	0.253	-0.174	0.018	0.187	0.086	-0.644	-0.785	-0.761	0.461	1.000

Note: Correlations greater than 0.75 have been shaded

Table 5-35
Spearman's correlation between different pollutants in a forested watershed

Prescott Peninsula Watersheds (MBDB and GRB)

	TKN	TP	TSS	TC	FC	*C. perf*	Rain	Q
TKN	1.000							
TP	0.729	1.000						
TSS	0.866	0.345	1.000					
TC	0.597	0.922	0.180	1.000				
FC	0.650	0.899	0.287	0.825	1.000			
C. perf.	-0.321	-0.243	-0.261	-0.384	-0.148	1.000		
Rain	0.544	0.741	0.192	0.642	0.820	0.247	1.000	
Q	0.444	-0.092	0.490	-0.056	-0.119	-0.096	0.044	1.000

Note: Correlations greater than 0.75 have been shaded

Table 5-36
Spearman's correlation between different pollutants across all watersheds

All Watersheds

	TKN	TP	TSS	TC	FC	Q
TKN	1.000					
TP	0.850	1.000				
TSS	0.843	0.833	1.000			
TC	0.128	0.308	0.183	1.000		
FC	0.493	0.605	0.438	0.737	1.000	
Q	0.000	0.158	0.115	0.013	-0.096	1.000

Note: Correlations greater than 0.75 have been shaded

SUMMARY AND DISCUSSION

In this chapter, monthly and event results for nutrients, standard indicator organism, and alternative indicator organisms were presented and discussed. The events captured spanned a range of hydrologic and contaminant response. Typically higher concentrations of contaminants were observed during event flows than during baseflow monitoring. Runoff from some events was such that little change in concentration was observed. In addition, some monthly sampling followed significant rainfall and was thus potentially more indicative of wet weather conditions. The captured events represent study site watershed response to a wide range of hydrologic conditions. The resulting data are useful for identifying water quality impacts due to land use and for assessing relative load contributions during dry and wet weather.

In this chapter, it was noted that the highest TC and FC levels occurred in the summer while nutrient and TSS data were variable in terms of magnitude across the seasons. In addition, no strong seasonality in the alternative indicator organism data was observed. The results suggest that their potential as surrogate measures of protozoa occurrence is unreliable due to underlying differences in seasonal occurrence. This will be explored further in the next chapter.

The watersheds were ranked in terms of the overall impact of land use activities on standard water quality, based on monthly baseline data. The pristine wildlife sites, regardless of beaver activity, were the least impacted in terms of nutrients and standard indicator organisms TC and FC. The residential study site was the most impacted in terms of nutrients and fecal coliform levels. High coliform levels were likely related to failing septic systems. It is expected that levels should drop as land owners continue to connect with the newly installed sewer system and sources in the remnant septic field are flushed. The dairy pasture and fields also negatively impacted water quality, but to a lesser extent. Total coliform was the most elevated parameter in BB.

Event mean concentrations were determined and presented. Values for the residential site were low in comparison to values reported in major urban runoff studies for separated storm sewer catchments, particularly in terms of FC. Downstream decreases in water quality, as measured by increasing concentrations of nutrients and standard indicator organisms, were observable in the monthly, event, and spatial data for BB and GB. The data suggest that even relatively small (less than 15% of overall land use) low residential development can have a negative impact on nutrient water quality over a pristine watershed. Negative impacts over that of pristine watersheds as measured by standard microbial pollution, however, were not observed. A key factor in terms of impact appears to be percent imperviousness, as noted in other studies. The effect of beaver on water quality tended to be positive overall in terms of nutrients and FC (not TC) but exhibited strong variability between events and seasons. Event loads were calculated and presented.

Product-moment correlation coefficients were determined to identify parameters which co-vary. In this chapter, alternative indicator organism, *Cryptosporidium*, and *Giardia* were eliminated from the analyses to facilitate inclusion of all the events. Few strong correlations with parameters of interest (such as with rain, discharge, physical water quality parameters or alternative indicators) were noted. When present, they were not consistent across the study watersheds or the lumped analysis. Such correlations would be useful to volunteer based organizations and state agencies in terms of assessing nutrient loads and developing TMDLs for lakes and streams. The lack of correlations and observed level of inconsistency are most likely related both to the relatively small number of samples included in the analysis and to the underlying variability of the hydrologic data (wide range of rainfall and discharge values).

The results summarized in this chapter show that negative water quality impacts should be expected as land use changes from pristine forests to agriculture to urban uses. These effects are evident both in baseline and wet weather sampling data, but may be more pronounced during wet weather. Upstream – downstream wet weather monitoring provides utilities with a means of estimating relative contributions to annual watershed loads from various land uses and wet versus dry weather. Wet weather concentrations, in particular, appear to vary from watershed to watershed and event to event. Collection of at least a limited amount of watershed specific wet weather data is thus necessary for developing watershed management plans with a high potential for improving water quality. Such monitoring schemes should take into account variability in precipitation, runoff and watershed characteristics. Co-collection of concentration and discharge data are critical for understanding cumulative impacts on water quality and placing wet weather transport in context with annual loads.

Collection of data on the rising limb of the hydrograph appears particularly important in terms of capturing variability of water quality parameters during wet weather. The use of passive or automated samplers is promising for reducing personnel requirements for wet weather sampling and for reducing missed sampling of the first flush and peak concentration of the contaminant of interest, particularly for nutrient data. Further work is necessary to see if composite samples may reduce analysis cost without reducing the ability of the data to describe the actual transport characteristics.

Experience suggests that preliminary collection of spatially distributed water quality data during dry and wet weather should be conducted prior to selecting locations for temporally varied wet weather data collection. Such data is necessary to identify stream reaches characterized by losses and gains. Sampling locations for wet weather data collection should be located as far downstream as possible from the suspected source of negative impact, yet upstream from tributary confluences and the sensitive stream reach, lake, or reservoir to be protected. This will allow natural stream attenuation processes to be accounted for when setting management goals and will avoid specific targeting of land owners. Bed sediments have been identified as a potential source of many contaminants. While events which result in scour of the beds may result in severe downstream water quality degradation, monitoring of such events will be difficult due to their comparatively low recurrence interval in most watersheds.

It is thus extremely important to systematically document wet weather events in terms of their hydrologic and environmental conditions. Co-location of sampling sites with long-term USGS stream gauges would facilitate this. However, relatively few USGS stream gauges are located within small water supply watersheds. Utility commitment to discharge and precipitation measurement is strongly suggested. Even with the approximately two years of continuous discharge data collected as part of this study, defining the events in terms of their long-term hydrologic and annual load characteristics (such as probability of occurrence) is difficult. This is largely due to inter-annual variability characteristic of the underlying hydrology. The existence of in-house precipitation and discharge data would facilitate development of watershed models based on collected data.

In terms of gaining a better watershed-wide understanding of wet weather effects on water quality, collection of a reduced number of samples over the rise and fall of the hydrograph but at a greater number of spatial locations may be beneficial. Locations of importance in terms of monitoring appear to differ between water quality parameters of interest. In addition, the time of year of most importance for sampling is not consistent across variables. Wet weather monitoring schemes on a per parameter basis may thus be more efficient in terms of personnel, analysis costs, and usefulness

CHAPTER 6
CRYPTOSPORIDIUM AND *GIARDIA* RESULTS

BACKGROUND AND OBJECTIVES

Microbial, nutrient, and physical water-quality data were collected at six sampling locations over a period of three years and under a variety of meteorological conditions. Three distinct land uses were isolated by pairing the sampling locations, Tables 6-1 and 6-2 (modified from Chapter 4 for clarity). Data collection frequency for *Cryptosporidium* and *Giardia* was storm-event based. Of particular interest was identification of meteorological and hydrologic conditions under which sampling and analysis of surface waters for traditional microbial organisms, emerging microbial organisms and non-bacterial pathogens are critical to ensure the integrity of surface-water drinking supplies. The final data base consists of nutrient, microbial, and pathogen data for twenty-one distinct storm events and three dry weather spatial events, in addition to monthly baseline nutrient and traditional microbial data, and continuous physical data. The focus of this chapter is the *Cryptosporidium* and *Giardia* results.

Table 6-1
Summary of sampling site locations

Influence	Sampling Location	Description	Code
Wildlife	Middle Branch Dickey Brook	Ranging wildlife	MBDB
	Griswold Brook	Beaver influenced	GRB
Agricultural	Burrow Brook Woods Road	Upstream of cow pasture	BBWR
	Burrow Brook Farm	Downstream of cow pasture	BBFM
Residential	Gates Brook Woodland Street	Upstream of residential and commercial area	GBWD
	Gates Brook Brayton Road	Downstream of residential and commercial area	GBBR

Table 6-2
Summary of land use characteristics

Site Name	Total Area (km^2)	Land Use Type				
		Forest	Agriculture	Urban	Water	Other
MBDB	1.15	95%	--	--	1%	4%
GRB	0.73	87%	--	--	3%	10%
BBFM	8.96	74%	10%	11%	2%	3%
BBWR	7.82	76%	7%	12%	2%	3%
Intervening	1.14	60%	30%	5%	0%	5%
GBBR	5.9	36%	8%	52%	--	4%
GBWD	4.3	41%	10%	43%	--	5%
Intervening	1.6	23%	0%	76%	--	0%

The main objectives of the project were to develop a strategy for selection of sampling locations, frequencies and methods to accurately depict pathogen occurrence and variability in relation to various sources within watersheds during and after weather (e.g., precipitation), hydrologic (e.g., flood or drought), or land-use (e.g., development) events. Of particular interest was monitoring for occurrence of *Cryptosporidium* oocysts and *Giardia* cysts. This chapter focuses on the following questions:

- Can storm-event sampling for emerging pathogens be optimized in terms of the sampling location, the number of samples collected over the hydrograph, timing of sample collection (including sample initiation), constituents measured, volumes analyzed, and monetary and personnel costs?
- Can occurrence of *Cryptosporidium* or *Giardia* be correlated with season, weather, hydrologic conditions, land-use, existence of potential sources?
- Do changes in standard water quality parameters provide an accurate indication of elevated emerging pathogen contamination?
- If storm-event monitoring is beneficial, what types of events are most important to monitor, particularly in terms of magnitude, antecedent conditions, and season?

Sampling locations and methods were previously described in Chapter 4. In this chapter, the conditions under which *Cryptosporidium* oocysts and *Giardia* cysts were detected will be presented. A discussion of the findings and suggestions for the most effective watershed monitoring plans will follow.

SUMMARY OF RESULTS

Cryptosporidium oocysts and *Giardia* cysts were detected in only a limited number of events out of the twenty-one wet weather and three dry-weather events monitored. Table 6-3 summarizes the location, season, date, total rainfall accumulation and *Crypto*/*Giardia* status of the twenty-four events monitored. Rainfall totals for the spatial events are listed as not available (NA) as these were dry-weather events.

Cryptosporidium oocysts were detected once at the beaver influenced wildlife watershed and once at both the up- and downstream sampling locations in the agricultural watershed, but during different events. They were not detected at either the up- or downstream sampling locations within the residential watershed nor at the wildlife site restricted to ranging wildlife. *Giardia* was detected at all of our study watershed (wildlife, agricultural, residential) sampling locations but not during all events and not always at both sampling locations during a given event. For example, *Giardia* was not detected at the upstream agricultural site (BBWR) during the 28-May-03 event but was detected downstream (BBFM). During the dry weather spatial sampling events neither *Cryptosporidium* oocysts nor *Giardia* cysts were detected.

The total number of samples analyzed for *Cryptosporidium* and *Giardia* are summarized in Table 6-4. Three samples per location were committed to a dry weather sampling event. The dry weather event, however, was spatial in character. Thus one dry weather sample was collected at each sampling location and an additional four samples were collected at other locations throughout the watershed. Because MBDB and GRB are distinct watersheds, two additional dry weather analyses for *Cryptosporidium* and *Giardia* were conducted at other locations within these watersheds. Table 6-4 thus lists the total number of dry weather samples taken both within the study watershed (see rows denoted wildlife, agricultural, and residential) and the one dry weather sample taken at each regular sampling location.

During wet weather events the goal at each sampling location was to conduct twenty-one analyses. This aim fell short for some locations. At GRB, only two analyses were completed for the 11-June-03 event due to holding time errors. During the 28-May-03 event at BBWR, the small cooler containing the filters for the site was stolen in the field during pre-event setup. Sufficient extra filters were available from the BBFM site to facilitate conducting two analyses rather than three. At the residential sites, GBBR and GBWD, field errors lead to the loss of one analysis at each site during the 11-December-03 event. In addition, during the 28-July-02 event monitored at the wildlife sites, the decision was made to only run two analyses due to the small precipitation accumulation (promising thunderstorm which dissipated). During a large precipitation accumulation event occurring at the site on 26-September-02, four analyses were conducted at GRB and MBDB to make up for the earlier event.

The percent positive detections observed over the course of the project are also summarized in Table 6-4. No column for dry weather positive detections is provided as neither *Cryptosporidium* nor *Giardi*a was detected during the spatial dry weather sampling. The percent positive detections during wet weather ranged from 0% to 10% for *Cryptosporidium* and from 10% to 15% for *Giardia.* The percent positive detection results are compared across sites on Figure 6-1. Percent recoveries at the more impacted, downstream site were typically around 5% higher, although *Giardia* percent positive detections were similar at both wildlife sampling locations (beaver influenced and no beaver). Sufficient data are not available to determine the statistical significance of these differences.

The reported QA/QC recoveries for the laboratory during the weeks of our sampling events are provided in Table 6-5. Cells for the events where positive detections occurred are shaded grey. Recoveries were not available for the 27-Oct-03 event at the wildlife study locations. The available data do not suggest any relationship between the laboratory QA/QC recovery results and the study results.

Table 6-3
Summary of events, noting the associated rainfall accumulation and detection of *Cryptosporidium* oocysts or *Giardia* cysts

Site/Season	Date	Rain Total (mm)	*Crypto* Detected		*Giardia* Detected	
Wildlife			*MBDB*	*GRB*	*MBDB*	*GRB*
Spring	11-Jun-03	11.68	No	No	No	No
	9-Jun-04	13.72	No	No	No	No
Summer	28-Jul-02	2.03	No	No	No	No
	26-Sep-02	55.12	No	No	No	No
Fall	27-Oct-03	49.58	No	No	No	No
	15-Oct-04	32.77	No	No	No	No
Winter	26-Mar-02	19.05	No	**Yes**	**Yes**	**Yes**
(spatial)	10-Mar-04	NA	No	No	No	No
Agricultural			*BBWR*	*BBFM*	*BBWR*	*BBFM*
Spring	28-May-03	5.33	No	**Yes**	No	**Yes**
	18-June-03	7.62	No	No	No	No
Summer	12-Aug-03	46.99	No	No	No	No
	19-Sept-03	10.16	**Yes**	No	**Yes**	**Yes**
(spatial)	14-Jul-04	NA	No	No	No	No
Fall	12-Nov-03	45.08	No	No	No	No
	17-Dec-03	58.42	No	No	No	No
Winter	31-Mar-04	69.85	No	No	No	No
Residential			*GBWD*	*GBBR*	*GBWD*	*GBBR*
Spring	2-May-03	3.81	No	No	No	No
	12-April-04	82.04	No	No	No	No
Summer	1-Aug-03	7.87	No	No	No	No
	2-Sept-03	18.54	No	No	No	No
(spatial)	4-Aug-04	NA	No	No	No	No
Fall	6-Nov-02	18.29	No	No	No	No
	11-Dec-03	29.21	No	No	No	No
Winter	4-Feb-03	9.65	No	No	**Yes**	**Yes**

Table 6-4
Number of samples per event type and percent positive detections for *Cryptosporidium* and *Giardia*

	Cryptosporidium			*Giardia*		
Site	No. Dry	No. Event	Event % detects	No. Dry	No. Event	Event % detects
Wildlife	***3 and 3***			***3 and 3***		
MBDB	1	21	nd	1	21	14.3 %
GRB	1	20	10.0 %	1	20	15.0 %
Agricultural	***6***			***6***		
BBFM	1	21	9.5 %	1	21	14.3 %
BBWR	1	20	5.0 %	1	20	10.0 %
Residential	***6***			***6***		
GBBR	1	20	nd	1	20	15.0 %
GBWD	1	20	nd	1	20	10.0 %

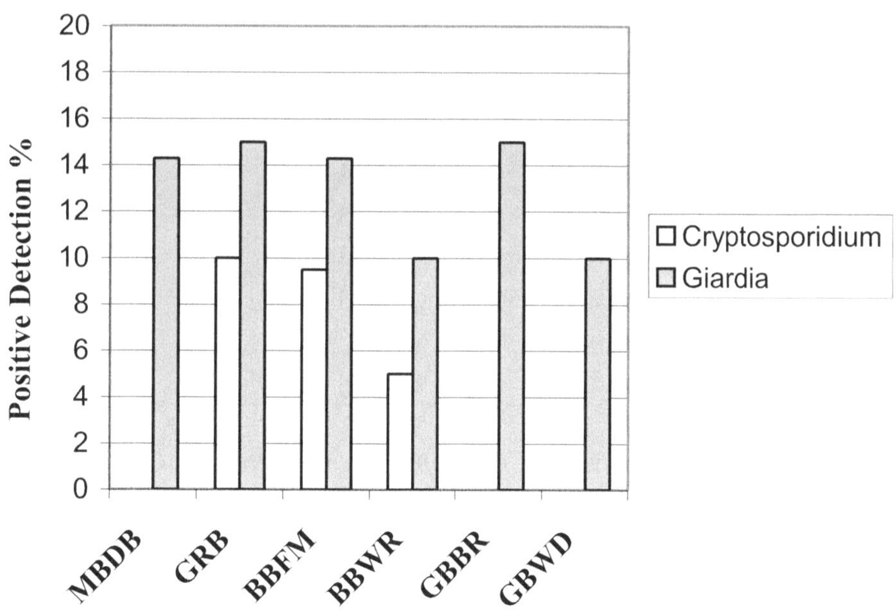

Figure 6-1 Percent positive detection results compared across study sites

Table 6-5
Laboratory QA/QC recovery results for the week when event samples were analyzed

Site/Season	Date	% *Crypto* Recovery	% *Giardia* Recovery
Wildlife			
Spring	11-Jun-03	44.4	53
	9-Jun-04	59.6	59.6
Summer	28-Jul-02	46.5	40.4
	26-Sep-02	47.5	49.5
Fall	27-Oct-03	NA	NA
	15-Oct-04	52.5	48
Winter	26-Mar-02	56.6	50
(spatial)	10-Mar-04	44	43.4
Agricultural			
Spring	28-May-03	51.5	61
	18-Jun-03	46.5	52
Summer	12-Aug-03	62.6	63
	19-Sep-03	51.5	58
(spatial)	14-Jul-04	63.6	45.5
Fall	11-Nov-03	48	54.5
	17-Dec-03	44	47.5
Winter	31-Mar-04	46.5	45.5
Residential			
Spring	2-May-03	44.4	52
	12-Apr-04	41.4	51.5
Summer	1-Aug-03	43.4	52
	2-Sep-03	45.5	48
(spatial)	4-Aug-04	41.4	45.5
Fall	6-Nov-02	52.5	50.5
	11-Dec-03	42	46.5
Winter	4-Feb-03	62.6	48

The observed enumerations for *Cryptosporidium* oocyst and *Giardia* cyst are summarized in Table 6-6. When both *Cryptosporidium* and *Giardia* were detected, *Giardia* counts tended to be at least slightly higher. Enumerations were also greater at the paired sampling locations considered to be more impacted. *Cryptosporidium* enumerations ranged from non-detect to 2 oocysts/100 L at the beaver influenced wildlife site and were all non-detect at the pristine wildlife site; they ranged from non-detect to 10 oocysts/100 L and 100 oocysts/100 L at the upstream and downstream agricultural sites, respectively; and they were all non-detect at the residential sites. *Giardia* enumerations ranged from non-detect to >30 and 300 cysts/100 L at the pristine and beaver influenced wildlife sites, respectively; they ranged from non-detect to 5 and 700 cysts/100 L at the upstream and downstream agricultural sites, respectively; and they ranged from non-detect to 20 and >200 cysts/100 L at the upstream and downstream residential sites, respectively. These ranges will be compared to reported literature values in a later section.

Table 6-6
***Cryptosporidium* oocysts or *Giardia* cysts results for events where detected. Values are reported as number detected per 100 L**

			Rising	Near Peak	Falling
26-Mar-02 Wildlife	*Crypto*	GRB	2	2	<2
	Giardia	MBDB	14	32	16*
		GRB	256	160	36
28-May-03 Agricultural	*Crypto*	BBFM	20†	100‡	<10
	Giardia	BBFM	10†	698‡	<10
19-Sept-03 Agricultural	*Crypto*	BBWR	<5	10	<5
	Giardia	BBWR	5	5	<5
		BBFM	70	<5	<5
4-Feb-03 Residential	*Giardia*	GBWD	19	10	<5
		GBBR	64	94	208

Notes: * Stage still rising; † Stage falling; ‡ Break in recession curve induced by additional rainfall.

Source: Rees *et al.*, 2005. Transport of *Cryptosporidium* and *Giardia* in Massachusetts Watersheds During Storm Events. In *Managing Watersheds for Human and Natural Impacts: Engineering, Ecological, and Economic Challenges*. Copyright ©2005 ASCE. Reproduced with permission.

RELATIVE RISK OF *CRYPTOSPORIDIUM* AND *GIARDIA* PRESENCE

Although limited in numbers, the results from this study may be used to infer the relative risk of emerging pathogen contamination to downstream surface waters by land use type, Table 6-7. This inference is feasible due to the nature of the study; relatively small areas (GRB: 0.75 km^2, MBDB: 1.15 km^2, BBFM – BBWR: 1.14 km^2, and GBBR – GBWD: 1.6 km^2, Table 5-2) of distinct land use types (100% ranging wildlife with beaver, 100% ranging wildlife, 30% agriculture / 60% forest, and 76% urban / 24% forest) were isolated to determine their distinct pathogenic occurrence and variability characteristics.

The apparent relative risk of *Cryptosporidium* contamination by land use was greatest for the isolated farm and lowest for the heavily urbanized area. Mixed results were obtained for the beaver influenced and ranging wildlife sites. While no *Cryptosporidium* was observed at MDBD, the solely ranging wildlife site, *Cryptosporidium* was detected at the upstream farm location, BBWR. Land use above BBWR is mostly forested (76%) with some agricultural fields (7%) and suburban areas (12%) which are characterized mainly by distributed large acre (greater than two) homes and at least one sub-development with houses on approximately half-acre lots. Because no *Cryptosporidium* was observed at the highly residential site (GBBR) and because the agricultural land use was mostly crops or fallow fields, it is theorized that the source of *Cryptosporidium* at BBWR was ranging wildlife. In addition, *Rhodococcus coprophilus* was detected in fairly high numbers (>500 pfu/100L) during the 19-Sep-03 event at BBWR, when *Cryptosporidium* was detected. This suggests that ranging wildlife were the likely source of the *Cryptosporidium.* In addition, stream walks did not identify any beaver dams, although large stretches of the stream above BBWR were not accessible due to terrain and thus the presence of beaver upstream of BBWR cannot be ruled out. While *Cryptosporidium* was observed at GRB, the beaver influenced ranging wildlife (no urban or agriculture) site, the observed levels were less than at BBWR (see 26-March-02 event). In addition, it cannot be determined if the source was specifically beaver or ranging wildlife (*Rhodococcus coprophilus* results are not available for this event). Ranging wildlife and beaver influenced wildlife are thus assigned the same relative risk in terms of *Cryptosporidium* contamination.

The data suggest that apparent relative risk of *Giardia* contamination by land use is highest for beaver influenced watersheds. Direct evidence is available from the 26-March-02 event at Prescott Peninsula, where high levels of *Giardia* were observed at GRB (beaver influenced) while much lower (order of magnitude) levels were observed at MBDB (ranging wildlife only). The highest enumerations, however, were observed at BBFM (cow influenced wildlife site) during the 28-May-03 event. However, beaver are known to inhabit the stream between BBWR and BBFM and evidence suggests that they, rather than the cows, were the likely source for the *Giardia* during the event. First, no *Rhodococcus coprophilus* was observed in the BBFM samples collected at the same time as filters for *Giardia* analysis were run, although *Rhodococcus coprophilus* was detected at the upstream BBWR site. Secondly, only 5.3 mm (0.21 inches) of rain fell during the event and the highest *Giardia* counts occurred at the hydrograph peak; surface runoff from the southern pasture appeared to be minimal, and by the peak of the hydrograph, runoff was likely dominated by upstream response, including flow through the beaver impoundments, rather than near stream zone response (both surface and baseflow). Based on this evidence, beaver influenced watersheds are given the highest risk category.

Table 6-7
Inferred relative risk of contamination to downstream surface waters by land use type, based on study results

Cryptosporidium		*Giardia*	
Level of Risk	Land Use	Level of Risk	Land Use
High	Agriculture	High	Beaver Influenced
Moderate	Ranging wildlife	Moderate High	Residential
Moderate	Beaver Influenced	Moderate Low	Agricultural
Low	Residential	Moderate Low	Ranging wildlife

Giardia was also observed in fairly high (but an order of magnitude lower) numbers at BBFM during the 19-September-03 event. As noted above, beaver inhabit the stream between BBWR and BBFM and are a potential source for the *Giardia* observed during the 19-September-03 event. However, during this event the cows cannot be ruled out as a contributing factor. Precipitation was modest (10 mm or 0.4 inches) but approximately double that which occurred during the 28-May-03 event. *Giardia* was observed only early in the event, suggesting its origin was either from near-stream surface runoff, the initial baseflow response, or first spillage rather than extensive flushing of the beaver impoundments. High levels of *Rhodococcus coprophilus* (763, 282, and 255 pfu/100 L) were detected in the samples collected at the same time as filters for *Giardia* analysis were run, although even higher levels were detected upstream and thus the *Rhodococcus coprophilus* may be due to upstream sources rather than the cows. While further data are needed to refine the risks associated with this land use category, results suggest that agricultural (cow influenced) watersheds are associated with moderate to low risk level for *Giardia*.

Residential land use is assigned a risk level of moderate to high based on the data. When detected at GBBR, levels begin to approach those observed at GRB (beaver influenced). In addition, levels at GBBR were a minimum of three times higher than at the upstream, GBWD, sampling location. The intervening area between GBWD and GBBR is 75% urban, mainly dense residential. While residential areas exist upstream of both GBWD and BBWR, these areas are less influenced both in terms of overall land use percentage and in terms of the density of the residential areas. In addition, these sites had low *Giardia* (less than 20/100 mL) detections when observed. The urban risk thus seems to be most associated with dense residential areas.

The lowest detection levels for *Giardia* were observed at the sites dominated by ranging wildlife, in particular MBDB. The upstream farm site, BBWR, which may also be considered largely dominated by ranging wildlife sources, also had fairly low detection levels. However, because observed levels were still greater than 10 cysts/100 L, a moderate to low risk category was assigned.

In summary, the studied land uses were assigned high to low risk in terms of *Cryptosporidium* oocyst contributions and high to moderate-low risk in terms of *Giardia* cyst contamination. No low risk land use areas were identified for *Giardia* because of its detection at levels equal or greater than 10 pfu/100 L at all sampling locations other than BBWR, which is best classified as a mixed land use area dominated by ranging wildlife. The presence of beaver between the up- and downstream sampling locations BBWR and BBFM at the agricultural study site was a confounding factor for fully identifying the relative risk of contamination due to cows relative to beaver. Ancillary data suggest that cows were the greater source in terms of *Cryptosporidium* while beaver were the greater source in terms of *Giardia*. Ancillary data and factors influencing the presence of *Cryptosporidium* and *Giardia* are more fully explored in the next section.

POSITIVE DETECTION EVENT SUMMARIES

26-March-02 Wildlife Site Event

The response at the wildlife sampling sites during the 26-March-02 event is shown in Figures 6-2 through 6-4. During this event, approximately 19 mm (0.75 inches) of rain fell on ground saturated from snowmelt and, in some locations, the remaining snowpack. The storm total rainfall is estimated from a National Acid Deposition Program (NADP) monitoring site located halfway between GRB and MBDB, which monitors daily rainfall totals on Prescott Peninsula. Precipitation data at GRB was lost due to a malfunction. At GRB, runoff response for this event was very modest, with the difference between the initial discharge and peak differing by only approximately 30%. Flows at MDBD, on the other hand, increased five fold. As previously noted, moderate and high *Giardia* levels were detected at MBDB and GRB, respectively, during this event. Low levels of *Cryptosporidium* were also detected at GRB. Despite the small increase in discharge observed at GRB, *Giardia* levels were in general an order of magnitude higher than at MBDB. The timing and magnitude of response at GRB suggest that the source of contamination more likely came from the watershed area below the beaver impoundment (first hydrograph response) and was mainly flushed out as spillage from the impoundment (second hydrograph response) reached the sampling location.

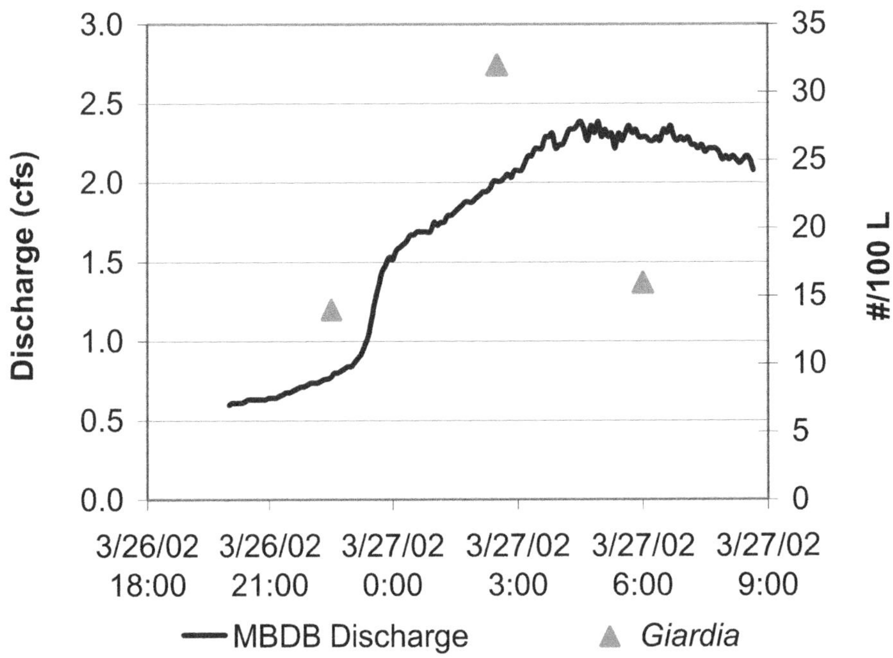

Figure 6-2 Detection of *Giardia* cysts with relation to runoff response at MBDB during the 26-March-02 event

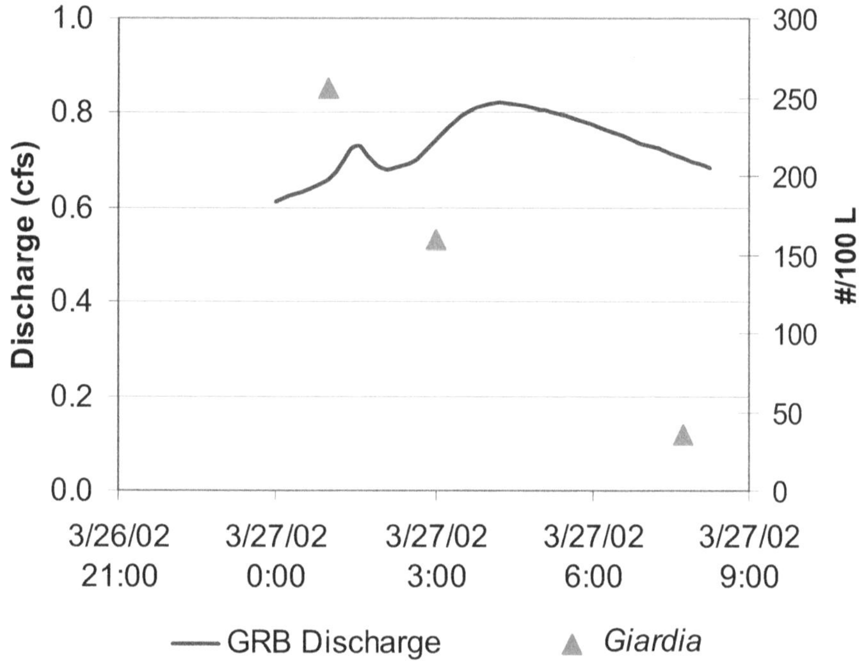

Figure 6-3 Detection of *Giardia* cysts with relation to runoff response at GRB during the 26-March-02 event

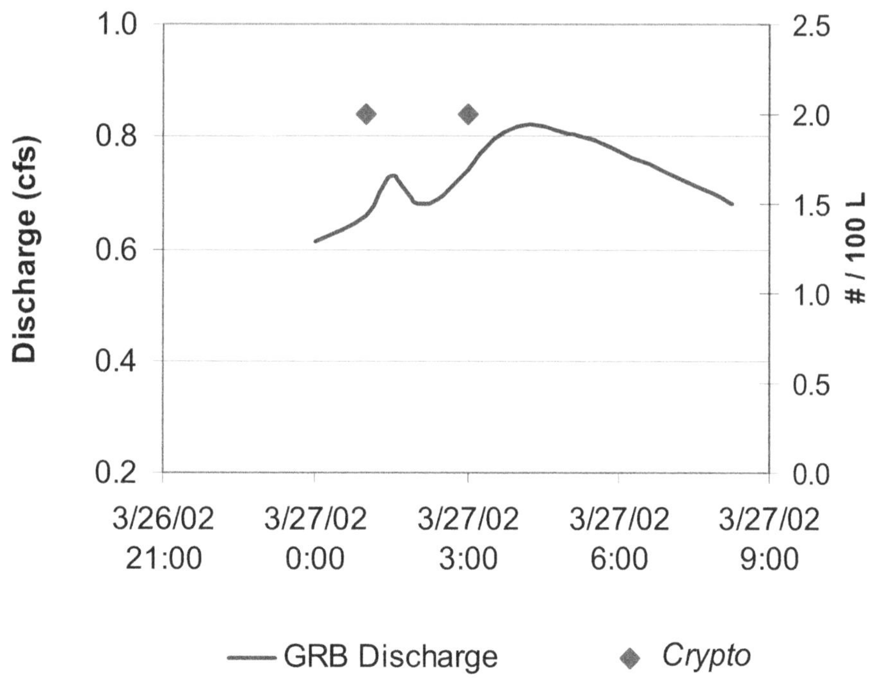

Figure 6-4 Detection of *Cryptosporidium* oocysts with relation to runoff response at GRB during the 26-March-02 event

At MBDB, the initial *Cryptosporidium* and *Giardia* sample occurred just prior to a very sharp rise in the hydrograph while the second sample occurred near the peak discharge, but on a less steep portion of the rising limb. Suspended sediment data, Figure 6-5, suggest that peak *Giardia* levels may have been missed by the sampling strategy; a secondary peak in suspended sediment concentration corresponds to the peak *Giardia* level but significantly higher suspended sediment concentration occurred at the top of the steepest portion of the rising limb hydrograph, unsampled for *Giardia* and *Cryptosporidium*. Peaks in total coliform and fecal coliform levels also occurred at the stop of the steepest portion of the hydrograph rising limb, Figure 6-6. Higher levels of *Giardia* and *Cryptosporidium* may have been detected if a sample had been collected during this period. Total coliform and fecal coliform levels rose again later in the event while suspended sediment concentrations continued to fall. Without additional *Cryptosporidium* and *Giardia* data points, firm conclusions cannot be drawn, but it appears that suspended sediment data have the potential to serve as a surrogate for determining the relative magnitude of *Cryptosporidium* and *Giardia* during storm events. Total coliform may be an additional surrogate.

At MBDB, no similarities between total Kjeldahl nitrogen or total phosphorus and *Cryptosporidium* or *Giardia* (Figure 6-7) were detected. *Rhodococcus coprophilus* were not detected during this event. While F+ and somatic phages were detected, they were found in very low levels, Figure 6-8. *Clostridium perfringens* were detected at higher levels, but their relative magnitude was characterized by the inverse shape of the *Giardia* data at MBDB, Figure 6-9.

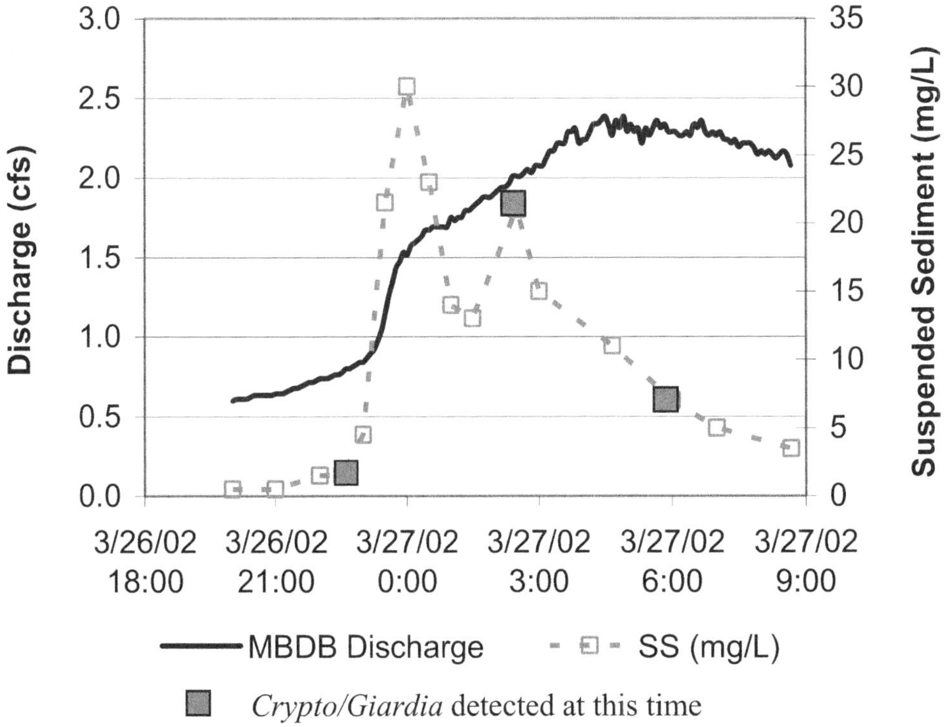

Figure 6-5 Suspended sediment concentrations at MBDB during the 26-March-02 event

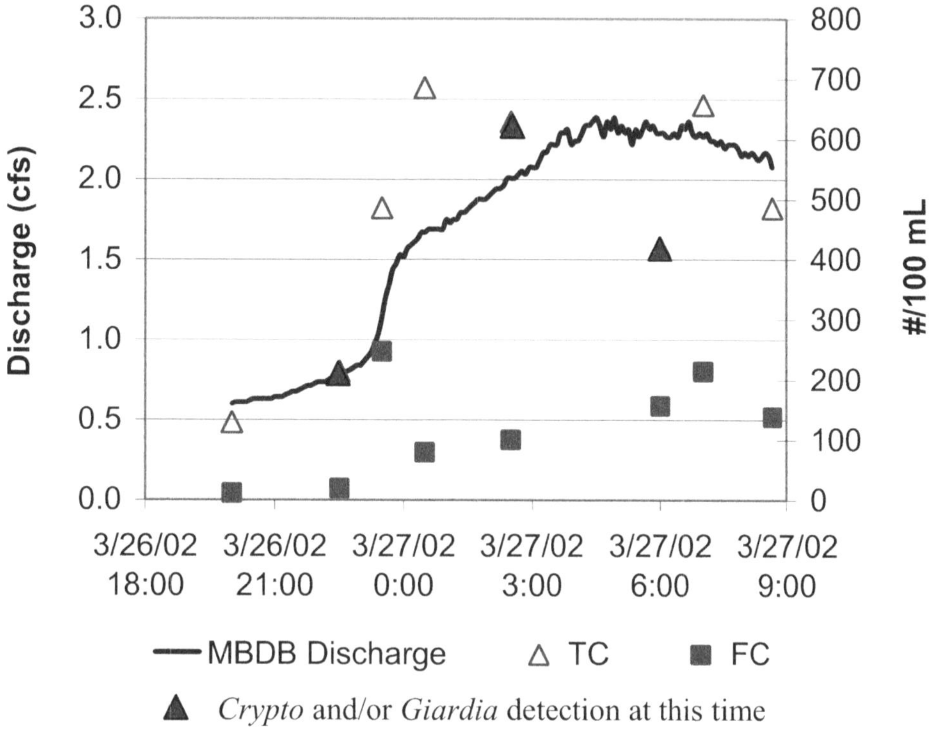

Figure 6-6 Total coliform and fecal coliform levels at MBDB during the 26-March-02 event

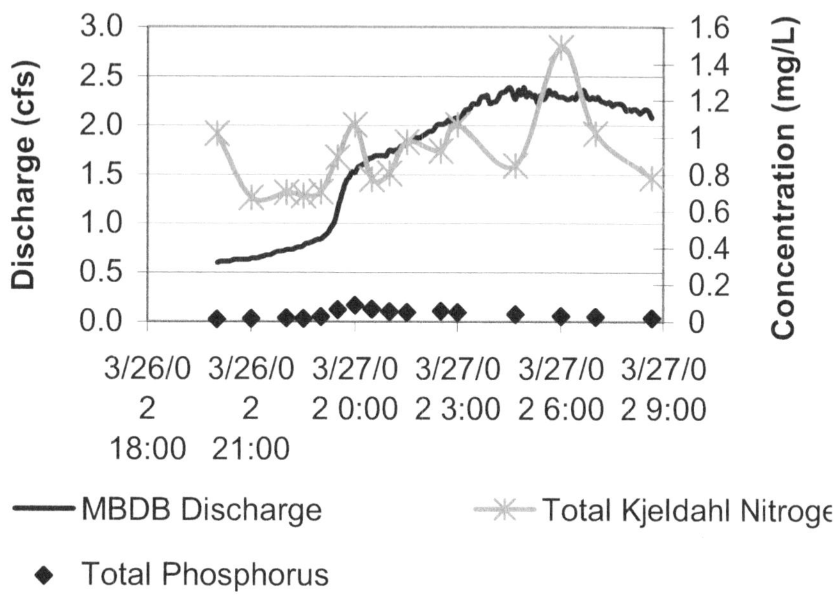

Figure 6-7 Total Kjeldahl nitrogen and total phosphorus concentrations during the 26-March-02 event at MBDB

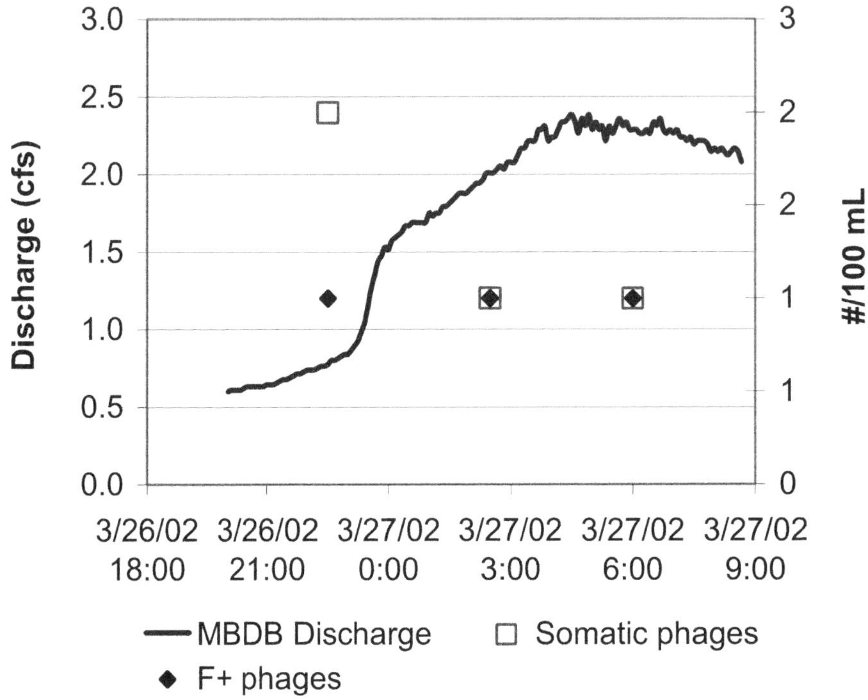

Figure 6-8 Somatic and F+ phage concentrations during the 26-March-02 event at MBDB

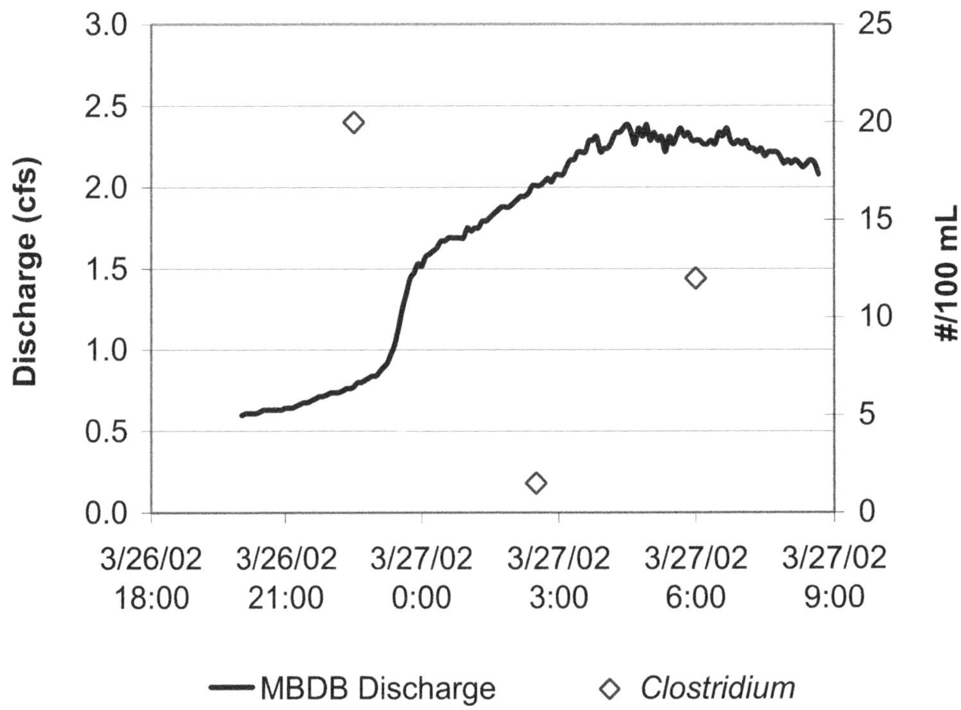

Figure 6-9 *Clostridium perfringens* concentrations during the 26-March-02 event at MBDB

At GRB, the initial *Cryptosporidium* and *Giardia* sample, with the highest detected *Giardia* levels, occurred during an initial rise in the hydrograph while the second sample occurred during the main hydrograph rise. The drop in *Giardia* levels over the course of the event suggests that the peak level may have been missed, either occurring prior to the initial sample or between the second and third samples. Suspended sediment data, Figure 6-10, suggest that peak *Giardia* levels may have occurred earlier. The peak in total coliform levels, however, corresponded with the *Giardia* data, Figure 6-11. Although this cannot be confirmed, ancillary suspended sediment and total coliform data suggest that the peak *Giardia* levels were either captured by the sampling scheme or occurred just prior.

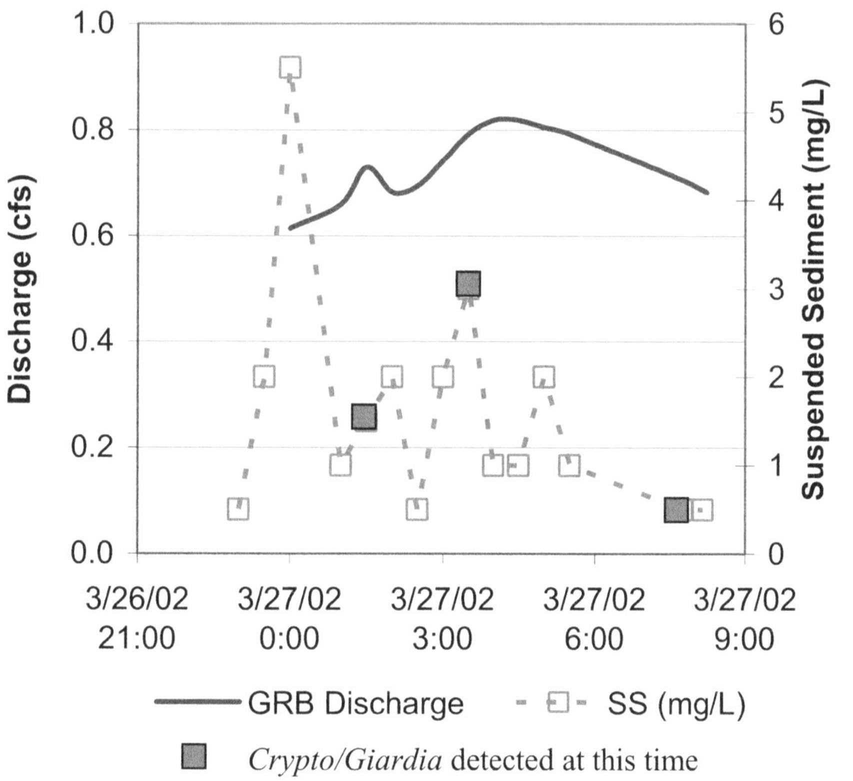

Figure 6-10 Suspended sediment concentrations at GRB during the 26-March-02 event. The samples corresponding to the time steps when *Giardia* and *Cryptosporidium* were detected are shaded grey.

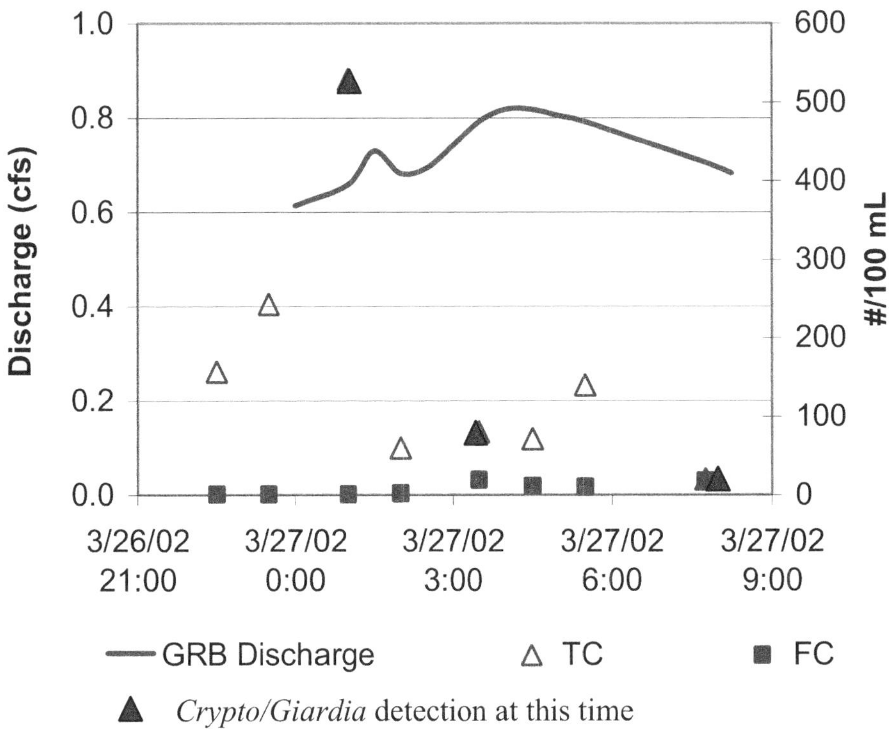

Figure 6-11 Total coliform and fecal coliform concentrations during the 26-March-02 event. The samples corresponding to the first *Giardia* and *Cryptosporidium* detection are shaded for total coliform. No total coliform data correspond exactly with the second *Giardia* sample.

As observed for MBDB for this event, there were no apparent similarities between total Kjeldahl nitrogen or total phosphorus and *Cryptosporidium* or *Giardia* (figure not shown). In addition, no *Rhodococcus coprophilus* or F+ phages were detected and only one somatic phage was detected. *Clostridium perfringens* were detected at higher levels, but (as for MBDB) their relative magnitude was characterized by the inverse shape of the *Giardia* data at GRB (figure not shown).

As *Rhodococcus coprophilus* were not detected, the most likely source for the *Giardia* contamination detected at GRB is beaver. However, at MBDB no known beaver exist, suggesting the potential for multiple sources during the event from within the two watersheds. Beaver live in family units and defend a common territory, which includes a series of connected ponds (Sun *et al.*, 2000). Each family unit typically consists of an adult pair and their offspring of the current and previous year. The two breeding adults normally give birth to 3 to 4 kits each year in May or June (Mass Wildlife website, 2003). Both male and female offspring disperse from their native family at the age of two (Jenkins and Busker, 1979). Hodgdon (1978) found beaver in Massachusetts disperse primarily in April to early May, although the presence of temporary beaver colonies in the summer suggests that it may take several months for dispersers to finally settle (Sun *et al.*, 2000). Dispersal patterns are associated with the availability of water and suitable food sources.

Based on the seasonal timing and timing relative to the hydrograph, it seems likely that the source of *Giardia* during the 26-March-02 event at GRB was from juvenile beavers dispersing from their native family. The existence of smaller beaver impoundments and partial impoundments between the main beaver impoundment at the headwaters of Griswold Brook and the sampling location suggest that such dispersal has happened at Griswold GRB, although the timing in correspondence with this event cannot be confirmed. Because dispersing beaver do not have well established impoundments and dens, cysts that they shed may be more available for transport due to less opportunity for settling and storage. Higher risk of contamination may thus be associated with late winter through early spring months as dispersal occurs.

4-February-03 Residential Site Event

The response at the residential sampling sites during the 4-February-03 event is shown in Figures 6-12 through 6-15. During this event, approximately 9.7 mm (0.38 inches) of rain fell. As previously noted, moderate *Giardia* levels were detected at both sampling locations during this event, with higher levels detected at the downstream sampling site. No *Cryptosporidium* were detected.

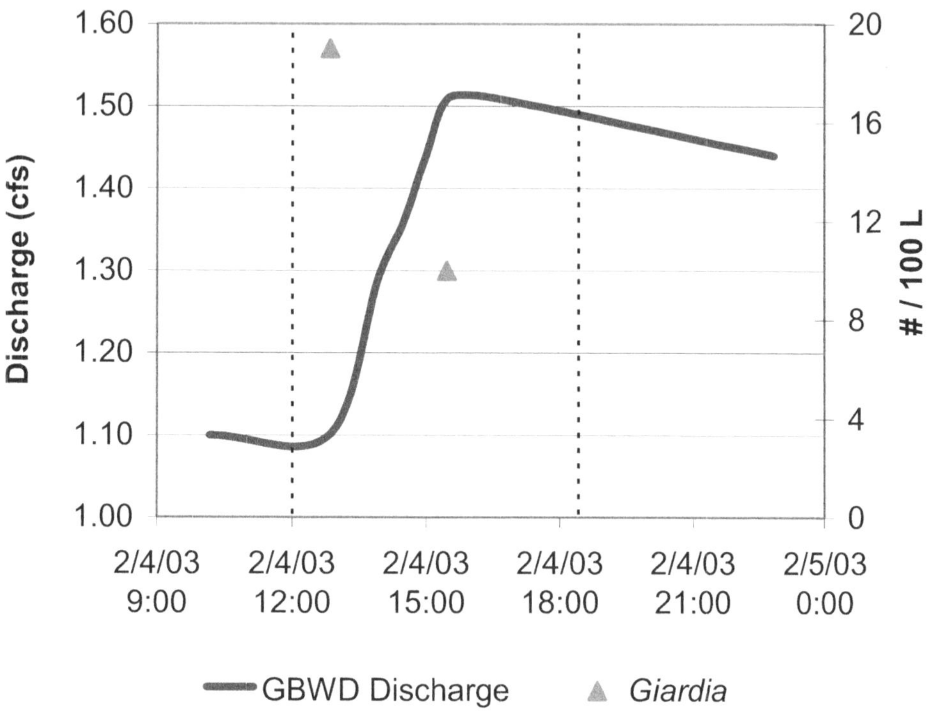

Figure 6-12 Detection of *Giardia* cysts with relation to runoff response at GBWD during the 4-February-03 event. The start and stop of rain are indicated by the dashed vertical lines.

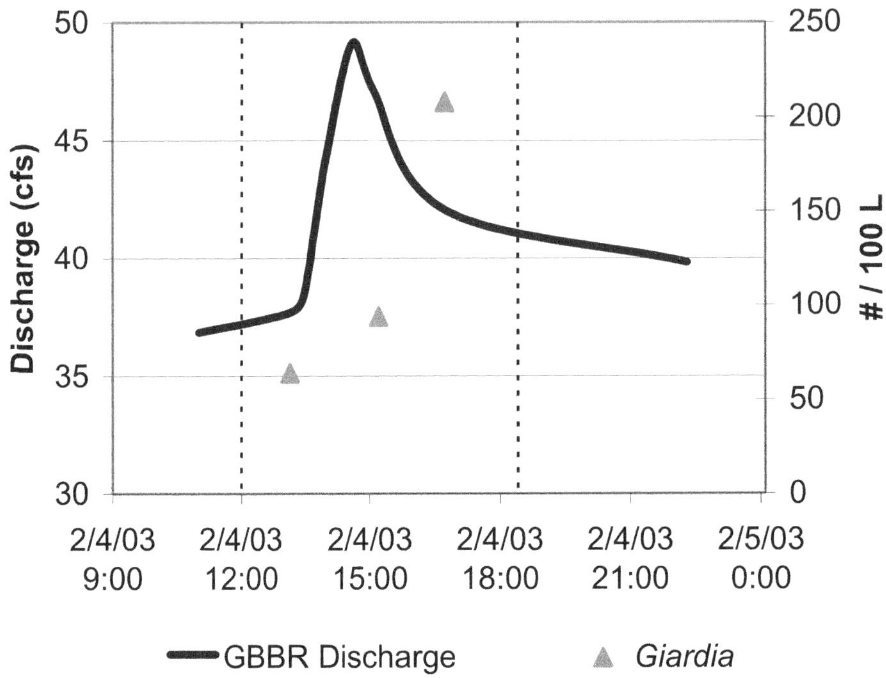

Figure 6-13 Detection of *Giardia* cysts with relation to runoff response at GBBR during the 4-February-03 event. The start and stop of rain are indicated by the dashed vertical lines

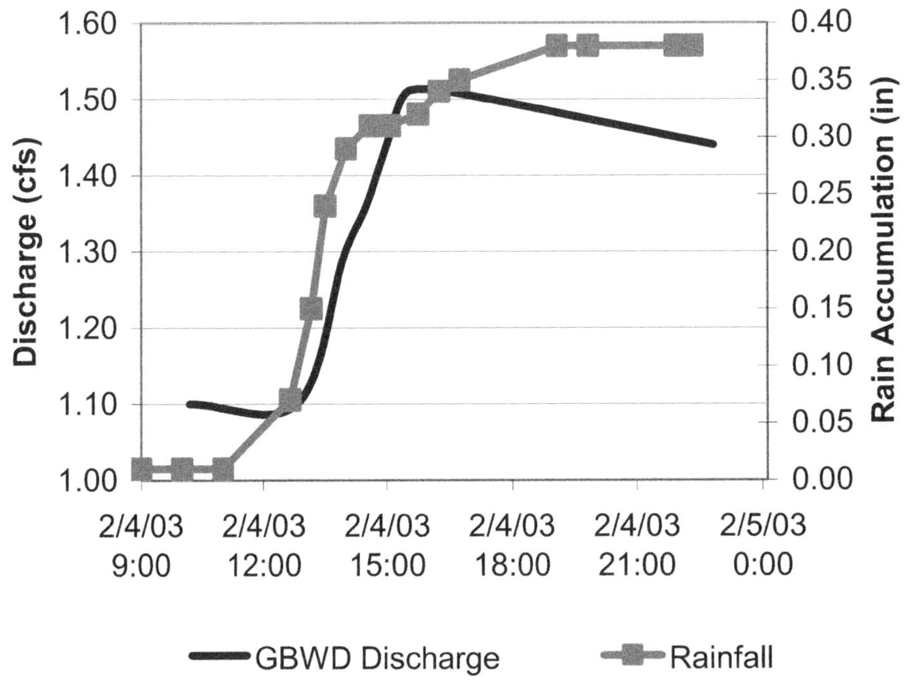

Figure 6-14 Runoff response to rainfall at GBWD during the 4-February-03 event

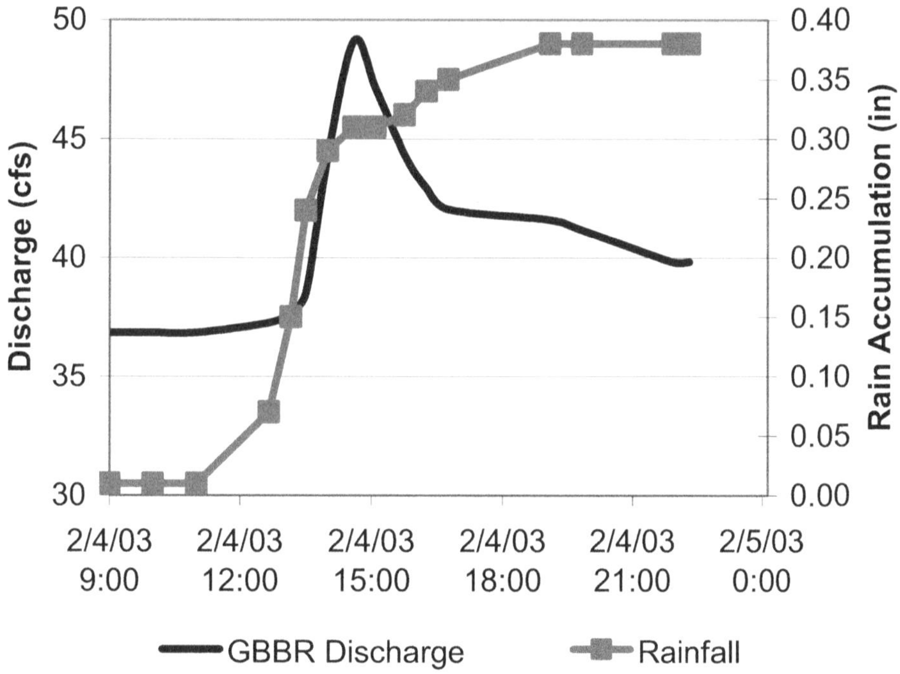

Figure 6-15 Runoff response to rainfall at GBBR during the 4-February-03 event

At the residential sites (4-February-03 event) *Giardia* enumerations at the upstream site decreased as the hydrograph rose (Figure 6-12) but were still increasing as the hydrograph fell (Figure 6-13) at the downstream sampling location. Runoff response to rainfall at both locations was rapid, suggesting saturated or frozen ground conditions. Although flow at GBWD is only a fraction of that observed at GBBR, the event peak discharge at both locations was approximately 35% greater than the pre-event flows. No major tributaries occur between the two sites, thus the increase in flow is generally associated with the larger watershed size for the downstream sampling point.

The highest detection levels of *Giardia* at GBWD occurred immediately after rainfall began, Figures 6-12 and 6-14. This dynamic suggests that the upstream sampling site was located in close hydrologic proximity to its source of *Giardia*, which was flushed quickly. *Clostridium perfringens* were detected at low numbers during this event and decreased in a similar fashion with respect to the hydrograph, Figure 6-16. Total coliform, fecal coliform, and dissolved organic carbon (not shown) were variable across the hydrograph (Figure 6-17). No sorbitol-fermenting *Bifidobacteria* were detected at this site during the event.

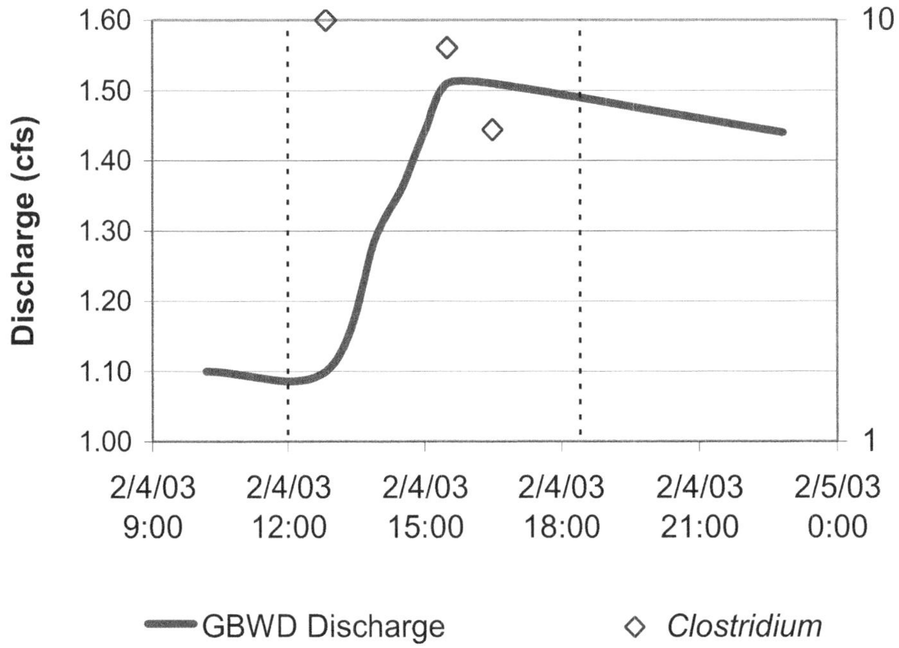

Figure 6-16 ***Clostridium perfringens*** **detections at GBWD during the 4-February-03 event**

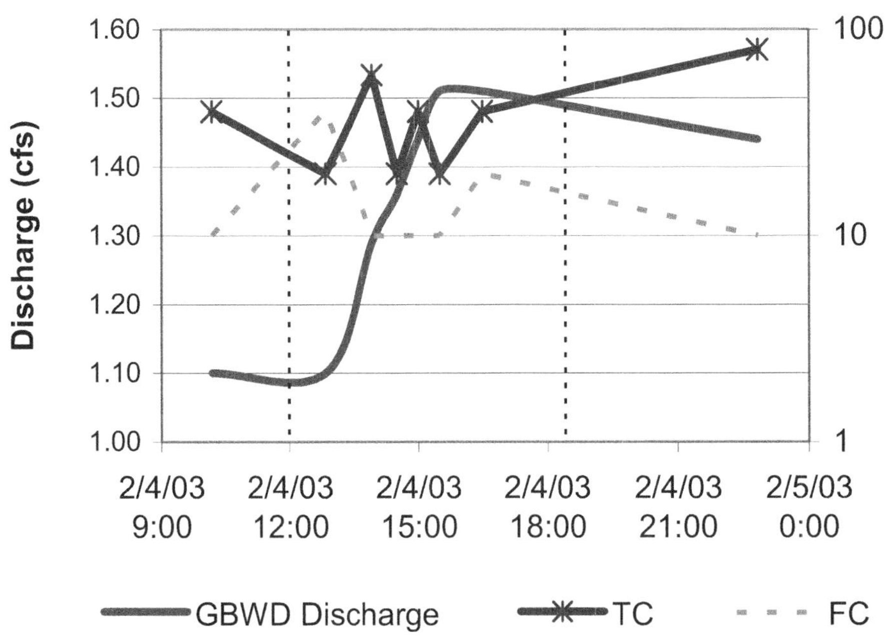

Figure 6-17 Total Coliform and Fecal Coliform detections at GBWD during the 4-February-03 event

At the downstream site, the rise in *Giardia* was a continuation of the upstream counts but was delayed in comparison to both the start of rainfall and the hydrograph rise. The peak in *Giardia* concentration rather appears to be associated with the peak of baseflow response, if baseflow and stormflow response are separated using the inflection point on the hydrograph recession (McCuen, 1998). The main *Giardia* source at GBBR appears hydrologically removed from the sampling location and is perhaps associated with subsurface flow. Although the potential for a subsurface *Giardia* source may not be directly confirmed, indicator organism results provide additional insight. Sorbitol-fermenting *Bifidobacteria* were detected in only small quantities during the event but had a similar response with respect to the hydrograph, Figure 6-18; somatic phages and *Clostridium perfringens* behaved similarly. Total and fecal coliform, however, exhibited the highest concentrations during the period of peak flow, Figure 6-19. These combined results suggest that the *Giardia* may have a human source, potentially from leaking septic or sewer systems.

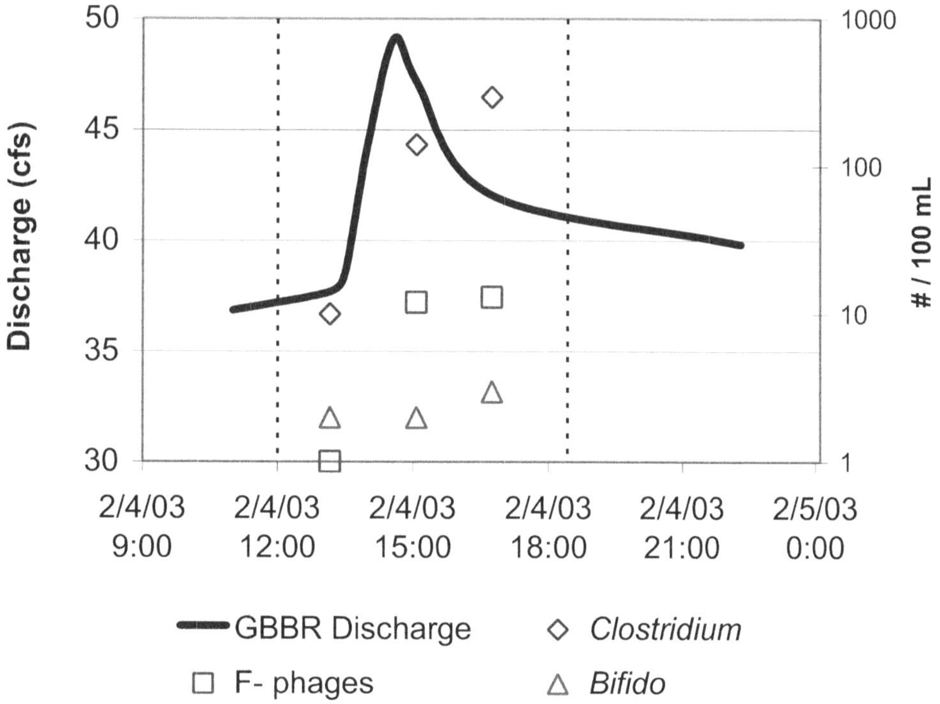

Figure 6-18 Alternative indicator organism detections at GBBR during the 4-February-03 event

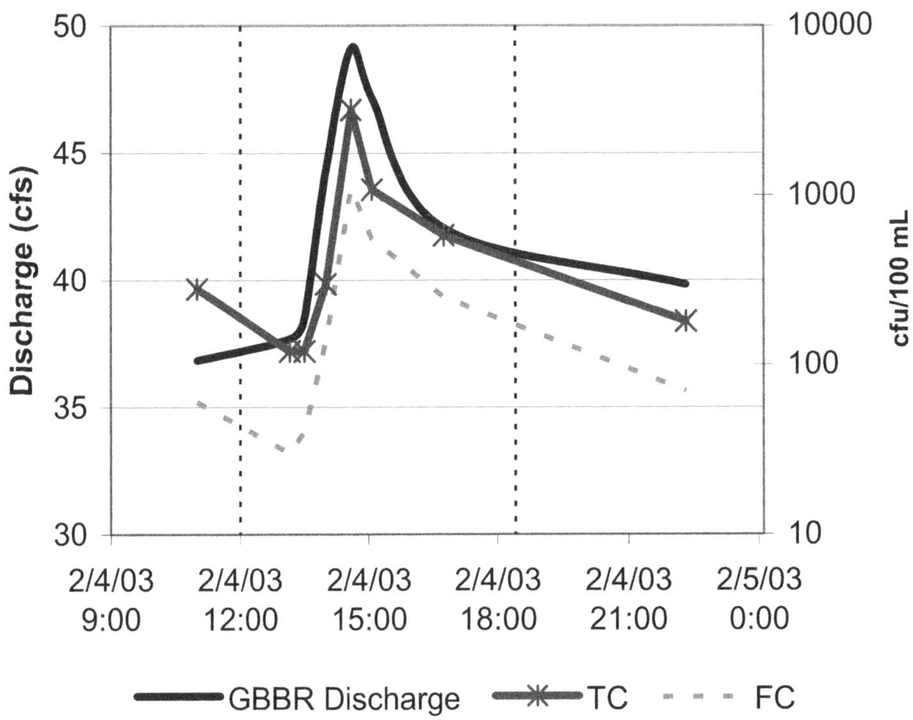

Figure 6-19 Total Coliform and Fecal Coliform detections at GBBR during the 4-February-03 event

28-May-03 Agricultural Site Event

The response at the agricultural sampling sites during the 28-May-03 event is shown in Figures 6-20 and 6-21. As previously noted, high levels of *Cryptosporidium* and *Giardia* were detected at the downstream sampling location (BBFM) during this event but not at the upstream location. Severe thunderstorms were forecast for the evening and overnight. Field work to install the weir at BBFM had been planned for several weeks to occur during the daytime; the decision was made to coordinate field work with capturing an event. Discharge at BBFM was thus inferred from discharge monitored upstream at BBWR, where the weir was already installed. Sampling for nutrients and standard indicator organisms was initiated after rainfall had started, when the initial hydrograph rise was detected at the sites. This delay was due to the anticipated afternoon/evening thunderstorms and the light nature of the rainfall. No sampling for *Cryptosporidium* and *Giardia* nor the alternative indicator organisms was conducted until late in the afternoon when the weir installation was almost complete and a wave of more intense rain blew through. Although the system dissipated, the spent samples were not discarded even though rainfall accumulation was only 5.3 mm (0.21 inches) total and 3 mm (0.1 inches) during the period samples were collected, resulting in a peak discharge at BBFM that was approximately 11% higher than the preceding baseflow. The results returned were the highest levels detected throughout the three-year study.

The highest detected levels of *Cryptosporidium* or *Giardia* during the event were observed approximately one and one-half hours after the last pulse of rain had occurred, Figures

6-20 through 6-22, and approximately two and one-half hours after the inferred peak discharge. It is possible that higher levels occurred between the two sample collections. The actual timing of the main hydrograph response and peak discharge at BBFM may have occurred later, due to lag effects, than the hydrograph inferred from BBWR suggests, and thus corresponded more closely with the peak in *Cryptosporidium* or *Giardia* detection levels. However, review of stage data collected upstream from the weir installation site at BBFM, not shown, suggest that the inferred timing is reasonably accurate. The peak detection levels appear to correspond with a leveling off of the recession curve induced by rainfall-runoff response.

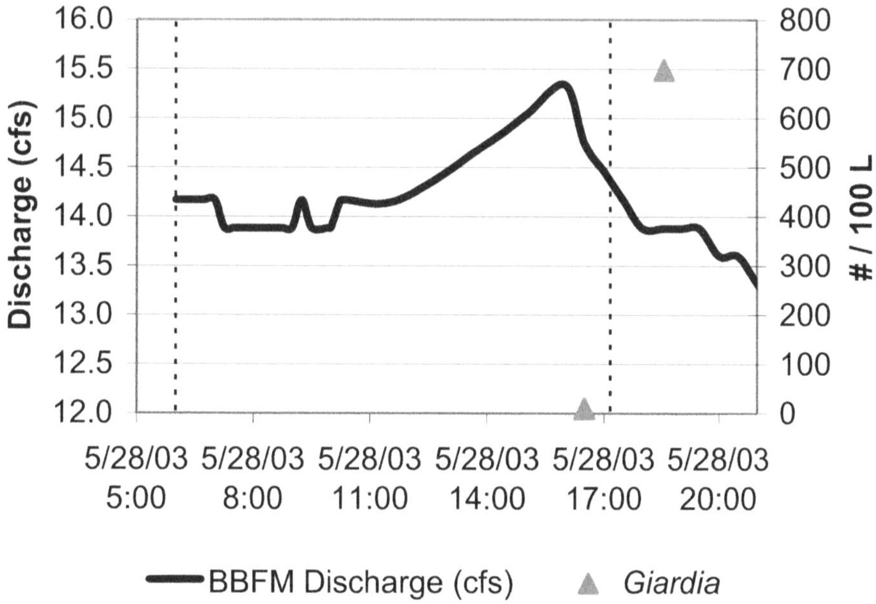

Figure 6-20 Detection of *Giardia* cysts with relation to runoff response at BBFM during the 28-May-03 event. The start and stop of rain are indicated by the dashed vertical lines. Note that discharge is inferred from BBWR.

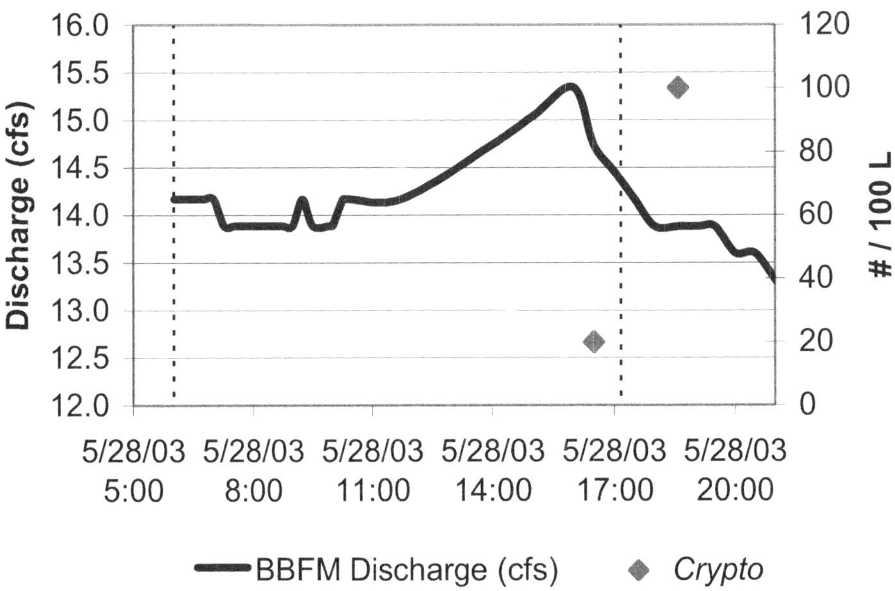

Figure 6-21 Detection of *Cryptosporidium* oocysts with relation to runoff response at BBFM during the 28-May-03 event. The start and stop of rain are indicated by the dashed vertical lines. Note that discharge is inferred from BBWR.

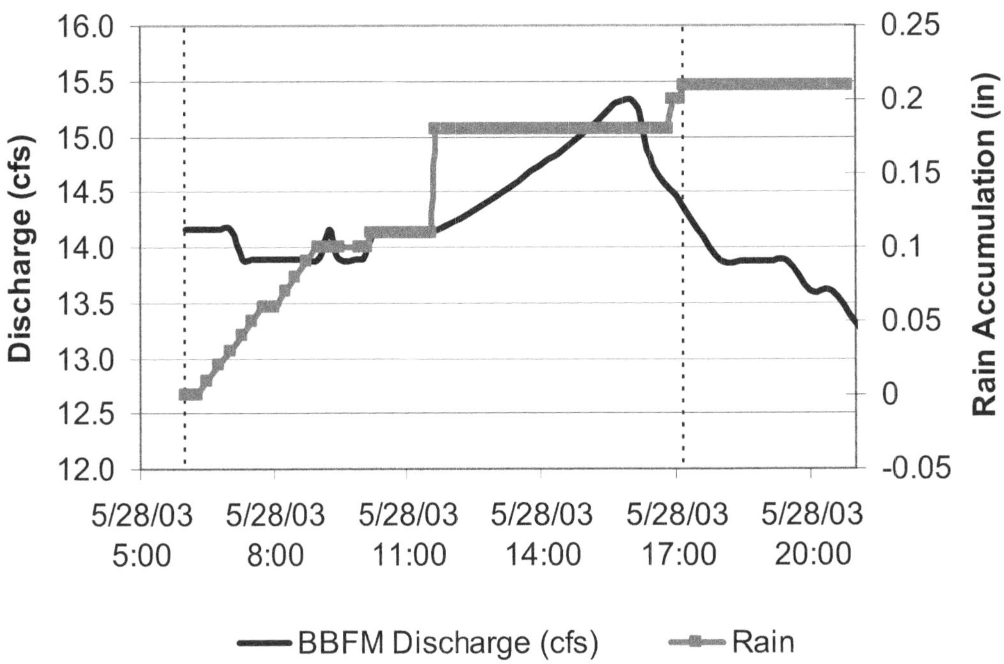

Figure 6-22 Rainfall and inferred discharge at BBFM during the 28-May-03 event

Due to the small rainfall accumulation and runoff response associated with this event, the large *Cryptosporidium* and *Giardia* levels were not anticipated. Ancillary data were thus reviewed to determine the likely source. Somatic coliphages, F+ coliphages and *Clostridium perfringens* were detected at both sites at approximately the same levels. At BBFM, the timing of response of the F+ coliphages and *Clostridium perfringens* matched that of *Cryptosporidium* and *Giardia,* Figure 6-23. Fecal coliform levels tended to decrease over the course of the event (not shown) while total coliform levels rose twice. *Cryptosporidium* and *Giardia* were not analyzed for during the first peak in TC. The second peak in TC preceded the peak in known *Cryptosporidium* and *Giardia* levels by one hour, Figure 6-24. Total coliform levels were much lower in the sample corresponding in time with the peak *Cryptosporidium* and *Giardia* levels. *Rhodococcus coprophilus* were not detected at BBFM during the event, although they were detected in two samples upstream at BBWR. The indicator microbial data thus cannot verify that cows were the source of the observed *Cryptosporidium* and *Giardia*.

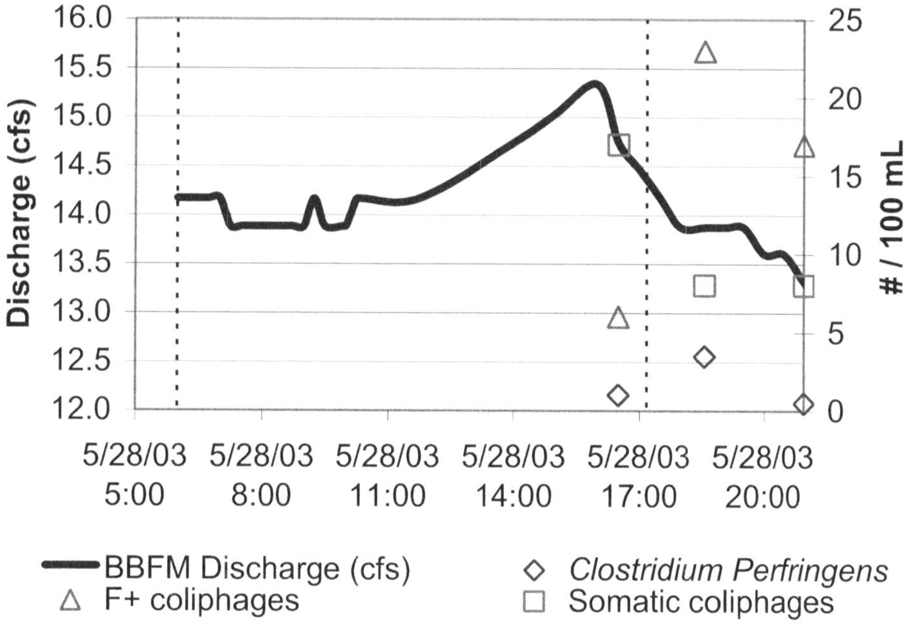

Figure 6-23 Alternative indicator organism levels detected at BBFM during the 28-May-03 event

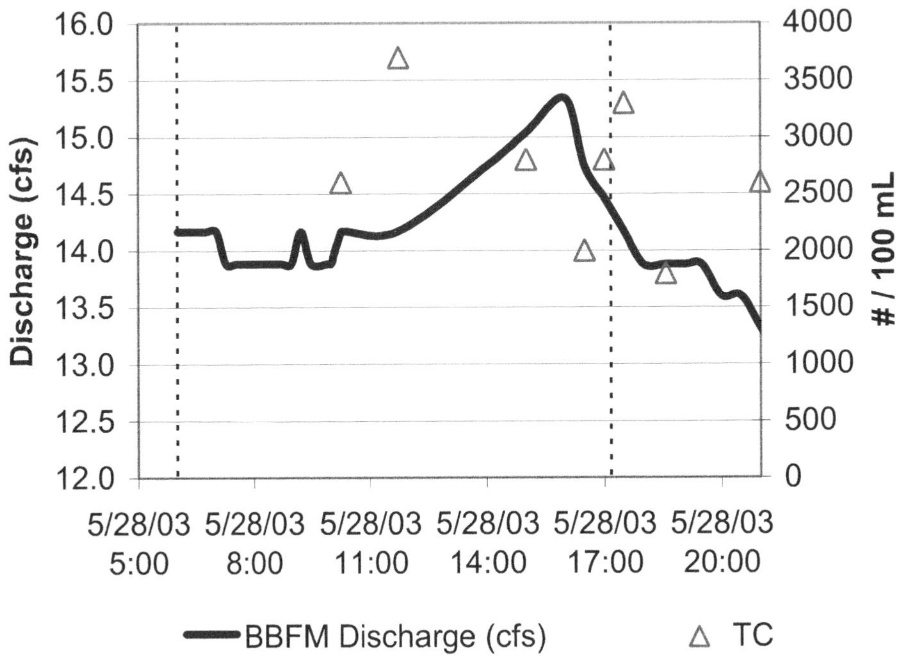

Figure 6-24 Total coliform levels detected at BBFM during the 28-May-03 event

If disturbed streambed sediments (due to installation of the weir) were the primary source of contamination, the likelihood of associated *Rhodococcus* detection would be decreased (due to die-off). Total coliform data (Figure 6-24) suggest, however, two distinct sources of contamination, the first likely associated with weir installation and the second likely associated with rainfall. Because *Cryptosporidium* and *Giardia* data are not available for the first period, the potential for the bed sediments to act as a source cannot be determined. The most likely source of *Cryptosporidium* and *Giardia* during the second peak in TC is from runoff. However, due to installation of the weir, the sampling location for this event was located a bit further downstream than typical. This location included an additional field not owned by the farmer who agreed to participate in the study and on whose property the weir was built. The field immediately upstream from this unique event sampling location was steeply sloped and contained beef cattle, including juveniles, indicating that beef cattle may also be a source of *Cryptosporidium* and *Giardia*. We had planned to continue sampling at this location throughout the study, however the participating farmer would not agree; he understood that his downstream neighbor had less efficient stream management practices and understandably wanted to know what his land was directly contributing.

19-September-03 Agricultural Site Event

The response at the agricultural sampling sites during the 19-September-03 event is shown on Figures 6-25 through 6-27. A total of 10.2 mm (0.4 inches) of rain fell during this event. At BBWR, discharge increased approximately 50% while at BBFM discharge increased approximately 24%. The timing and shape of the response was similar at both sites, as expected, because there are no major tributaries in between. Both *Cryptosporidium* and *Giardia* were detected at the upstream sampling location (BBWR), but at low levels. While only *Giardia* was detected downstream (BBFM), it was detected at higher levels than upstream at BBWR. At the downstream site, *Giardia* was detected just as the hydrograph began to rise but was not detected again over the course of the event. This sample occurred during a period of intense (~18 mm hr^{-1}) rainfall, Figure 6-28 (BBWR discharge shown). *Cryptosporidium* was detected near the peak discharge at BBWR, but at no other time during the event. This sample occurred at the end of a period of relatively low (~ 3 to 6 mm hr^{-1}) rainfall rates. *Giardia* cysts were detected at constant, although low, levels over the rise and peak of the hydrograph and then were not detectable. Detection of *Cryptosporidium* and *Giardia* for this event appeared to be associated with periods of high rainfall rate.

Due to turn around issues between events, somatic and F+ coliphages were excluded from the analyses during this event. *Rhodococcus coprophilus* and *Clostridium perfringens* were detected at both sampling sites, Figure 6-29. At the downstream BBFM site, the sample with the highest levels of *Rhodococcus coprophilus* corresponded with *Giardia* detection. However, although *Rhodococcus coprophilus* was detected in later samples, *Giardia* was not. Although *Rhodococcus coprophilus* and *Clostridium perfringens* were both detected upstream at BBWR, there appeared to be no correspondence in terms of relative magnitude with either *Cryptosporidium* or *Giardia* detection, Figures 6-29 and 6-30. Similarly, total coliform and fecal coliform concentrations did not appear to be correlated with *Cryptosporidium* or *Giardia* positive detections, Figure 6-31 (BBWR) and Figure 6-32 (BBFM). At the downstream site, BBFM, suspended solids, total Kjeldahl nitrogen, and total phosphorus concentrations all spiked in the sample corresponding to the time of the positive *Giardia* hit, Figure 6-33. No similar spike was seen at BBWR; nutrient and suspended sediment concentrations were relatively flat throughout the event at that site.

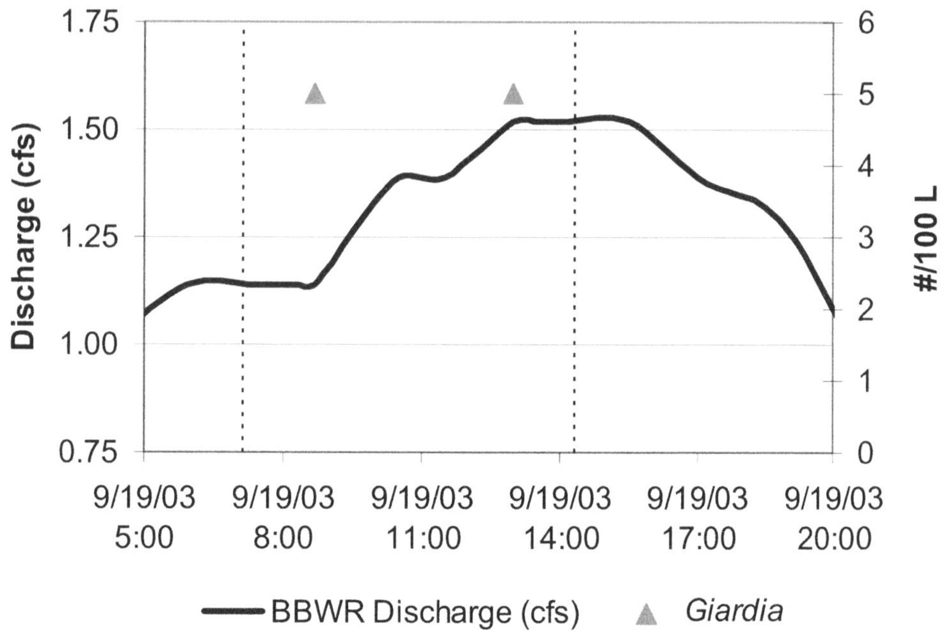

Figure 6-25 Detection of *Giardia* cysts with relation to runoff response at BBWR during the 19-September-03 event. The start and stop of rain are indicated by the dashed vertical lines.

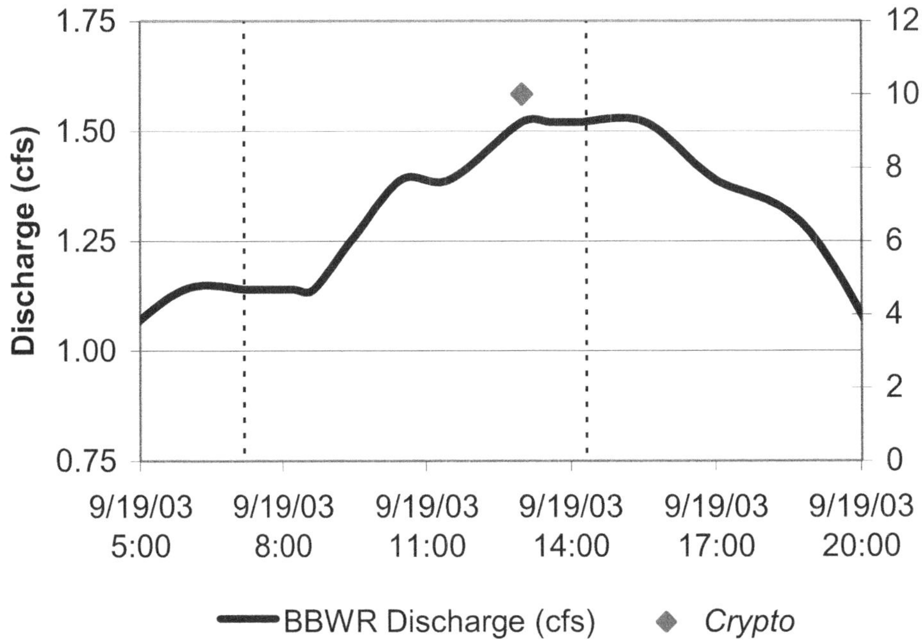

Figure 6-26 Detection of *Cryptosporidium* oocysts with relation to runoff response at BBWR during the 19-September-03 event. The start and stop of rain are indicated by the dashed vertical lines.

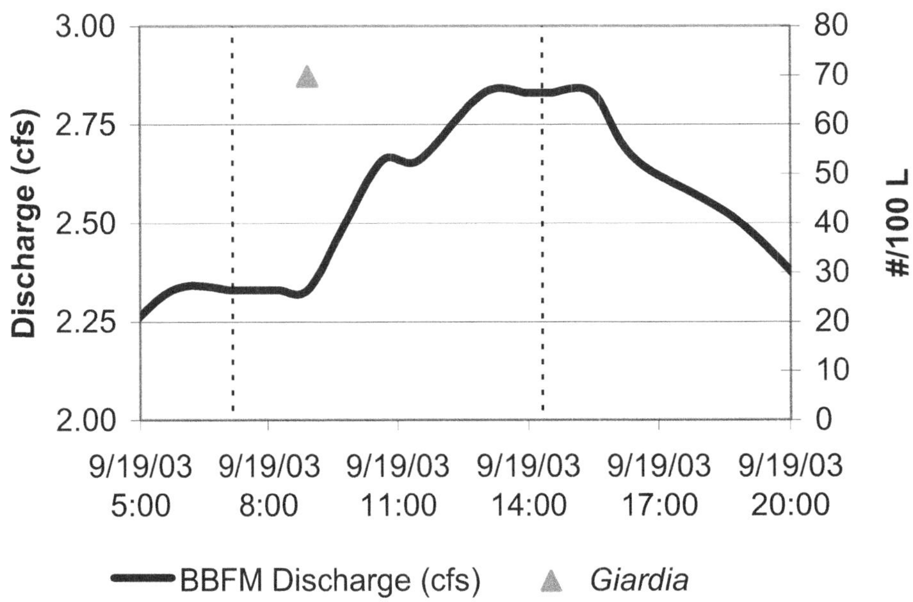

Figure 6-27 Detection of *Giardia* cysts with relation to runoff response at BBFM during the 19-September-03 event. The start and stop of rain are indicated by the dashed vertical lines.

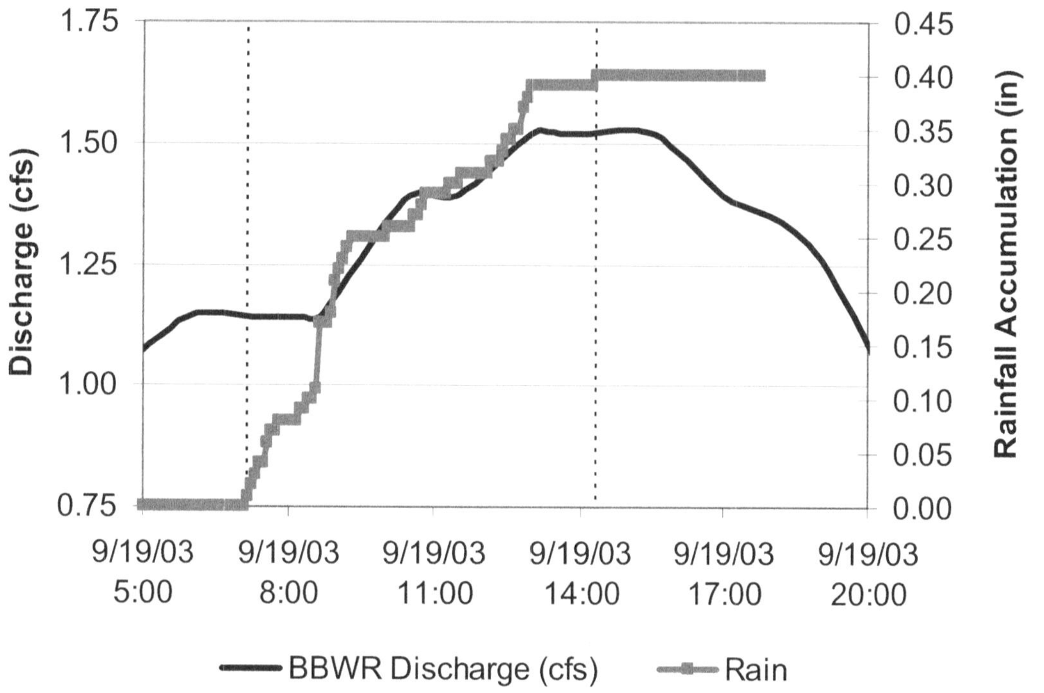

Figure 6-28 Rainfall and runoff response at BBWR during the 19-September-03 event

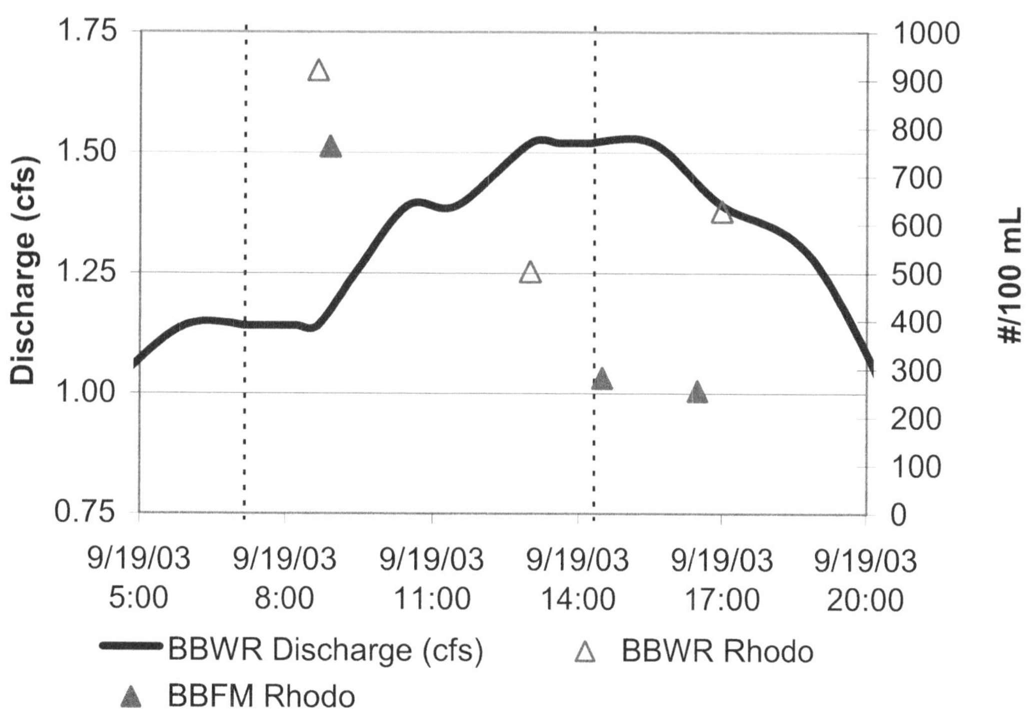

Figure 6-29 *Rhodococcus coprophilus* detection at BBFM and BBWR and corresponding runoff response (shown only at BBWR) during the 19-September-03 event

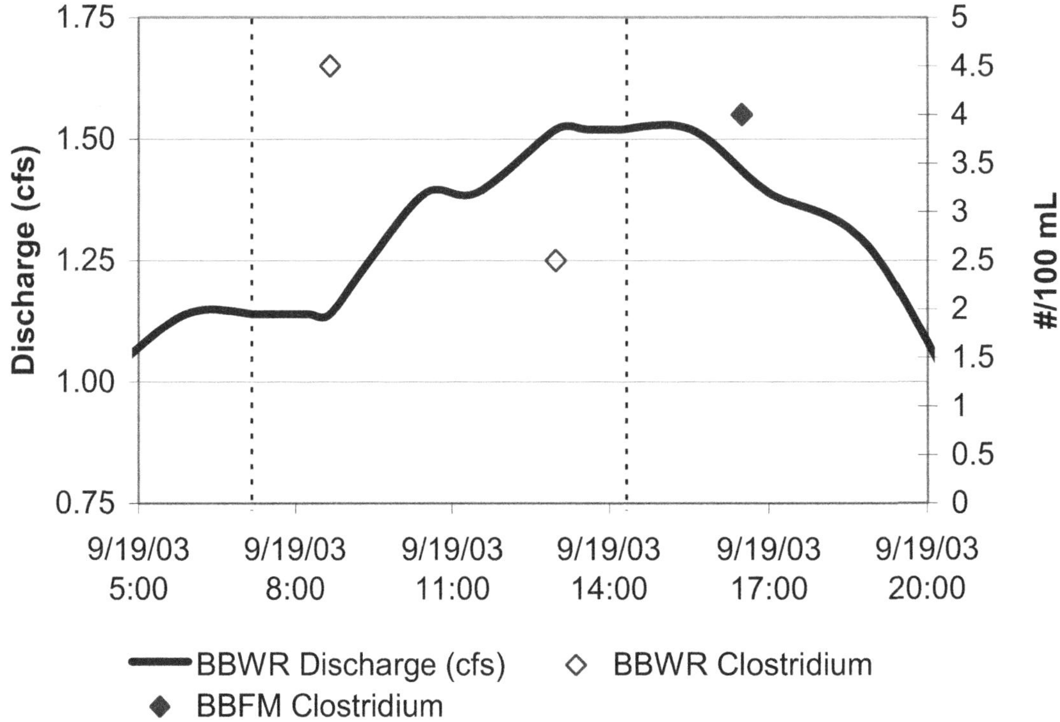

Figure 6-30 ***Clostridium perfringens*** **detection at BBFM and BBWR and corresponding runoff response (shown only at BBWR) during the 19-September-03 event**

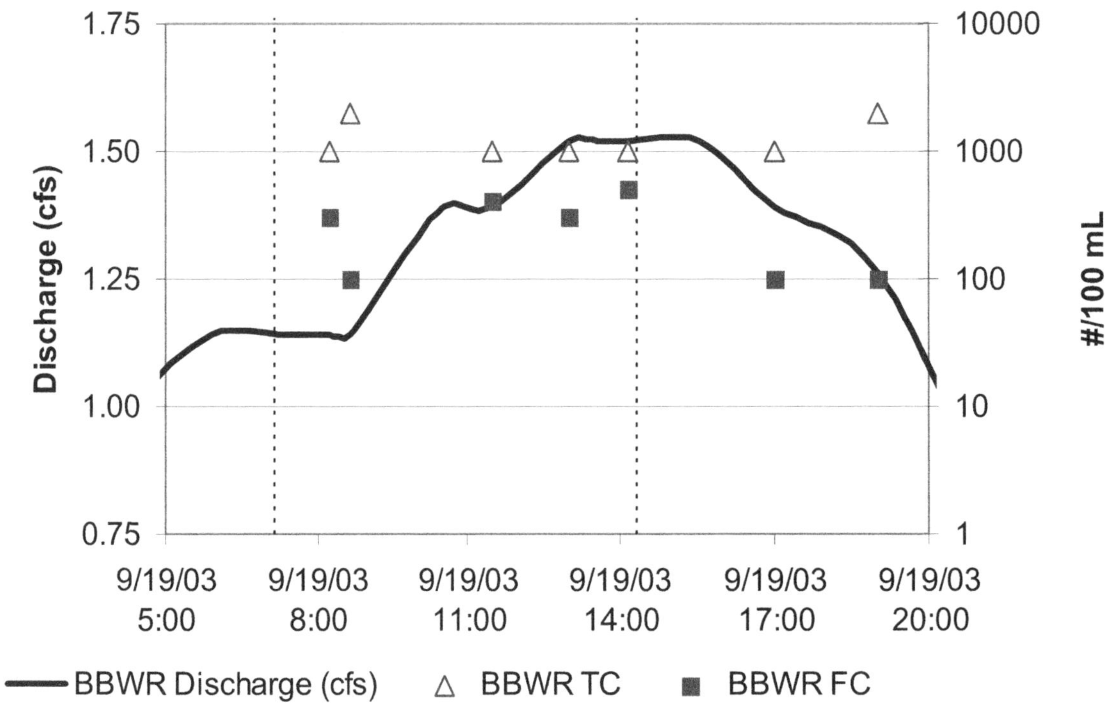

Figure 6-31 Total coliform and fecal coliform enumerations at BBWR during the 19-September-03 event

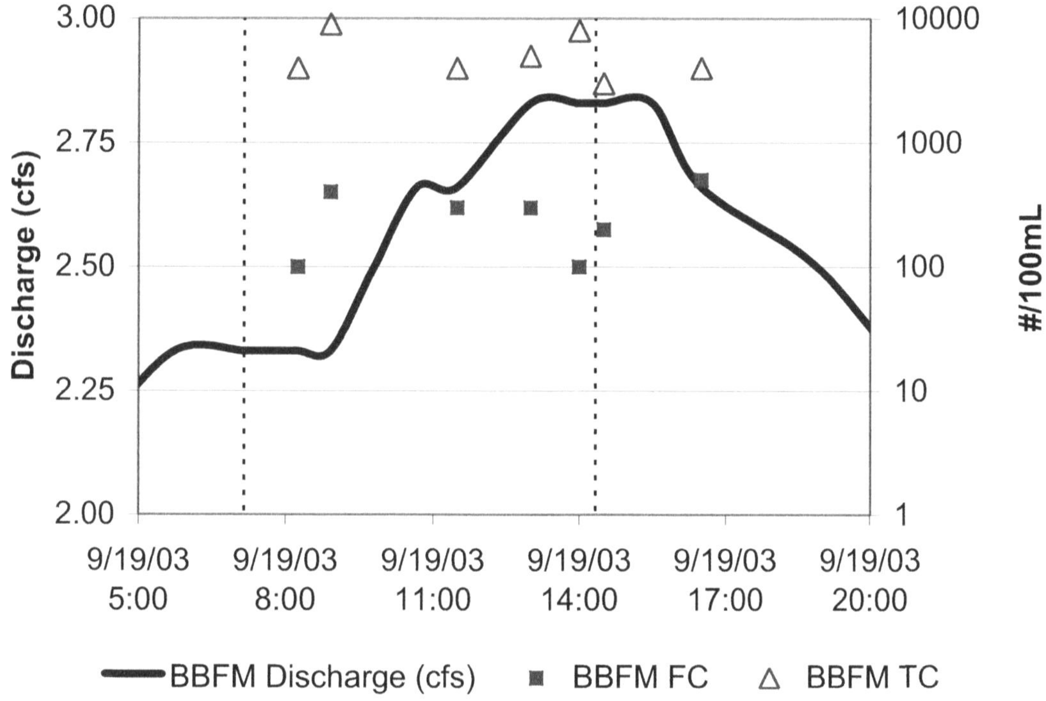

Figure 6-32 Total coliform and fecal coliform enumerations at BBFM during the 19-September-03 event

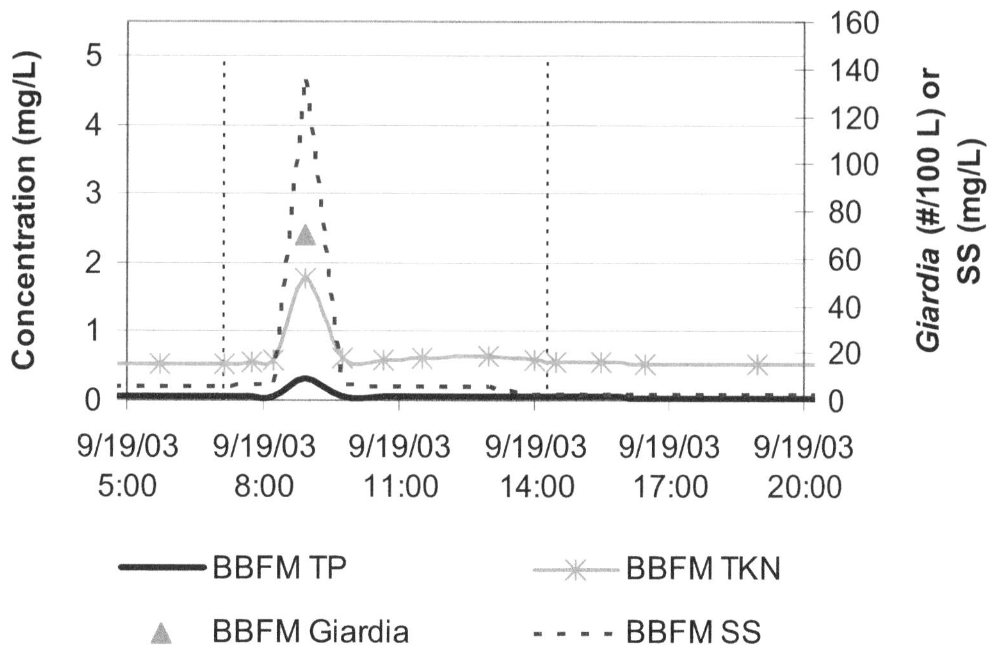

Figure 6-33 Total phosphorus, total Kjeldahl nitrogen, and total suspended solids spike corresponding to *Giardia* detection at BBFM during the 19-September-03 event

Ancillary data suggest that the source of *Giardia* at BBFM during the 19-September-03 event was grazing animals, likely the cows in the southern farm pasture. High rainfall rates may have washed a plug of contaminated sediment, also containing elevated levels of phosphorus and Kjeldahl nitrogen, into the stream just upstream of the BBFM sampling site. While it is clear that upstream contamination sources also existed during this event, transport did not appear to be linked as strongly to surface runoff, although it was correlated with moderate to high rainfall rates and the rise of the stream hydrograph. Along stream sinks of *Giardia* and *Cryptosporidium* appear to exist between BBWR and BBFM, although apparent losses could be due in part to the sampling scheme. The beaver impoundments between the sites are potential sinks due to enhanced conditions for settling.

Event Summary

Detailed review of the events during which *Cryptosporidium* and *Giardia* were detected provides insight into the occurrence and transport of these pathogens, including:

- *Cryptosporidium* and *Giardia* tend to exhibit strong flushing characteristics in non-urbanized areas, with highest levels occurring during the rising limb of the hydrograph or near the peak.
- In agricultural and wildlife (forested) areas, the highest probability for transport of *Cryptosporidium* and *Giardia* to the stream appears to be during periods of high intensity rain, accompanied by sharp rises in the stream hydrograph. This may in part be an affect of our sampling strategy.
- Although not discussed explicitly in the above event summaries, a change in the cattle herd at the agricultural site occurred during the fall of 2003 and spring of 2004. During this period, the dairy cows were sold and replaced with a reduced number of beef cattle. While no *Cryptosporidium* and *Giardia* hits were detected after this switch, only one wet-weather sampling event occurred after this point in time. This event was characterized by the largest event rainfall accumulation but low antecedent rainfall accumulation (see next two sections). Results from the 28-May-03 event suggest that beef cows are a probable source of *Cryptosporidium* and *Giardia*, suggesting that the absence of dairy cattle was not the cause of the non-detects during the 31-March-04 event. Rather, the non-detects were likely due to the absence of cows (beef or dairy) from the field (barn housed during winter).
- At the downstream agriculture site, BBFM, identification of the source of *Cryptosporidium* and *Giardia* is complicated by the existence of both beaver and cows within the intervening area. Beaver cannot be eliminated as a potential source during the 28-May-03 event because lower *Rhodococcus* levels were detected at the downstream (compared to upstream) sampling location, although the source of *Cryptosporidium* and *Giardia* was apparently the intervening area; recall that *Rhodococcus* has been observed in the feces of cows but not of beaver (Chapter 4). Large numbers of *Rhodococcus* were, however, observed during the 19-September-03 event at both the upstream and downstream sampling locations, suggesting grazing animals (rather than beaver) as the likely potential source of *Cryptosporidium* and *Giardia*. However, higher concentrations of *Rhodococcus* were again observed upstream. Sufficient data are not available to distinguish if this negates cows as the source of *Cryptosporidium* and *Giardia* or has no relation. In summary, an

insufficient number of positive detects were obtained to isolate cows as a source versus the beaver at the agricultural site.

- In heavily populated residential settings, baseflow appears to be an important transport mechanism of *Giardia*. No data are available to determine if this also holds for *Cryptosporidium*. This is likely limited to areas where septic systems or combined septic/sewer systems are in use.
- Data suggest that when *Cryptosporidium* and *Giardia* have been detected during an event, more frequent sampling of suspended sediment (and possibly total coliform) has the potential to serve as a surrogate for determining the relative magnitude of *Cryptosporidium* and *Giardia* over the course of the event.
- There appears to be a higher potential for transport of *Cryptosporidium* and *Giardia* during late winter and spring. It is hypothesized that this may be related to young beaver and other wildlife dispersal timing.
- Ponds, lakes, and beaver impoundments may serve as an along-stream sink of *Cryptosporidium* and *Giardia* from upstream sources, depending on the relative velocity and flow through patterns for a particular event.
- The stream bed near and downstream from agricultural areas may be an important source of *Cryptosporidium* and *Giardia*. Events which result in high stream velocities and erosion in these areas may be particularly important in terms of monitoring. Data from this study, however, were insufficient to confirm this as a potential source.
- Additional data are necessary to further elucidate the occurrence and transport of *Cryptosporidium* and *Giardia*, in addition to the usefulness of common water quality and standard indicator organism data for inferring pathogenic contamination.

FACTORS INFLUENCING PRESENCE AND DETECTION

Data from the seventeen wet weather events where *Cryptosporidium* oocyst or *Giardia* cyst were not detected are also useful for gaining insight into the occurrence and transport of these pathogens. In this section, data from all twenty-one wet weather events are combined and analyzed to determine a broader picture of the factors that influence apparent *Cryptosporidium* and *Giardia* levels in the environment, including the potential lack of detection when pathogens are actually present.

Discharge

The correspondence between *Cryptosporidium* oocyst or *Giardia* cyst detection and the hydrograph rise or fall was summarized in a simplified manner within Table 5-6, along with the observed enumeration levels for the events where positive detections occurred, and in more detail in the individual event descriptions. For clarity, the observed relationships with discharge are reviewed in this section.

When present at the wildlife and agricultural sampling locations, the number of oocysts or cysts detected varied with flow, typically showing flushing characteristics with peak detection occurring either near the peak in flow or preceding it. At the residential sites (4-February-03 event) *Giardia* enumerations at the upstream site decreased as the hydrograph rose but were still increasing as the hydrograph fell at the downstream sampling location. The timing of peak *Giardia* levels corresponded with the peak in flow.

Typically enumerations at the downstream sampling location, or the beaver influenced site in the case of the wildlife watersheds, exhibited higher values, indicating that sources existed within the intervening area. During one event at the agriculture watershed, occurring on 19-September-03, *Cryptosporidium* oocysts were detected at the upstream site but not downstream of the farm. Impoundments between the two reaches may have served as a sink. Alternatively, *Cryptosporidium* may have been present at the downstream sampling site but not detected by our sampling. During this event, *Giardia* was detected at both sites, but exhibited a strong "first flush" at the sampling location downstream of the farm.

Additional insight may be gained by analyzing the exceedence probability, perhaps on a weekly basis, of peak discharges for events where positive *Cryptosporidium* oocyst or *Giardia* cyst detections occurred. Such analyses, however, have not yet been carried out. In addition, peak discharges for the positive and negative detection events have not yet been compared, although the data have been used to make load estimations. It is anticipated that this work will be completed for a future publication.

Rainfall

Due to the challenges inherent in capturing wet weather event monitoring data, the captured events were characterized by a wide range of rainfall accumulation and runoff response. Event rainfall accumulations and antecedent precipitation are summarized in Table 6-8 and were also provided in Table 6-3 (event rainfall accumulation) and Table 5-5 (antecedent precipitation). Events during which *Cryptosporidium* oocysts or *Giardia* cysts were detected are shaded light gray. Rainfall accumulations for the wet-weather events ranged from 2.0 to 82.0 mm (0.02 to 3.23 inches) across all sites. The range in magnitude of events sampled at the individual study sites was similar (wildlife – 2.0 to 55.1 mm, agricultural – 5.0 to 70.0 mm, and residential 3.7 to 82.0 mm). However, a higher number of events with rainfall accumulations under 15 mm were monitored at all study sites (wildlife – 3 of 7, agriculture – 4 of 7, residential – 5 of 7). With the exception of the May 28, 2003 and September 19, 2003 events at the agricultural site and the February 4, 2003 event at the residential site, the rainfall accumulations during the event far exceed the rainfall accumulation occurring during the previous 72 hours. It should be noted that wet-weather sampling plans often stipulate a 72-hour antecedent period of no or low rainfall. No such pre-requisite was set for this study.

Table 6-8
Summary of rainfall accumulations before and during monitored wet-weather events

Season	Date	Rain (mm)	Antecedent Rain Previous 24hrs (mm)	Previous 48hrs (mm)	Previous 72hrs (mm)
Wildlife					
Spring	11-Jun-03	11.68	0	0.51	0.51
	9-Jun-04	13.72	0	0	0.25
Summer	28-Jul-02	2.03	0	0	0
	26-Sep-02	55.12	0	0	0
Fall	27-Oct-03	49.58	0.76	0.76	3.05
	15-Oct-04	32.77	3.81	3.81	3.8
Winter	26-Mar-02	19.05	0	0	0
Agriculture					
Spring	28-May-03	5.33	0.76	36.07	36.32
	18-Jun-03	7.62	1.52	1.52	2.29
Summer	12-Aug-03	46.99	0	4.57	6.60
	19-Sep-03	10.16	0	0.25	21.08
Fall	11-Nov-03	5.08	0	0	0
	17-Dec-03	58.42	0.51	0.51	9.14
Winter	31-Mar-04	69.85	0.76	0.76	0.76
Residential					
Spring	2-May-03	3.56	0.76	NA	NA
	12-Apr-04	82.04	0	0	0
Summer	1-Aug-03	7.87	NA	NA	NA
	2-Sep-03	18.54	0	0	0
Fall	6-Nov-02	18.29	0.76	1.02	1.02
	11-Dec-03	29.21	0	2.54	2.54
Winter	4-Feb-03	9.65	5.33	21.84	22.35

Cryptosporidium oocysts or *Giardia* cysts were detected even when very modest runoff occurred in response to less than 6 mm (0.24 inches) of rainfall. The relationship between the detection of *Cryptosporidium* oocysts or *Giardia* cysts and rainfall accruing during the event is more clearly illustrated on Figures 6-34 and 6-35. In these plots, all data are shown by plotting the level of detection, based on the volume of water filtered, for samples where no *Cryptosporidium* or *Giardia* was found. Less than 20 mm of rain fell during all of the events where detection occurred for *Cryptosporidium* oocysts and *Giardia* cysts. However, rainfall accumulations ranging between 20 to 40 mm and 60 to 80 mm are not well represented by the data.

The events for which detections occurred were all very modest to low rainfall accumulation events with significantly higher rainfall accumulations accruing during the previous 72 hours than during the event itself, except in the case of the March 26, 2002 event at the wildlife site. During this event, our rainfall monitoring equipment was down. However, the National Acid Deposition Program station located on Prescott Peninsula indicated no rain had fallen during the previous three days. It should be noted that the March 26, 2002 event was characterized by rain falling on an existing snowpack. The relationship between the detection of *Cryptosporidium* oocysts or *Giardia* cysts and rainfall accruing during the preceding 72-hours, ending at 7:00 AM, plus any rainfall occurring on that day prior to sampling the event is plotted on Figures 6-36 and 6-37 for *Cryptosporidium* and *Giardia*, respectively. In these plots, all data are shown by plotting the level of detection, based on the volume of water filtered, for samples where no *Cryptosporidium* or *Giardia* was found. The significance of the 72-hour antecedent precipitation is more clearly illustrated in Table 6-8. This is due to the fact that during the highlighted events, *Cryptosporidium* and *Giardia* were not detected in ALL of the samples, although *Giardia* was detected in all but one; Figure 6-36 and 6-37 show both the positive and negative detects. In addition, *Cryptosporidium* was not detected during the February 4, 2003 event at Gates Brook.

In addition, analyses of the individual events where positive detections for pathogens occurred suggest that rainfall rate may be a more important indicator of *Cryptosporidium* or *Giardia* transport than total storm accumulation. The limited data suggest that streams experiencing rainfall rates in excess of 15 mm/h may contain significant wet weather volumes of *Cryptosporidium* oocysts and *Giardia* cysts prior to any significant rise in stage.

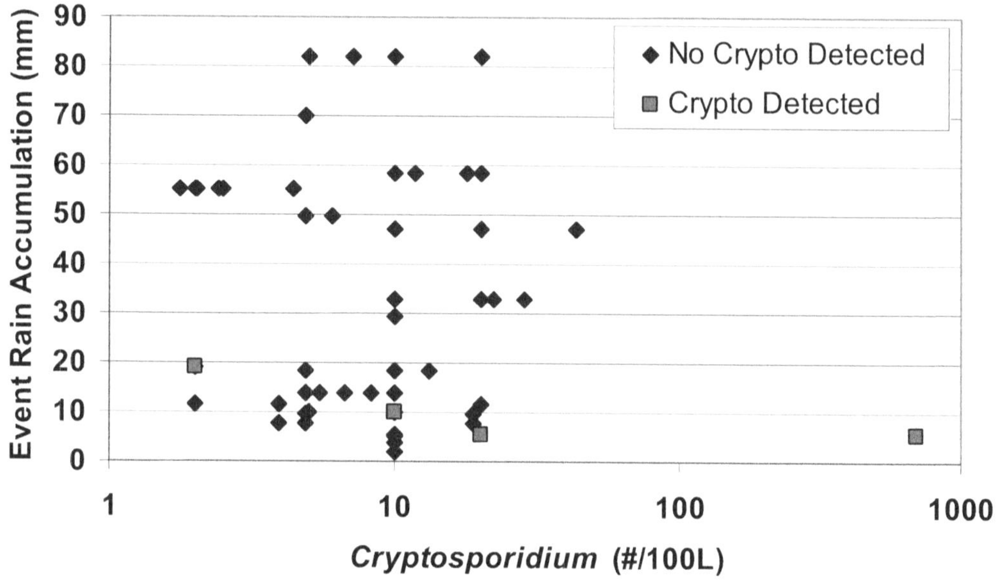

Figure 6-34 Event rainfall accumulation versus either *Cryptosporidium* oocysts detection limit based on volume of water filtered (no *Crypto* detected case) or the detected number of oocysts expressed per 100 L (*Crypto* detected case) *Source*: Rees *et al.*, 2005. Transport of *Cryptosporidium* and *Giardia* in Massachusetts Watersheds During Storm Events. In *Managing Watersheds for Human and Natural Impacts: Engineering, Ecological, and Economic Challenges*. Copyright ©2005 ASCE. Reproduced with permission.

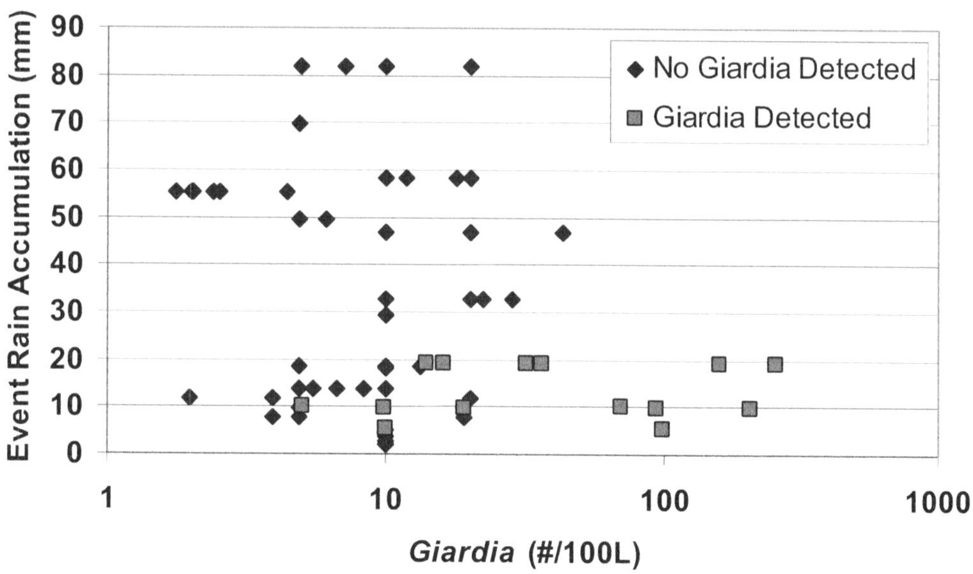

Figure 6-35 Event rainfall accumulation versus either *Giardia* cyst detection limit based on volume of water filtered (no *Giardia* detected case) or the detected number of cysts expressed per 100 L (*Giardia* detected case). *Source*: Rees *et al.*, 2005. Transport of *Cryptosporidium* and *Giardia* in Massachusetts Watersheds During Storm Events. In *Managing Watersheds for Human and Natural Impacts: Engineering, Ecological, and Economic Challenges*. Copyright ©2005 ASCE. Reproduced with permission.

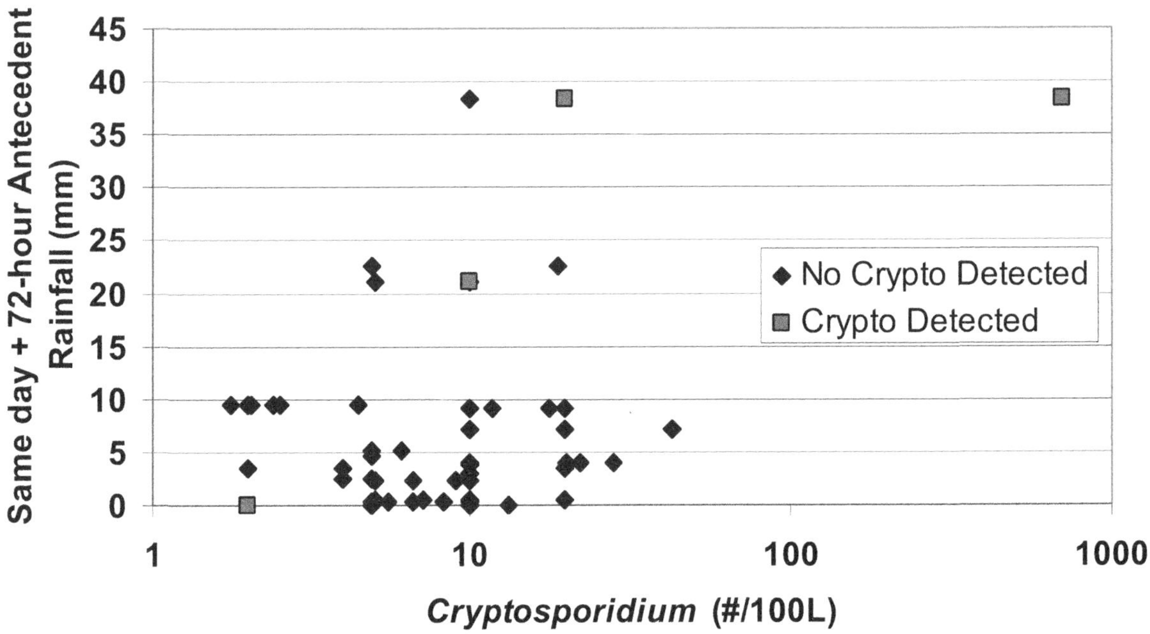

Figure 6-36 Antecedent rainfall occurring prior the same day plus over the preceding 72-hour period ending at 7:00 AM versus either *Cryptosporidium* oocyst detection limit based on volume of water filtered (no *Crypto* detected case) or the detected number of oocysts expressed per 100 L (*Crypto* detected case)

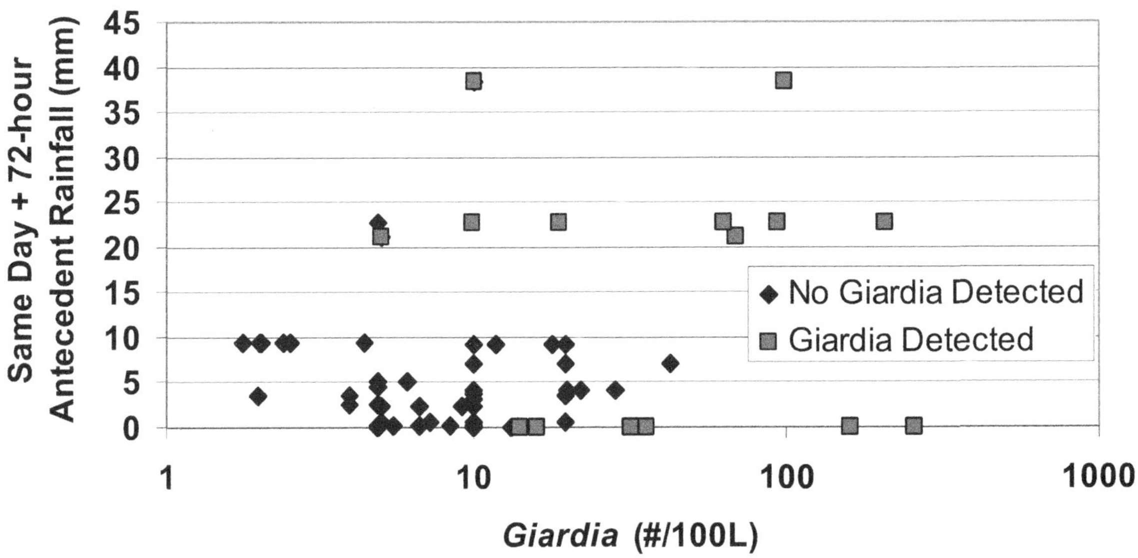

Figure 6-37 Antecedent rainfall occurring prior the same day plus over the preceding 72-hour period ending at 7:00 AM versus either *Giardia* cyst detection limit based on volume of water filtered (no *Giardia* detected case) or the detected number of oocysts expressed per 100 L (*Giardia* detected case)

Filtered Volumes and Suspended Sediment Concentrations

Based on the literature concerning known outbreaks of *Cryptosporidiosis*, higher detections during large rainfall accumulation events were expected. One confounding factor could be suspended sediment concentration and the resulting influence on the volume of water which can be pumped through the *Cryptosporidium* and *Giardia* filter. Larger events likely result in higher suspended sediment concentrations and thus reduced filter capacity. Although reporting limits for *Cryptosporidium* and *Giardia* are based on 100 L, the ability to pump this volume is typically limited to reservoirs or lakes with low suspended solids concentrations. We anticipated filter clogging issues due to sediment loads in our tributary watersheds. Additionally, we were concerned about the time required to pump such a large volume versus discharge rate within the tributaries and our sampling interval for other parameters. Thus we originally aimed to filter 50 L of water. We were able to filter this volume for several events at the wildlife sampling sites, but were never able to filter this large of a volume at either the agricultural or residential sites. Based on initial field experience, we modified our sampling strategy to filter 20 L of water at the wildlife sampling locations and 10 L of water at the agricultural and residential sampling locations.

The volume of water pumped per filter during events where detections occurred is shown in Table 6-9. Hits were observed for filtered volumes as low as 5.3 L to as large as 50 L. The distribution of volume filtered and detect versus no-detect is more clearly illustrated in Figures 6-38 and 6-39. Due to detections at relatively low filtered volumes, there is no clear indication that inadequate volume sampling has resulted in lack of detection of *Cryptosporidium* or *Giardia* when actually present. While filtering of larger volumes is desirable, it is not feasible for most wet weather sampling of New England streams. This problem may not be as pronounced in other regions of the country.

Table 6-9
Summary of volume, in Liters of raw water, filtered for analysis during events where *Cryptosporidium* or *Giardia* were detected

		Rising	Near Peak	Falling
26-Mar-02	GRB	50	50	50
Wildlife	MBDB	50	50	50
28-May-03	BBFM	10	10	10
Agricultural				
19-Sept-03	BBWR	20	20	20
Agricultural	BBFM	10	20	20
4-Feb-03	GBWD	5.3	20	20
Residential	GBBR	20	5.3	5.3

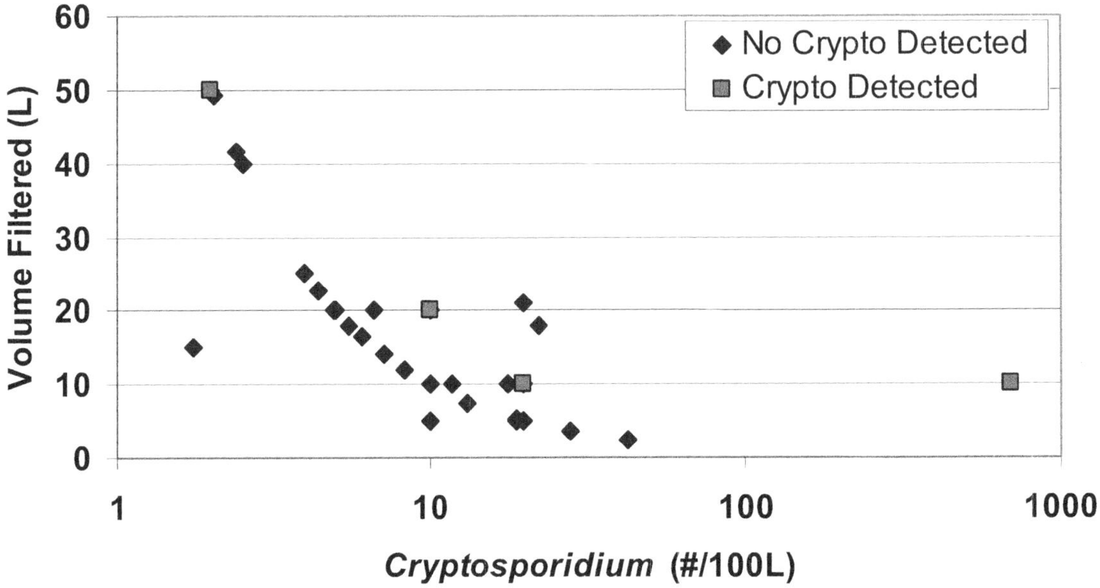

Figure 6-38 ***Cryptosporidium*** **oocysts detection based on the volume of water filtered**

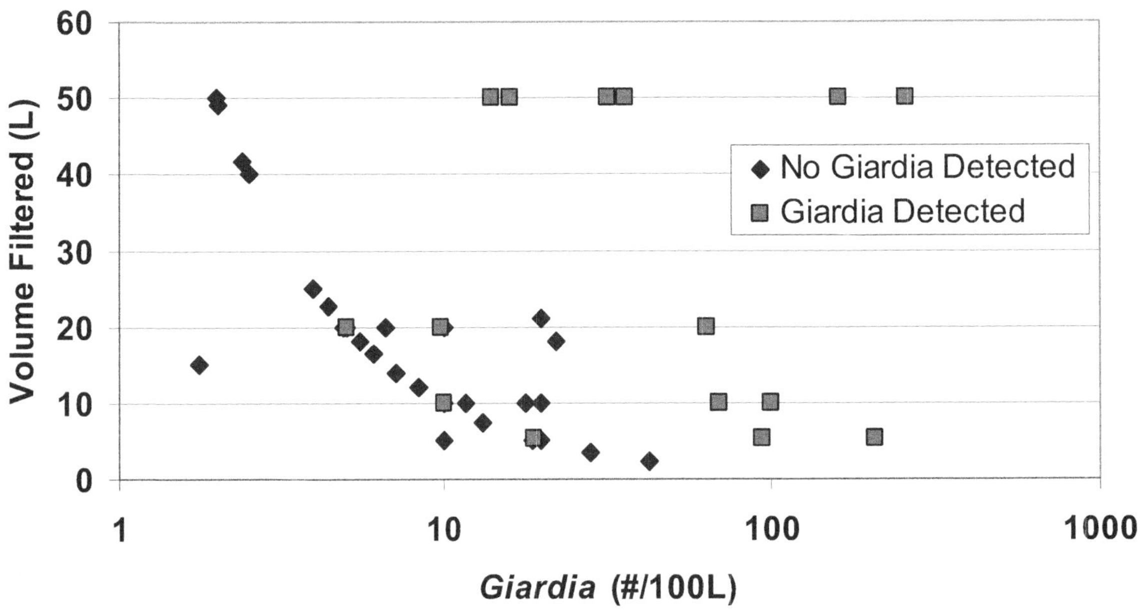

Figure 6-39 ***Giardia*** **cysts detection based on the volume of water filtered**

Figures 6-40 and 6-41 show the relationship between suspended sediment concentration in the corresponding sample and detection. Although no *Cryptosporidium* or *Giardia* was detected at the highest suspended sediment concentrations, positive detections were found at mid-range (100 – 150 mg/L; one *Cryptosporidium* and two *Giardia* detections) and very low (less than 25 mg/L; three *Cryptosporidium* and five *Giardia* detections) suspended sediment concentrations. Analysis of the individual events where positive detections occurred suggested that presence of *Cryptosporidium* or *Giardia* could be confirmed even during periods of high suspended sediment transport. A higher number of detections when suspended sediment concentrations were low, however, may suggest a masking effect for *Cryptosporidium* and *Giardia* detection when suspended sediment concentrations are high, despite the apparent lack of correlation between low filtered volumes and detection (Figures 6-38 and 6-39) and a possible correspondence between trends in suspended sediment concentration and *Cryptosporidium* and Giardia levels **for events when they are present**. If present, it is theorized that a masking effect would be caused by filter clogging, reducing the volume sampled and thus the detection limit for *Cryptosporidium* and *Giardia*. Additional data collection, using filters run in series until a larger sample volume is met, could provide further insight into the potential existence of false negatives due to high suspended sediment concentrations.

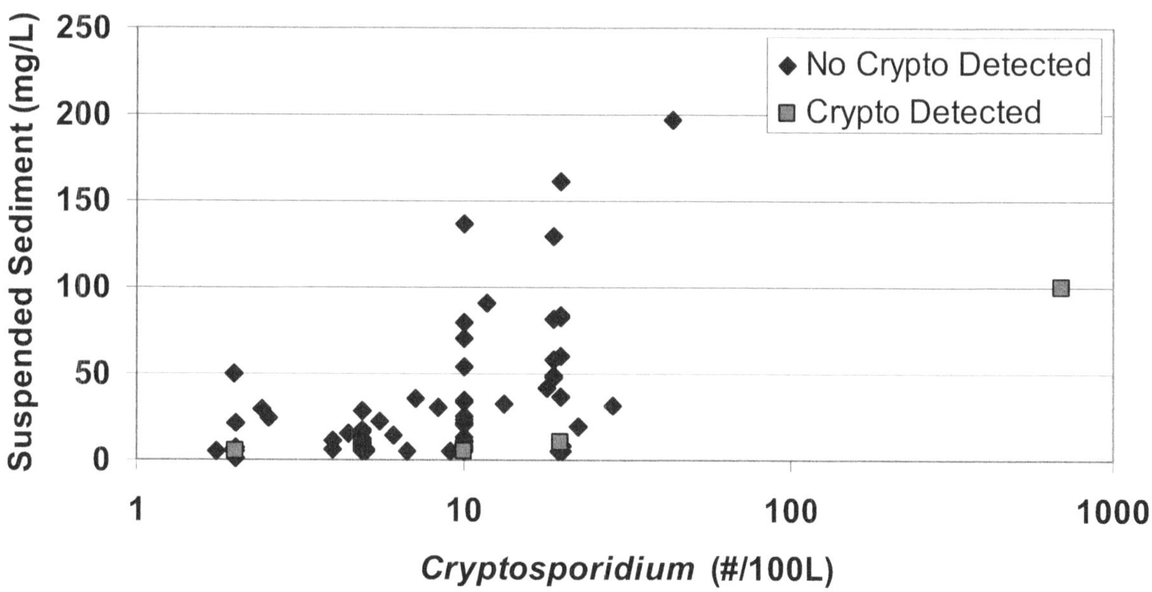

Figure 6-40 ***Cryptosporidium*** **oocysts detection based on suspended sediment concentrations**

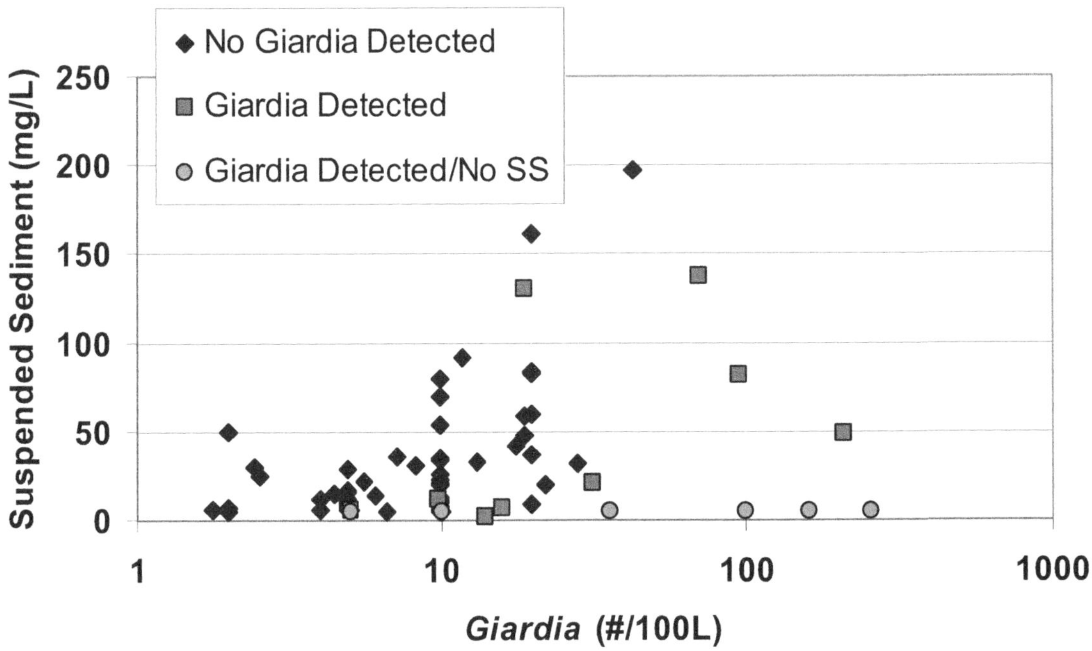

Figure 6-41 ***Giardia*** **cysts detection based on suspended sediment concentrations**

Based on analysis of events where detections occurred, suspended sediment concentrations appeared to be a potential surrogate indicator for relative magnitude over the duration of an event, as it exhibited flushing characteristics similar to that for *Cryptosporidium* and *Giardia* in non-urban watersheds. This was based on the fact that peak levels of pathogen detection often occurred during secondary peaks in suspended sediment concentration, and sampling for *Cryptosporidium* or *Giardia* often did not occur at the peak suspended sediment concentration. Inclusion of the non-detect data suggests that this relationship is likely only valid when a source for *Cryptosporidium* and *Giardia* exists within the stream and near-stream zone; during many events, *Cryptosporidium* and *Giardia* were not detected despite high suspended sediment levels (Figures 6-40 and 6-41).

Annual Variability and Antecedent Conditions

Three National Climatic Data Center (NCDC) precipitation monitoring locations and one National Acid Deposition Program (NADP) monitoring location near the sampling sites provide long-term annual precipitation data for the region, Figure 6-42. The wildlife study sites are located closest to the NADP site, but are also relatively close to the Amherst location, which has a longer period of record. The agricultural study site is located closest to the Barre Falls Dam precipitation gauge while the residential study site is located closest to the Worcester precipitation gauge. Water year 2002 was a very wet year, exceeding the 75th quartile of annual precipitation as measured at most locations since 1959 (the period of record for the NADP site was significantly shorter, extending back only to 1982). Water year 2003 was also wetter than average. The water year 2001, however, was a drought year. Snowfall data are also available at the three long-term NCDC locations, Figure 6-43. Water years 2002 and 2003 were high snowfall accumulation years for Amherst, but not for the other two locations.

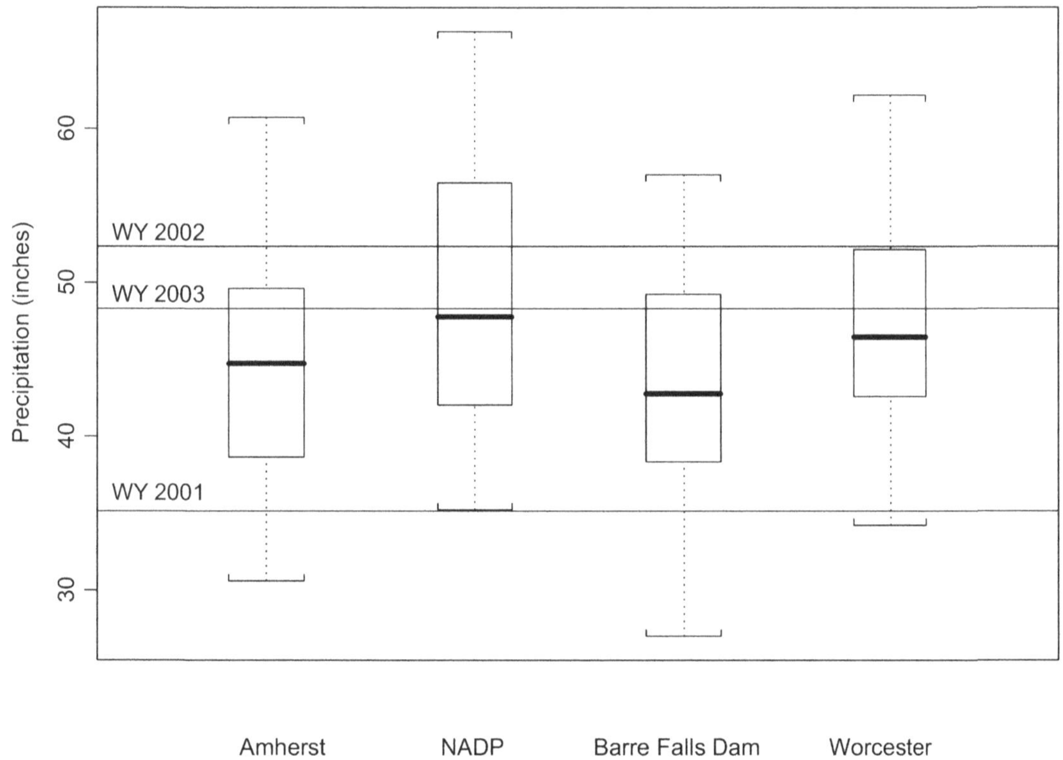

Figure 6-42 Annual precipitation accumulation inter-quartile ranges for four long-term precipitation monitoring locations near the study sites, by water year

Wet-weather data for this project were collected between March 2002 and October 2004. Three of the four wet-weather events where *Cryptosporidium* oocysts or *Giardia* cysts were detected occurred in the 2002 water year (October 1, 2002 to September 30, 2003), resulting in a 30% event detection rate. This water year was wetter than average and was the wettest of the study years overall, Figure 6-42. A total of ten of the twenty-one wet weather events were collected during the 2002 water year. No *Cryptosporidium* oocysts or *Giardia* cysts were detected during the 2003 water year despite extensive sampling (8 of 21 wet-weather events, if the October 2004 event is included, and the three spatial events) during this period and a fairly wet year. *Cryptosporidium* oocysts and *Giardia* cysts were detected in one of the three events monitored in the 2001 drought year.

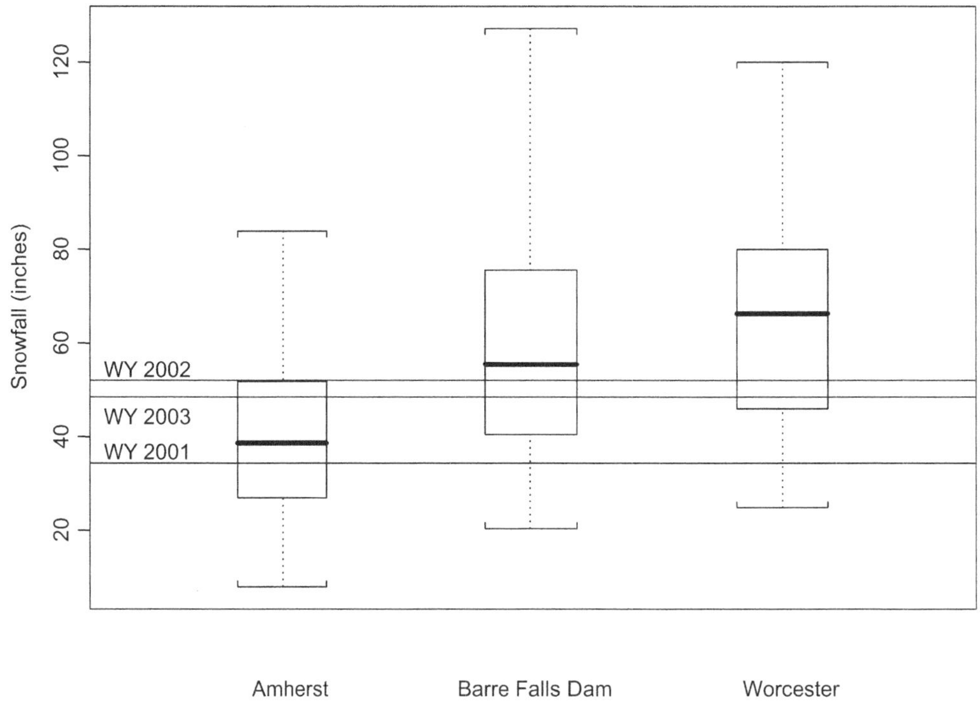

Figure 6-43 Annual snowfall total inter-quartile ranges for three long-term precipitation monitoring locations near the study sites, by water year

Analyses conducted of antecedent rainfall data collected at the study sites suggested that during events, *Cryptosporidium* and *Giardia* were more likely to be present if antecedent conditions were fairly wet. However, routine (e.g., bi-monthly or monthly) monitoring data for *Cryptosporidium* and *Giardia* are not available for the study. To further evaluate the importance of antecedent precipitation, the long-term station data were used to provide a regional picture of antecedent conditions in the month prior to events where positive detections occurred, Figures 6-44 through 6-46.

Although small in comparison to historical data, a total of 1.55 inches of water had fallen on Prescott Peninsula as snow during the ten days prior to the 26-March-2002 sampling event, Figure 6-44. Runoff from this event was the result of rain on snow. Significant levels of *Giardia* were detected at the residential study sites during the 4-February-03 wet weather event. Twenty two inches of snow (as snow) had fallen during the preceding month, Figure 6-45. Runoff from this event was also due to a combination of precipitation and snowmelt. Such events have significant potential to cause flooding. The data collected in this study suggest that they also have significant potential to transport *Cryptosporidium* and *Giardia* if sources within the watershed exist.

Historical monthly precipitation accumulation data for the Barre Falls Dam NCDC long-term precipitation gauge are shown on Figure 6-46. This site is the closest NCDC location to the agricultural site. *Cryptosporidium* and *Giardia* were detected at the agricultural site during the 28-May-03 event, when 5.3 mm (0.21 inches) of rain fell. A total of 104.1 mm (4.1 inches) of rain fell in May 03 which is a high accumulation for the month based on historical data. *Cryptosporidium* and *Giardia* were also detected at the agricultural site during the 19-September-03 event, when 10.2 mm (0.4 inches) of rain fell. A total of 145.5 mm (5.73 inches) of rain fell in September 03, which is a high accumulation for the month based on historical data.

One event is of particular note. Neither *Cryptosporidium* nor *Giardia* was detected during one winter event which was monitored, the 31-March-04 event at the agricultural site. This was one of the largest rainfall accumulation events monitored (70 mm or 2.75 inches) but occurred at the end of an extremely wet March compared to historical data (highest on record, with a total of 9.28 inches of rainfall, consisting of only 11 inches of snow). During the prior 72-hours, only 0.762 mm (0.03 inches) of rainfall had occurred. It is likely that either flushing had occurred or the ground was no longer saturated prior to the event.

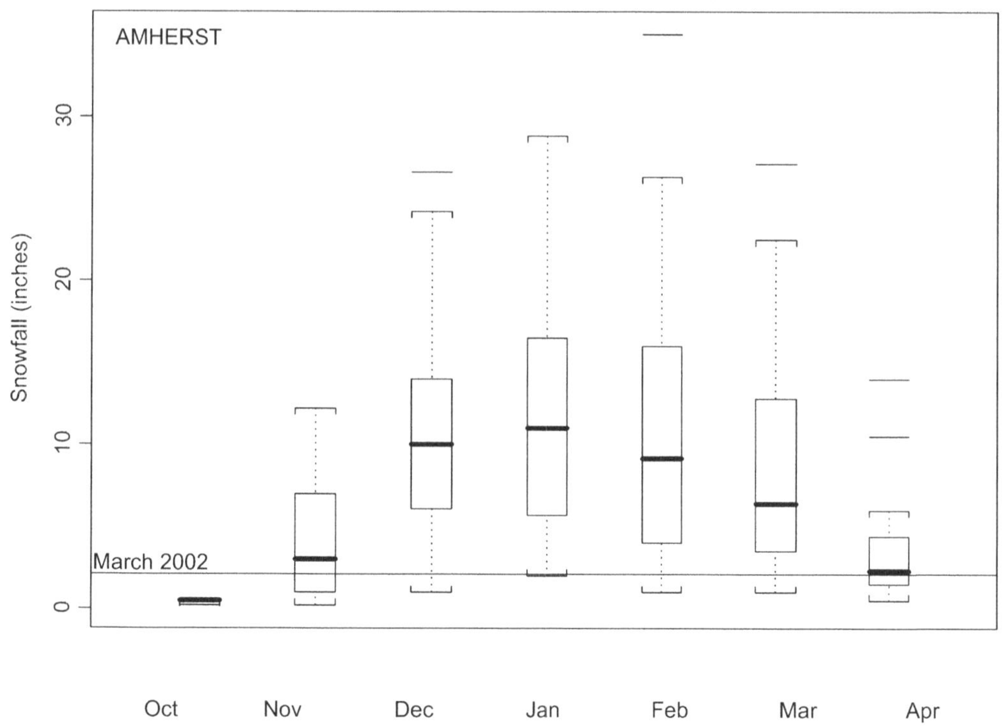

Figure 6-44 Summary of typical snowfall accumulations compared to snowfall occurring during March 2002 at the Amherst NCDC monitoring location. *Giardia* and *Cryptosporidium* were detected during the 26-March-02 event monitored at the wildlife site, in proximity to this gauge.

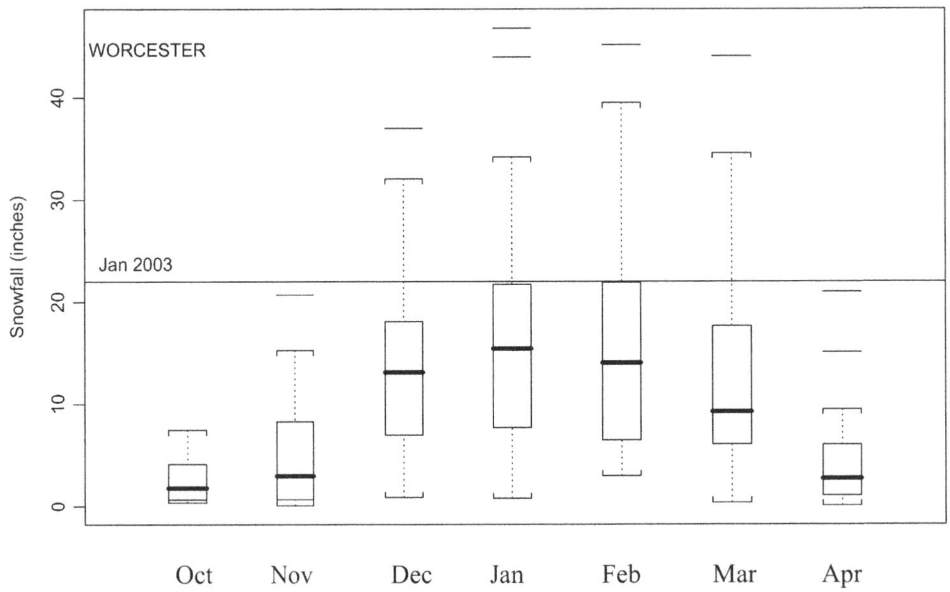

Figure 6-45 Summary of typical snowfall accumulations compared to snowfall occurring during January 2003 at the Worcester NCDC monitoring location. *Giardia* was detected during the 4-February-03 event monitored at the residential site, in close proximity to this gauge.

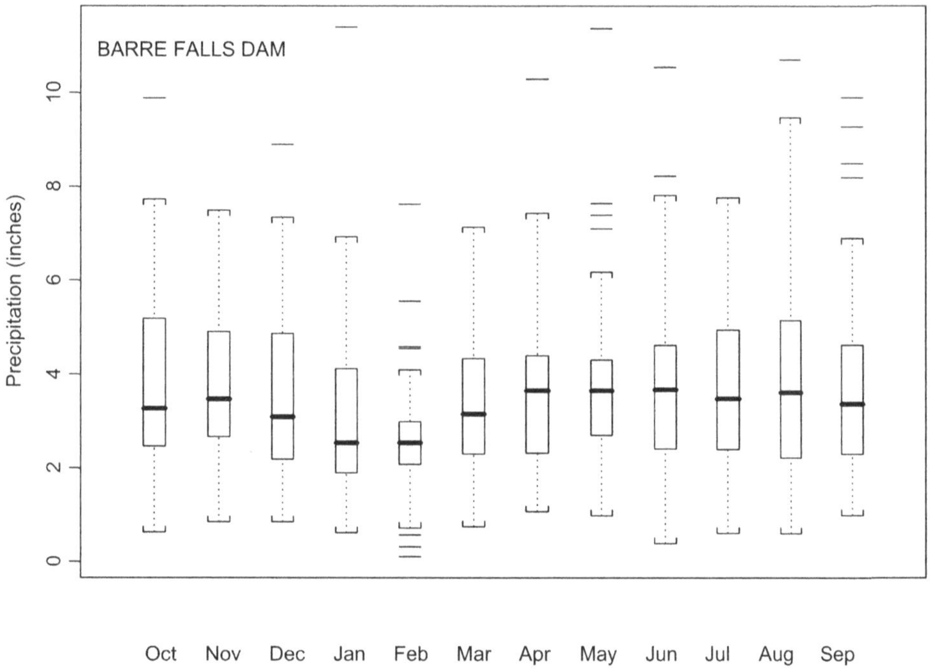

Figure 6-46 Summary of typical monthly precipitation at the Barre Falls Dam NCDC monitoring location

Review of long-term historical records suggests that a higher probability for detection of *Giardia* and *Cryptosporidium* exists during months of high rainfall accumulation in comparison to long-term historical records. In addition, data suggests a higher probability for detection for events characterized by even small rainfall accumulation if significant rainfall has occurred in the prior 72-hours. Events occurring in months where excessive (as opposed to simply high) rainfall occurs may result in a flushing of oocysts and cysts and a resulting lack of detection in later events. While further study is necessary, wet weather events occurring after monthly rainfall accumulations reach approximately the 75th percentile, in terms of magnitude in comparison to historical data, may warrant monitoring. Such a criterion, however, overlooks the need to also monitor storms occurring early in the month when soils are just saturated from precipitation during the prior month. Thus a monthly running average of precipitation may be a better gauge of transport potential. In addition, the 72-hour antecedent precipitation accumulation may provide additional guidance for when monitoring is warranted. However, sufficient data do not exist to confirm the importance of this measure, particularly for larger watersheds. Remotely sensed soil moisture data should be evaluated for its potential to provide guidance in terms of runoff and thus transport potential. This later approach is likely the better criterion. In addition, rainfall-on-snow events appear to have a particularly high probability of transport. Monitoring schemes that capture such events may yield more accurate results in terms of understanding the frequency and characteristics of *Cryptosporidium* and *Giardia* occurrence and transport.

Inter-annual variability in *Giardia* and *Cryptosporidium* occurrence in streams and rivers appears to be correlated to some extent with antecedent conditions. While annual precipitation accumulations provide some insight into such variability, this study suggests that they may not be the best metric to use for evaluating historical data and extrapolating to other years. Monthly precipitation and snowfall data may instead be a better overall metric. Long-term regional records, rather than site specific rainfall data, provide a better means of evaluating current conditions in a historical context. Data which have been collected should be evaluated in terms of historical moisture conditions to better understand their potential relation to long-term occurrence and transport.

Seasonal Variability

Trends in *Cryptosporidium* and *Giardia* detection with season are of interest. Seasons for this project were defined in terms of the hydrologic year: Fall (October 1 to December 31), Winter (January 1 to March 31), Spring (April 1 to June 30), and Summer (July 1 to September 30). Two wet-weather events were captured at each sampling site in spring, summer, and fall, typically during different years. Only a single winter wet-weather event was captured at each site due to frequent hazardous sampling conditions and the associated difficulty in capturing such events. Wet-weather events were thus monitored at least once in each month except for January. *Cryptosporidium* oocysts and *Giardia* cysts were detected during winter (February and March), mid-spring (May), and late summer (September) events. No *Cryptosporidium* or *Giardia* was detected during the fall or during June through August, Figures 6-47 and 6-48.

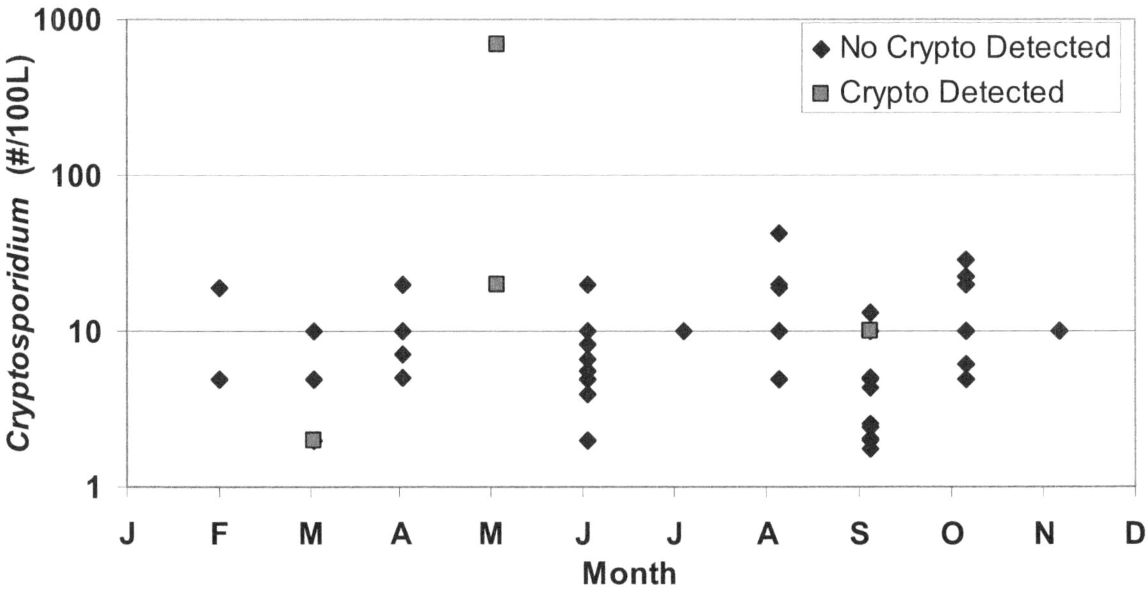

Figure 6-47 Seasonal patterns of *Cryptosporidium* detection

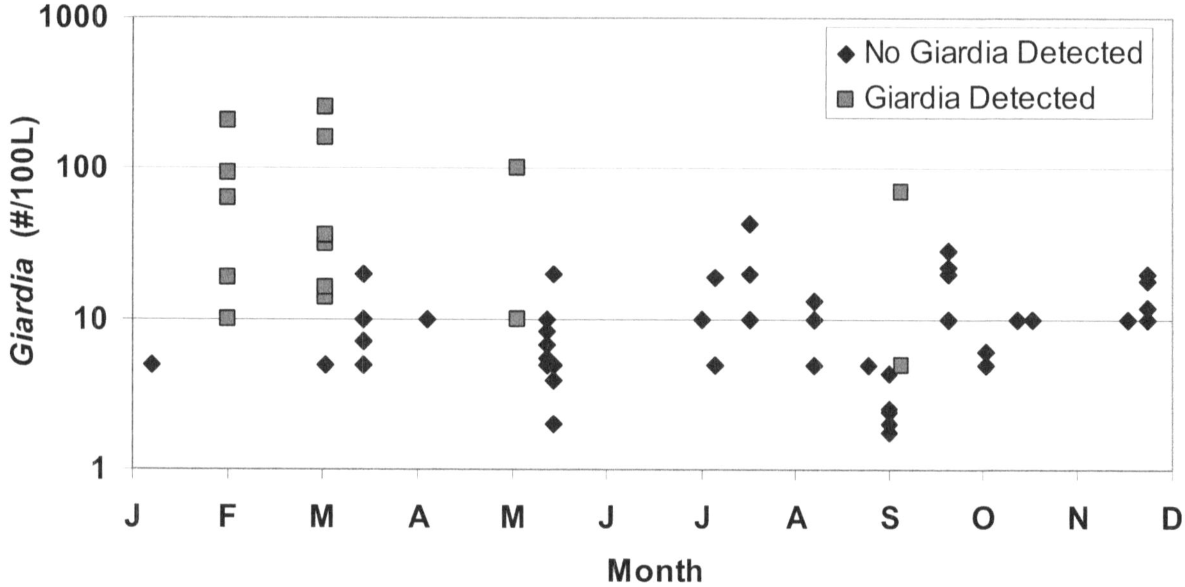

Figure 6-48 Seasonal patterns of *Giardia* detection

Previous sections of this chapter have incorporated discussions on seasonal effects. Results suggest that the probability of *Cryptosporidium* and *Giardia* occurrence in surface waters may be greater during winter months. Two potential underlying causes are suggested. One potential cause is the propensity of high runoff events to occur during this season as a result of saturated ground conditions and high baseflow. Rain-on-snow events appear to be particularly important. The second potential cause is related to wildlife activity, specifically the timing of juvenile beaver dispersion. In New England, this tends to occur during late winter through early spring. However, these beaver may not be fully established until the fall. It is hypothesized that *Giardia* and/or *Cryptosporidium* shed by dispersing beaver have a higher potential for transport, as dispersing beaver are more likely to deposit feces in flowing streams while they seek suitable locations and build their impoundments. Once established, impoundments provide an environment in which oocysts and cysts may settle rather than being transported to downstream locations.

Relation to Standard Indicator Organisms

One goal of the study was to evaluate the potential usefulness of standard indicator organisms as a surrogate for *Cryptosporidium* and *Giardia* testing. If a relationship between total or fecal coliform and *Cryptosporidium* oocysts or *Giardia* cyst levels in the environment exists, then combined monitoring strategies could be developed to provide better temporal variability information in terms of *Cryptosporidium* and *Giardia*. Individual event analyses suggested a possible correlation between total coliform and *Cryptosporidium*/*Giardia* levels. The potential relationship across events and for periods of positive and negative detection was evaluated by plotting either the level detected or the level of detection, based on the volume of water filtered, for samples where no *Cryptosporidium* or *Giardia* was found. Results are shown on Figures 6-49 and 6-50 for total coliform and on Figures 6-51 and 6-52 for fecal coliform.

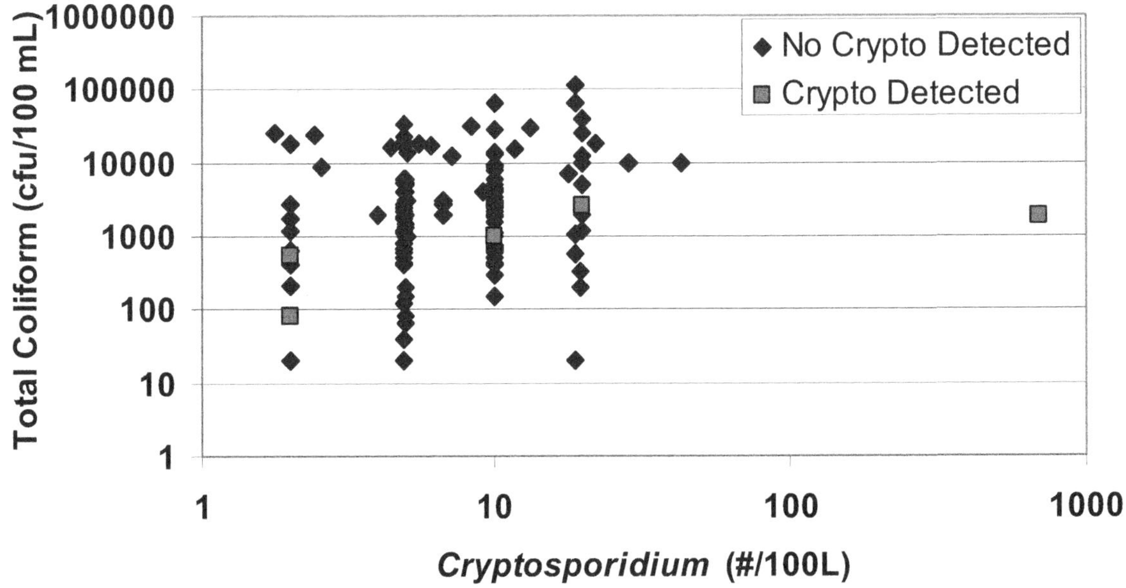

Figure 6-49 Relationship between total coliform and *Cryptosporidium* oocyst levels. Samples where no oocysts were detected are plotted as the detection limit

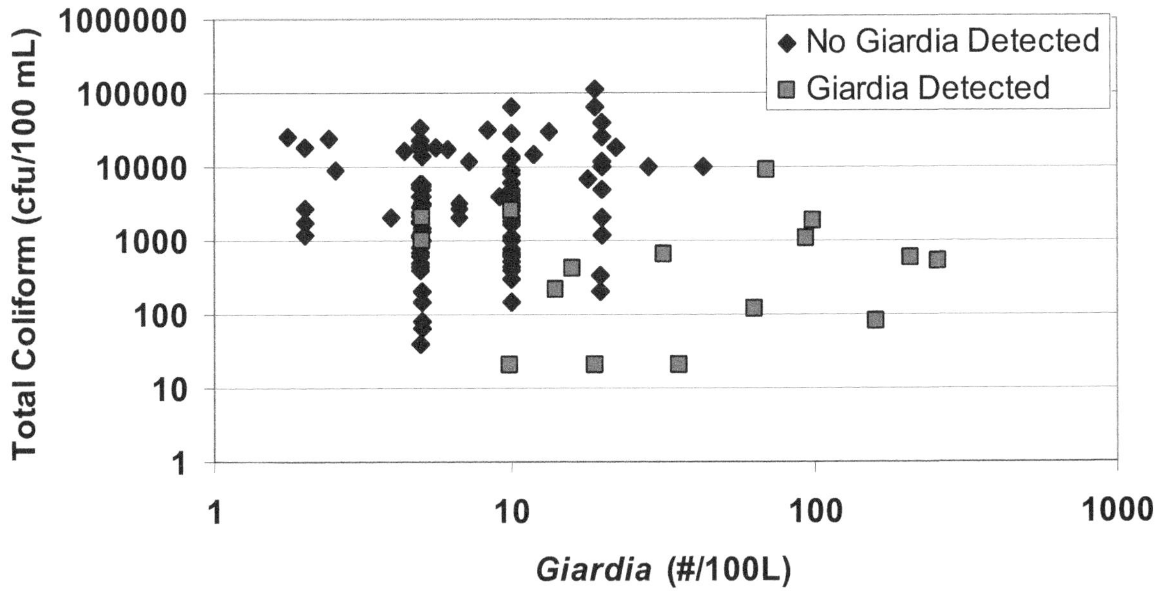

Figure 6-50 Relationship between total coliform and *Giardia* cyst levels. Samples where no oocysts were detected are plotted as the detection limit

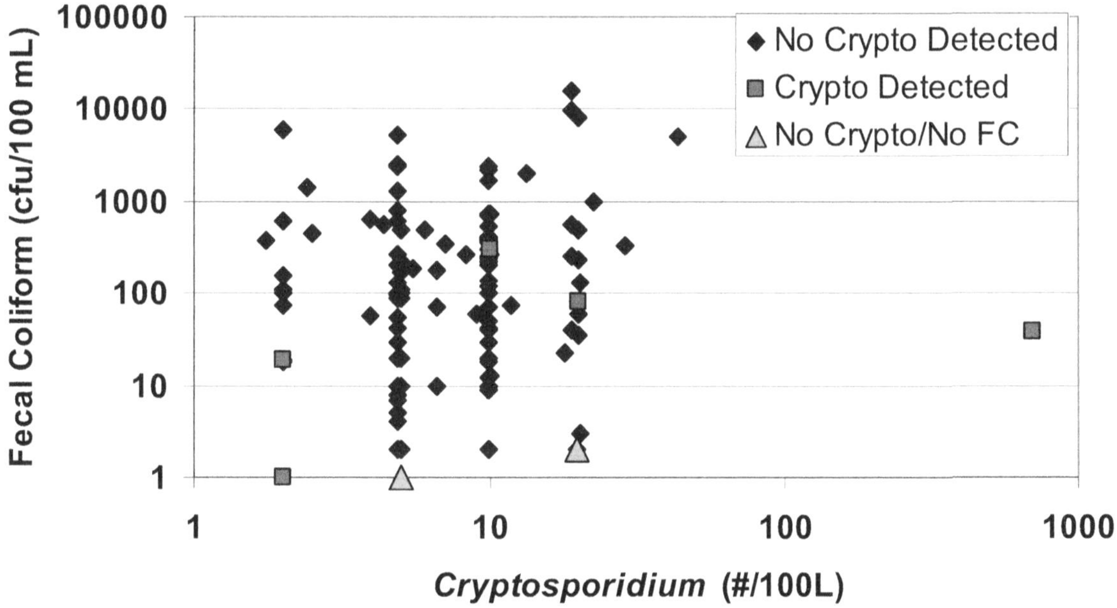

Figure 6-51 Relationship between fecal coliform and *Cryptosporidium* oocyst levels. Samples where no oocysts were detected are plotted as the detection limit

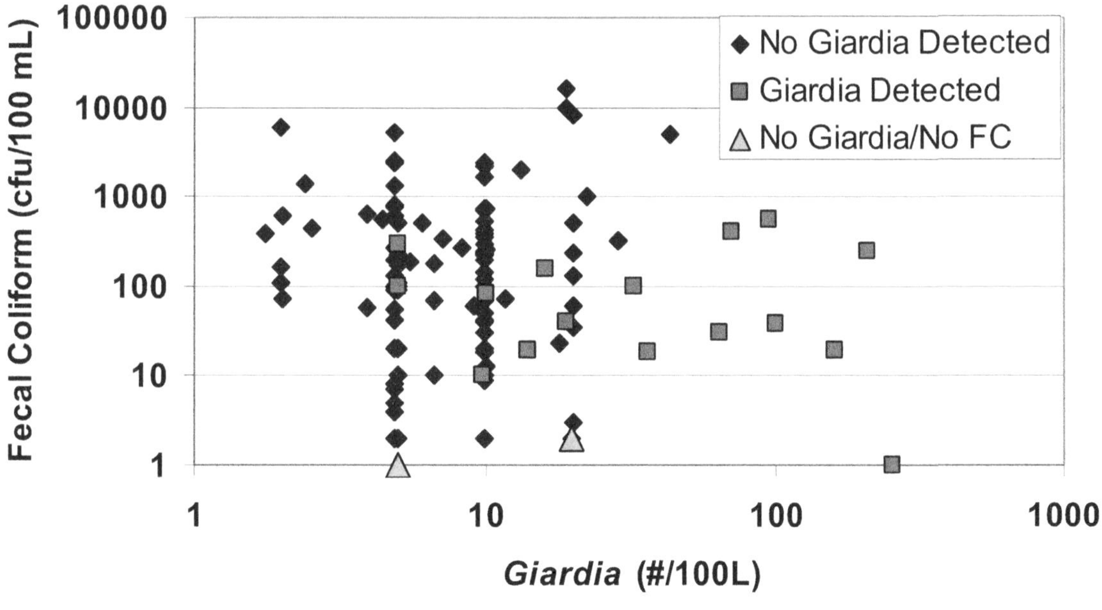

Figure 6-52 Relationship between fecal coliform and *Giardia* cyst levels. Samples where no oocysts were detected are plotted as the detection limit

The data suggests that there is no relationship between standard indicator organism (in this study represented as total and fecal coliform data) and either *Cryptosporidium* or *Giardia* levels. Similar findings have been found by other researchers (Atherholt *et al.*, 1998; LeChevallier *et al.*, 1991). High levels of both total and fecal coliform were detected in samples where no *Cryptosporidium* or *Giardia* was found. Conversely, low levels of total and fecal coliform were detected for some samples where *Cryptosporidium* or *Giardia* was detected. Analysis of individual events where positive *Cryptosporidium* or *Giardia* detections occurred suggest that there may be a significant within event correlation between total coliform and both *Cryptosporidium* and *Giardia*. This is explored further below. However, such a correspondence would not eliminate the need for *Cryptosporidium* or *Giardia* monitoring. Rather, it suggests that two to three samples collected in tandem with more refined total coliform (or suspended sediment) data collection may be adequate to characterize pathogen transport on a per event basis.

Relation to Alternative Indicator Organisms

Several alternative indicator organisms were included in the study in order to evaluate their potential usefulness both as a surrogate for *Cryptosporidium* and *Giardia* testing and for identifying potential sources of contamination. These included *Clostridium perfringens*, F+ phages, somatic or CN phages, *Rhodococcus coprophilus* (agricultural and wildlife sites) and soribitol-fermenting *Bifidobacteria* (residential). Due to low detection levels, phages were dropped from the analyses in the summer of 2003. For some events, some additional analyses were not run due to turn-around processing (analysis of previous event samples was still underway) or lab conflict issues (personnel not available or other project conflicts). Table 6-10 summarizes the analyses run per event and site. An additional event, that occurring on 14 December 2001, is included in the table. This event was a practice event where all parameters other than *Cryptosporidium* and *Giardia* were analyzed for.

The potential relationship across events and for periods of positive and negative detection was evaluated by plotting either the level detected or the level of detection, based on the volume of water filtered, for samples where no *Cryptosporidium* or *Giardia* was found. Results for *Clostridium perfringens* (Figures 6-53 and 6-54), F+ phages (Figures 6-55 and 6-56), somatic or F- phages (Figures 6-57 and 6-58), *Rhodococcus coprophilus* (Figures 6-59 and 6-60), and soribitol-fermenting *Bifidobacteria* (Figures 6-61 and 6-62) are presented. Recall that fewer analyses for phages were conducted.

No clear correlations between alternative indicator organisms and either *Cryptosporidium* or *Giardia* were found when the data across events was considered, although a few observations are noteworthy. High levels (greater than 100 cfu/100 mL) of *Clostridium perfringens* were typically associated with fairly high levels of *Giardia*. Both *Giardia* and *Clostridium perfringens* were more typically detected when *Giardia* levels exceeded 10 per 100 L. While only limited phage data were available, neither detection of F+ nor CN phages corresponded well with levels of *Cryptosporidium* and *Giardia*. Similarly, *Rhodococcus coprophilus* and sorbitol-fermenting *Bifidobacteria* were not useful across events for identifying potential occurrence of *Cryptosporidium* and *Giardia*. *Rhodococcus coprophilus* and sorbitol-fermenting *Bifidobacteria* were, however, useful for helping identify potential sources of contamination when *Cryptosporidium* and *Giardia* are detected during individual events. The lack of temporal trends in co-occurrence between alternative indicator organisms, *Cryptosporidium*, and *Giardia* were also noted by LeChevallier *et al.* (1991).

Table 6-10
Summary of alternative indicator analyses completed per event and sampling site

Site	Event Date	*C. perf* (cfu/100 mL)	F+ (pfu/100 mL)	F- (pfu/100 mL)	*Rhodococus*	*Bifidobacteria*
Burrow	5/28/2003	*x*	*x*	*x*	*x*	
Brook	6/18/2003					
	8/12/2003	*x*			*x*	
(Agriculture)	9/19/2003	*x*			*x*	
	11/12/2003	*x*			*x*	
	12/17/2003					
	3/31/2004	*x*			*x*	
	7/14/2003	*x*			*x*	
Gates	11/6/2002	*x*	*x*	*x*		*x*
Brook	2/4/2003	*x*	*x*	*x*		*x*
	5/2/2003	*x*	*x*	*x*		*x*
(Residential)	8/1/2003	*x*				*x*
	9/2/2003	*x*				*x*
	12/11/2003	*x*				*x*
	4/12/2004	*x*				*x*
	8/4/2004	*x*				*x*
Prescott	12/14/2001	*x*	*x*	*x*		
Peninsula	3/26/2002	*x*	*x*	*x*	*x*	
	7/28/2002	*x*	*x*	*x*	*x*	
(Wildlife)	9/26/2002	*x*	*x*	*x*	*x*	
	6/11/2003	*x*			*x*	
	10/27/2003	*x*			*x*	
	3/10/2004	*x*			*x*	
	6/9/2004	*x*			*x*	
	10/15/2004	*x*			*x*	

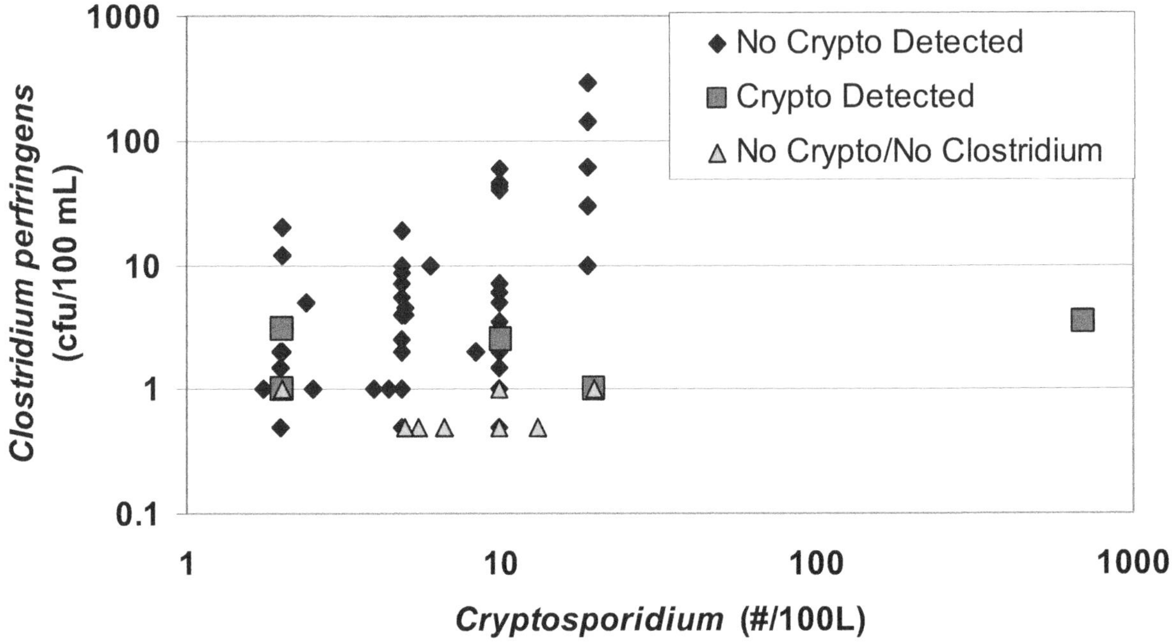

Figure 6-53 Relationship between *Clostridium perfringens* and *Cryptosporidium* oocyst levels. Samples where no oocysts were detected are plotted as the detection limit

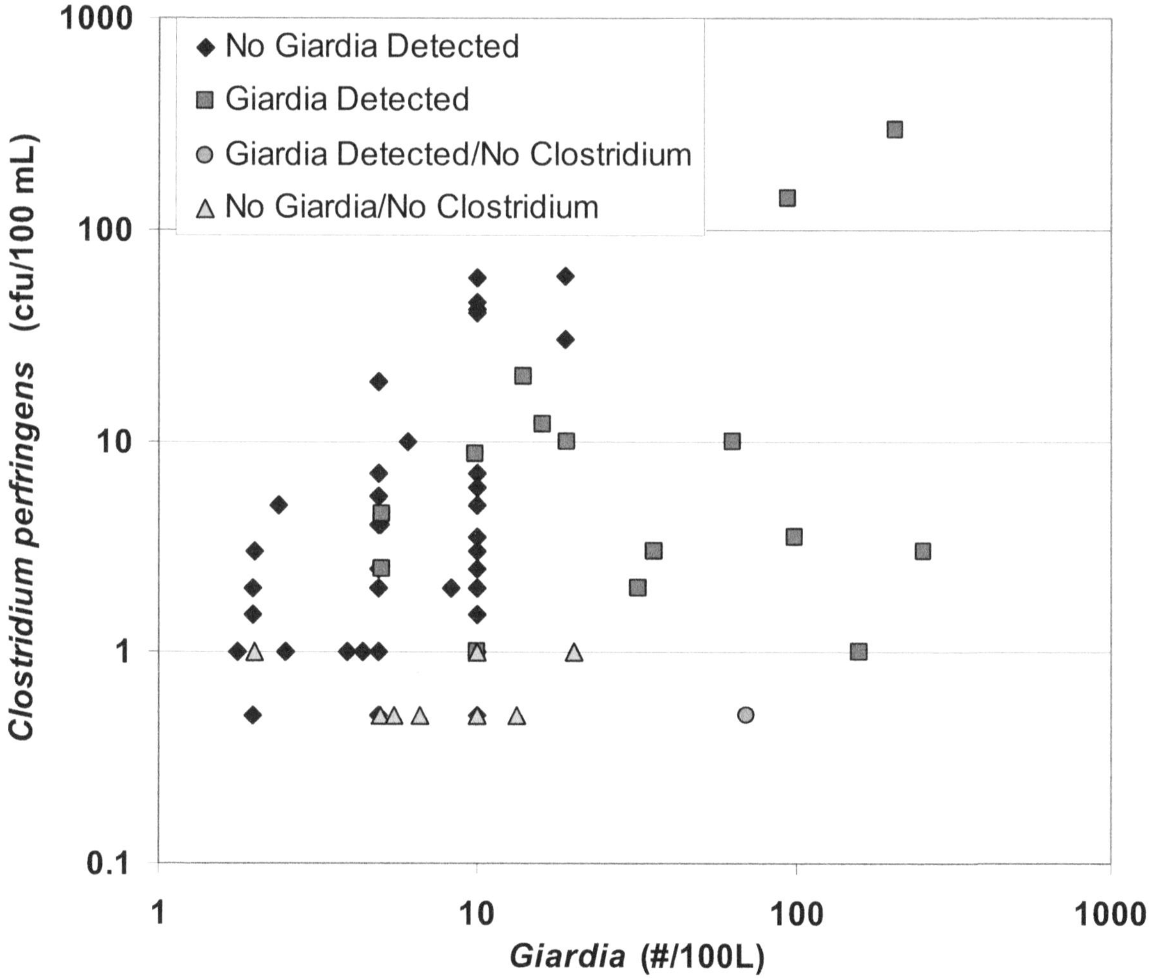

Figure 6-54 Relationship between *Clostridium perfringens* and *Giardia* cyst levels. Samples where no cysts were detected are plotted as the detection limit

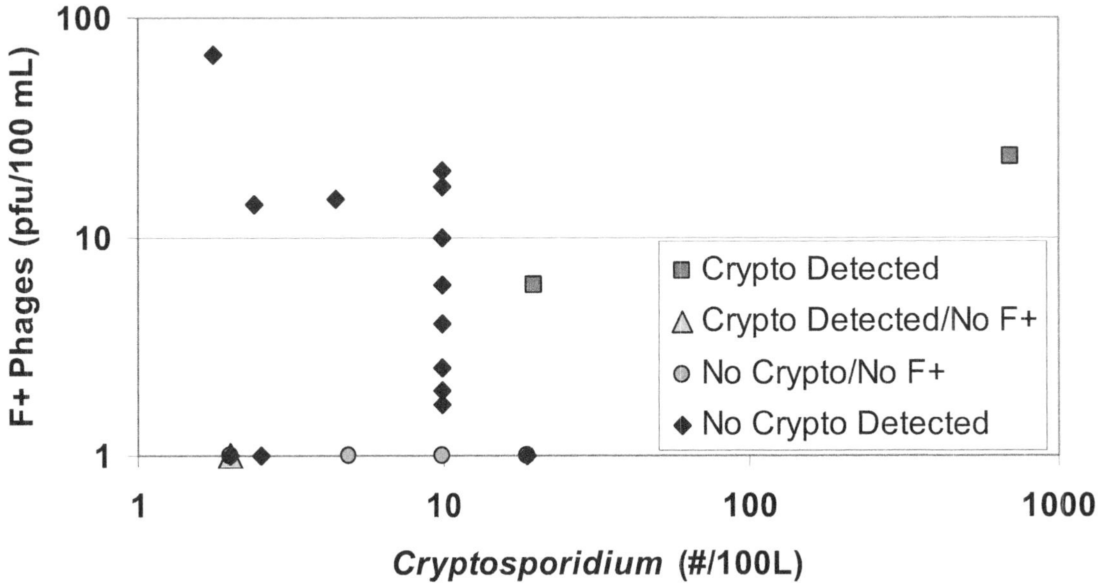

Figure 6-55 Relationship between F+ phages and *Cryptosporidium* oocyst levels. Samples where no oocysts were detected are plotted as the detection limit

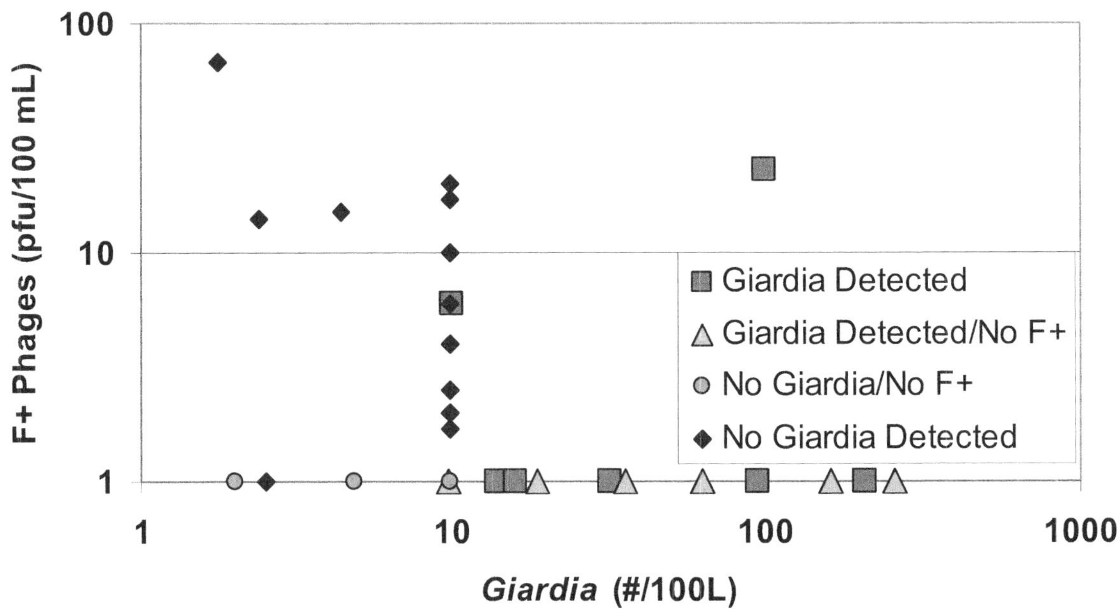

Figure 6-56 Relationship between F+ phages and *Giardia* cyst levels. Samples where no cysts were detected are plotted as the detection limit

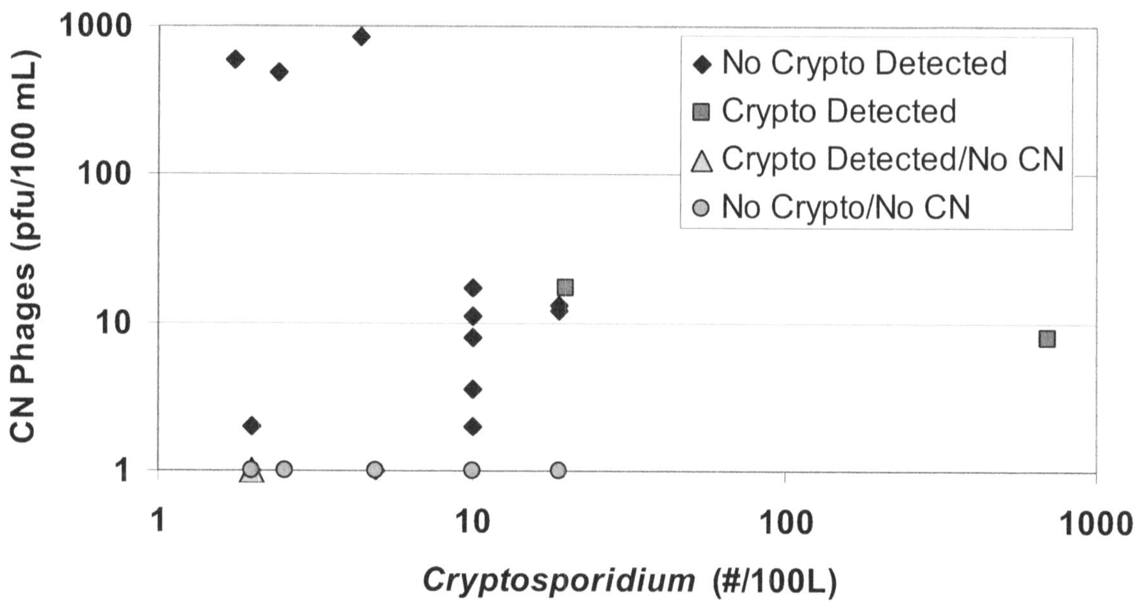

Figure 6-57 Relationship between CN phages and *Cryptosporidium* oocyst levels. Samples where no oocysts were detected are plotted as the detection limit

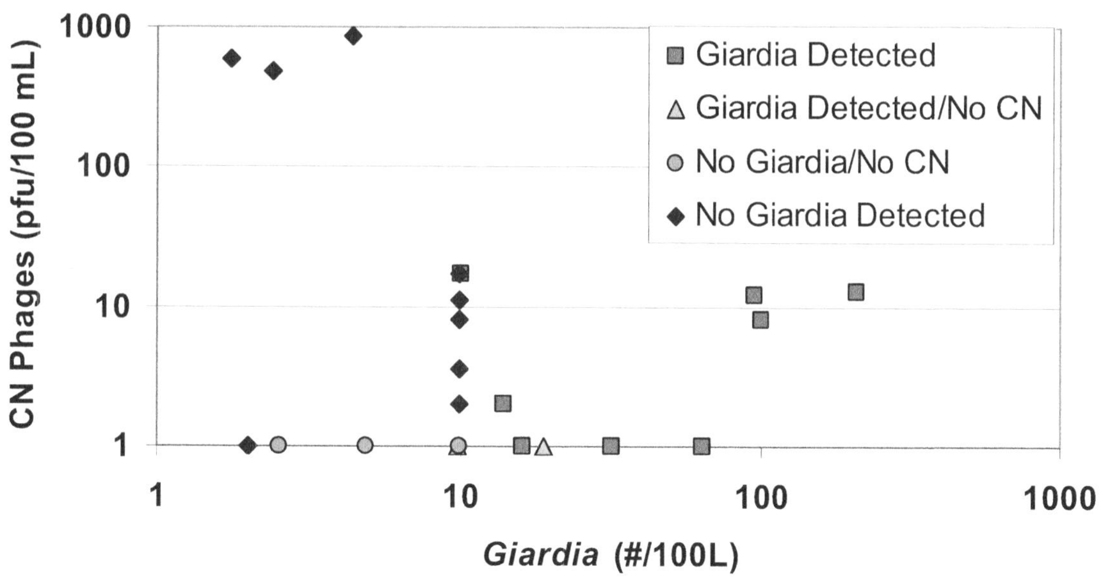

Figure 6-58 Relationship between CN phages and *Giardia* cyst levels. Samples where no cysts were detected are plotted as the detection limit

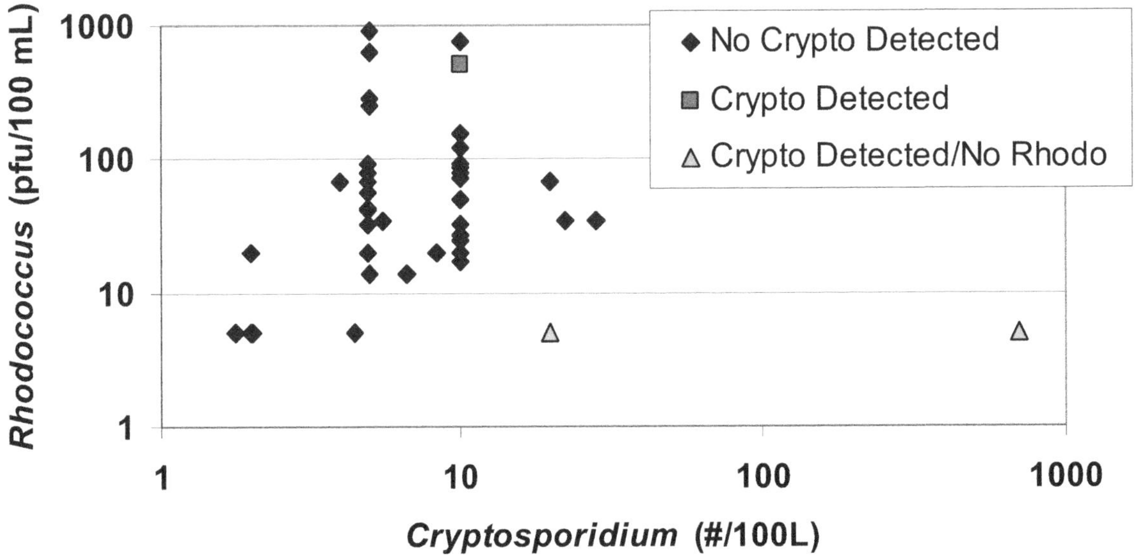

Figure 6-59 Relationship between *Rhodococcus* and *Cryptosporidium* oocyst levels. Samples where no oocysts were detected are plotted as the detection limit

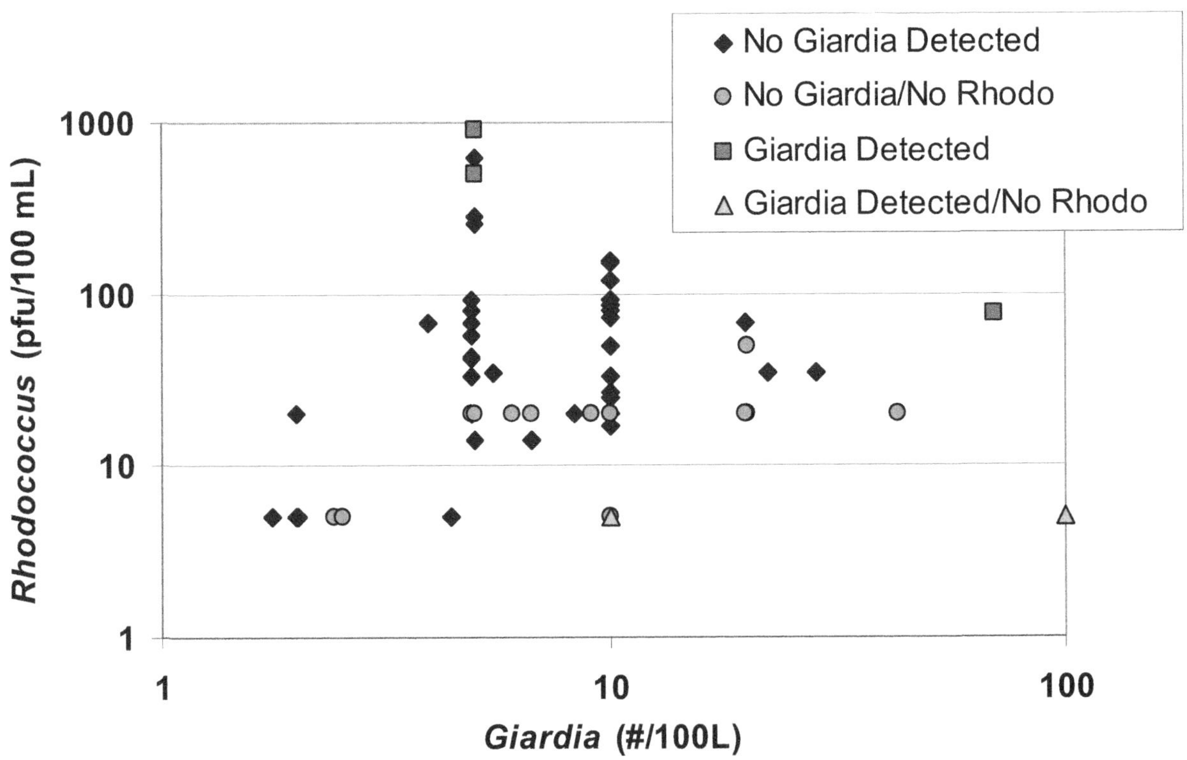

Figure 6-60 Relationship between *Rhodococcus* and *Giardia* cyst levels. Samples where no cysts were detected are plotted as the detection limit

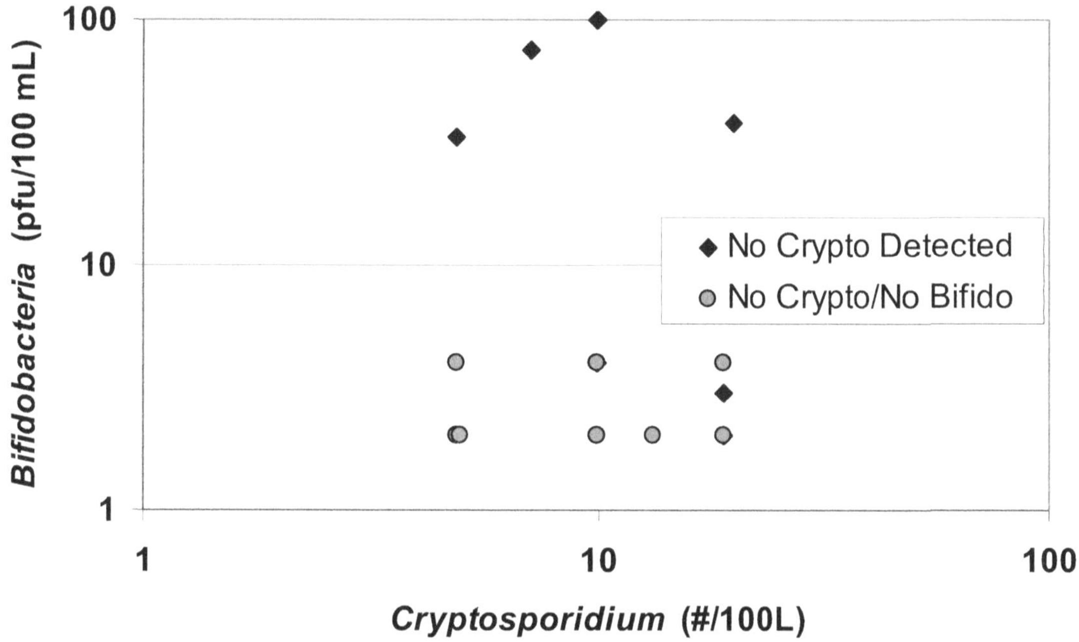

Figure 6-61 Relationship between *Bifidobacteria* and *Cryptosporidium* oocyst levels. Samples where no oocysts were detected are plotted as the detection limit

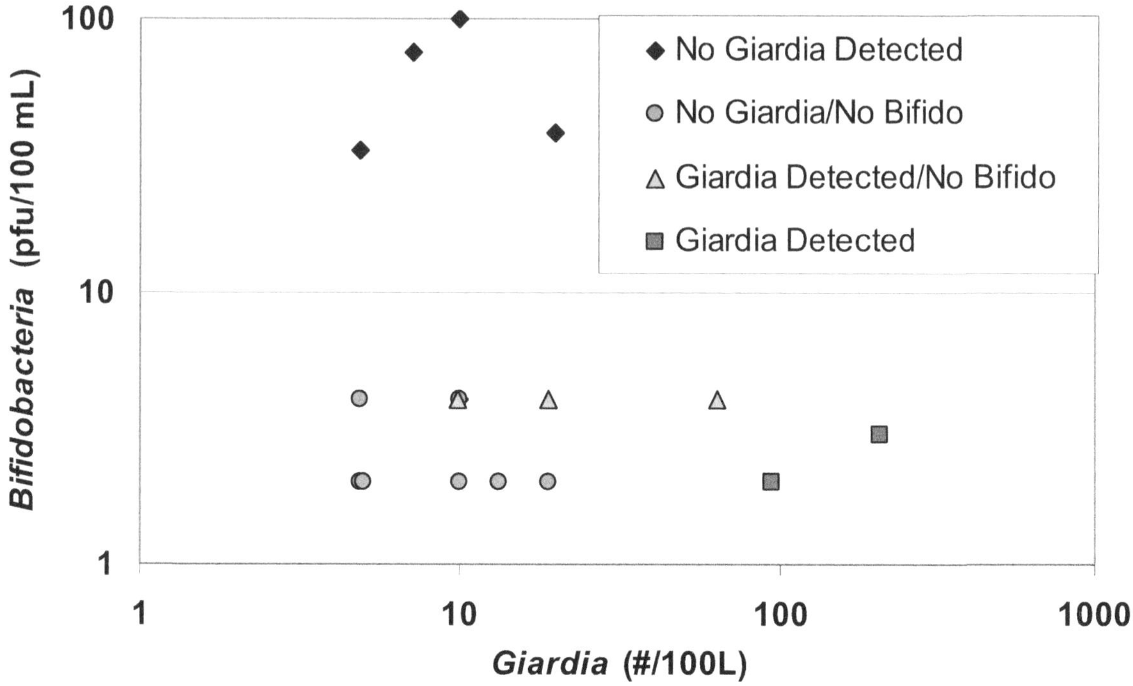

Figure 6-62 Relationship between *Bifidobacteria* and *Giardia* cyst levels. Samples where no cysts were detected are plotted as the detection limit

EVENT CORRELATIONS

For the events where *Cryptosporidium* and *Giardia* were detected, event flow weighted mean concentrations (EFWMCs) were used to determine the correlation between pollutants based on the product-moment correlation coefficient. Linear regression and logistic regression analyses could not be conducted because of the limited data set. Results for events where *Cryptosporidium* was detected are shown in Table 6-11; results for events where *Giardia* was detected are shown in Table 6-12. Physical water quality parameter data were included in the analysis. Alternative indicator organisms had to be excluded from the analyses due to their small EFWMC values in relation to the other parameters.

Cryptosporidium EFWMC values exhibited a strong (r=0.824) positive correlation with EFWMC values of the log TC values but were not highly correlated with other parameters. Weak negative correlations existed between *Cryptosporidium* and TKN (r=-.542), FC (r=-.781), conductivity (r=-.679), temperature (r=-.643), and rain (r=-.706). Other studies have observed greater numbers of oocysts at low water temperatures (LeChevallier, 2002) and seasonal analysis have suggested a higher probability of occurrence in winter months. The negative correlation with rainfall is an artifact of the underlying data; it should be viewed as an indication that even small amounts of rain are capable of transporting *Cryptosporidium* and *Giardia* to headwater streams when sources are present in the watershed. A physical basis for negative correlations with TKN and FC is not clear. Few reported studies have included nutrient analyses with *Cryptosporidium* and *Giardia* sampling and used the resulting data to look for correlations. Adsorption with a portion of the TKN fraction may occur and result in losses but cannot be substantiated. UV254 was excluded from the correlation, but should perhaps be revisited. In addition, total organic carbon may be a parameter worth monitoring; it was included in this study only during select events as part of a different DCR project. Other studies have noted positive correlations with turbidity (Atherholt, 1998). However, turbidity was excluded from these analyses due to inconsistencies of the in-situ data. The FC negative correlation is counterintuitive when one considers both the positive correlation with TC, evident in Table 6-11 and discussed previously in the individual event analyses, and other studies (Atherholt et al., 1998). The correlations found in this study thus must be used with caution due to the small sample set they are based upon.

Giardia EFWMCs exhibited significant correlations with nitrate, TP, TSS, and conductivity but not TC. Individual event analyses, however, suggested that TC could be useful for inferring *Giardia* levels between actual samples. The lack of a strong correlation between TC and *Giardia* may be due in part to their wide difference in EFWMCs; Spearman rank correlations will address this issue and the results are provided below in Table 6-14. The correlation with conductivity may be due to an ability to act as a marker for the arrival (at the sampling point) of water from different areas of the catchment, such as from the beaver impoundment or groundwater sources. The continuous conductivity and pH data collected for this study have not yet been fully utilized in this regard. Correlations with both nitrate and TP/TSS suggest that both "wash flow" and particulate transport mechanisms are important with respect to *Giardia* fate and transport.

While strong correlations suggest parameters which may assist as surrogate measures of contamination level during an event after positive pathogen levels have been noted, additional event based data are necessary to draw firm conclusions from this analyses.

Table 6-11

Correlations between monitored parameters when *Cryptosporidium* was detected

	NO3	TKN.	TP.	TSS.	Log.TC.	Log.FC.	*Crypto.*	Cond..	Temp..	Rain.
NO3	1.000									
TKN.	0.983	1.000								
TP.	0.988	0.943	1.000							
TSS.	0.888	0.788	0.948	1.000						
Log.TC.	0.213	0.030	0.360	0.638	1.000					
Log.FC.	0.873	0.948	0.788	0.551	-0.291	1.000				
Crypto.	-0.378	-0.542	-0.231	0.090	0.824	-0.781	1.000			
Cond..	0.936	0.985	0.871	0.670	-0.144	0.989	-0.679	1.000		
Temp..	0.952	0.992	0.894	0.705	-0.096	0.980	-0.643	0.999	1.000	
Rain.	0.922	0.978	0.852	0.642	-0.181	0.994	-0.706	0.999	0.996	1.00

Note: Correlations greater than 0.75 have been shaded; correlations between 0.4 and 0.75 have been underlined.

Table 6-12

Correlations between monitored parameters when *Giardia* was detected

	NO3	TKN	TP	TSS	Log.TC	Log.FC	*Giardia*	Cond	Temp	Rain
NO3	1.000									
TKN.	0.796	1.000								
TP.	0.998	0.833	1.000							
TSS.	0.964	0.929	0.979	1.000						
Log.TC.	-0.411	-0.879	-0.468	-0.640	1.000					
Log.FC.	-0.207	-0.757	-0.269	-0.461	0.977	1.000				
Giardia.	0.904	0.461	0.875	0.757	0.018	0.231	1.000			
Cond..	0.863	0.381	0.829	0.696	0.107	0.316	0.996	1.000		
Temp..	-0.836	-0.998	-0.870	-0.952	0.843	0.710	-0.522	-0.444	1.000	
Rain.	-0.855	-0.994	-0.887	-0.963	0.824	0.684	-0.552	-0.476	0.999	1.000

Note: Correlations greater than 0.75 have been shaded; correlations between 0.4 and 0.75 have been underlined.

In addition to the product-moment correlation coefficients discussed above, Spearman's rho were also calculated to determine the correlation between pollutants for events where *Cryptosporidium* and *Giardia* were detected. Results for events where *Cryptosporidium* was detected are shown in Table 6-13; results for events where *Giardia* was detected are shown in Table 6-14. Physical water quality parameter data were included in the analysis. Alternative indicator organisms had to be excluded from the analysis due to the incomplete data set. Antecedent precipitation was not included.

Cryptosporidium EFWMC values exhibited a strong positive correlation with EFWMC values of the log TC and discharge values (r=0.800 and 1.000, respectively). Strong negative correlations were exhibited between Cryptosporidium and TKN (r=-0.800) and rainfall (r=-0.949). Weak correlations existed between *Cryptosporidium* and FC (r=0.400) and TSS (r=0.400). The negative correlation with rainfall is an artifact of the underlying data; it should be viewed as an indication that even small amounts of rain are capable of transporting *Cryptosporidium* and *Giardia* to headwater streams when sources are present in the watershed. A physical basis for the negative correlation with TKN is not clear. The correlations found in this study thus must be used with caution due to the small sample set they are based upon.

Giardia EFWMCs exhibited no significant correlations with any of the evaluated water quality parameters. Individual event analyses, however, suggested that TC could be useful for inferring *Giardia* levels between actual samples. The lack of a strong correlation between TC and *Giardia* may be due in part to their wide difference in EFWMCs. Weak negative correlations are observed between *Giardia* and TP (r=-0.543), TSS (r=-0.429) and Log FC (r=-0.486).

Table 6-13
Spearman's correlations between monitored parameters when *Cryptosporidium* was detected

	TKN	TP	TSS	Log TC	Log FC	*Crypto.*	Rain	Q
TKN	1.000							
TP	-0.200	1.000						
TSS	-0.200	1.000	1.000					
Log TC	-0.400	0.800	0.800	1.000				
Log FC	-0.400	0.800	0.800	0.400	1.000			
Crypto.	-0.800	0.400	0.400	0.800	0.200	1.000		
Rain	0.949	-0.316	-0.316	-0.632	-0.316	-0.949	1.000	
Q	-0.800	0.400	0.400	0.800	0.200	1.000	-0.949	1.000

Table 6-14
Spearman's correlations between monitored parameters when *Giardia* was detected

	TKN	TP	TSS	Log TC	Log FC	*Giardia*	Rain	Q
TKN	1.000							
TP	0.371	1.000						
TSS	0.600	0.943	1.000					
Log TC	-0.600	-0.257	-0.314	1.000				
Log FC	-0.086	0.543	0.429	0.543	1.000			
Giardia	-0.371	-0.543	-0.429	0.200	-0.486	1.000		
Rain	0.441	-0.353	-0.265	-0.265	-0.088	-0.265	1.000	
Q	-0.086	-0.371	-0.257	0.486	-0.143	0.143	-0.265	1.000

SUMMARY AND CONCLUSIONS

Microbial, nutrient, and physical water-quality data were collected at six sampling locations over a period of three years and under a variety of meteorological conditions. Three distinct land uses (forested wildlife with and without beaver, agricultural cow pasture, and densely populated residential/commercial urban area) were isolated by pairing the sampling locations. The final data base consists of nutrient, microbial, and pathogen data for twenty-one distinct storm events and three dry weather spatial events, in addition to monthly baseline nutrient and microbial data, and continuous physical data. The focus of this chapter was presentation of the results with respect to *Cryptosporidium* and *Giardia* detection.

This study is unique from others not only in terms of monitoring over the rise and the fall of a storm-event hydrograph, but also in the nature of the watersheds studied. The focus watersheds for this study were all small (less than 10 km^2). Other studies have typically focused on much larger (order of magnitude) watersheds with mixed land uses. A limited number of studies have utilized upstream – downstream sampling techniques, as herein, to isolate suspect sources. The probability of presence, particularly of *Cryptosporidium* and *Giardi*a, is thus likely lower in watersheds included in this study while the probability of transport to the stream and outlet, when present, is higher. In addition, it should be noted that manure spreading for crop fertilization was not practiced in any of the watersheds.

Detection rates per study site ranged from non-detect (residential study site) to 10% when individual event samples (40 to 42 in total) were considered for *Cryptosporidium*. Detection rates per study site for *Giardia* ranged from 10 to 15% when individual event samples (40 to 42 in total) were considered. In all, *Cryptosporidium* was detected in 14% and *Giardia* in 19% of the events studied.

Literature detection rates (see Chapter 3) between no detect to 100% have been reported in the literature for *Giardia*. The detection rate for this study is on the low end of reported values, falling closest to values reported by Ong *et al.* (1998) for forested protected areas or LeChevallier *et al.* (1991) for routine sampling in much larger catchments. *Giardia* levels ranged from no detect to 256 cysts/100 L in the wildlife watershed (geometric mean of positive values, 46.85 cysts/100L), from no detect to 698 cysts/100 L in the agricultural watershed (geometric mean of positive values, 78.76 cysts/100 L), and from no detect to 208 cysts/100 L in the residential watersheds (geometric mean of positive values, 47.34 cysts/100 L). Comparison with literature values is complicated by uncertainty in the averaging method used for some reported values, in particular how no detects are treated. Although a lower detection rate was observed, when detected, *Giardia* levels tended to be on the high side of those reported for forested areas and exceeded the range of literature values reported for agricultural watersheds and cattle areas, approaching the values reported for much larger mixed land use watersheds. Literature values specific to residential areas are scarce and no direct comparison can be made.

Detection rates between no detect to 100% have been reported in the literature for *Cryptosporidium*. The detection rate for this study was low, most similar to the results for studies in Florida. *Cryptosporidium* levels ranged from no detect to 2 oocysts/100 L in the wildlife watersheds (geometric mean of positive values, 2 cysts/100 L), from no detect to 100 oocysts/100 L in the agricultural watershed (geometric mean of positive values, 27.14 oocysts/100 L) and were all no detect at the residential site. Values for the agricultural watershed were similar to those reported for the British Columbia study (Ong et al., 1996), but were less than values reported by Rose et al. (1988) and in Arizona. Literature values reported for pristine and protected watersheds vary greatly. Our values are on the low end of those reported.

Potential relationships across events and for periods of positive and negative detection were evaluated by plotting either the level detected or the level of detection, based on the volume of water filtered, for samples where no *Cryptosporidium* or *Giardia* was found. Results for *Clostridium perfringens* (Figures 6-53 and 6-54), F+ phages (Figures 6-55 and 6-56), somatic or F- phages (Figures 6-57 and 6-58), *Rhodococcus coprophilus* (Figures 6-59 and 6-60), and Soribitol-fermenting *Bifidobacteria* (Figures 6-61 and 6-62) are presented. Recall that fewer analyses for phages were conducted. No clear correlations between alternative indicator organisms and either *Cryptosporidium* or *Giardia* were found when the data across events was considered, although a few observations are noteworthy. High levels (greater than 100 cfu/100 mL) of *Clostridium perfringens* were typically associated with fairly high levels of *Giardia*. Both *Giardia* and *Clostridium perfringens* were more typically detected when *Giardia* levels exceeded 10 per 100 L. While only limited phage data were available, neither detection F+ nor CN phages corresponded well with levels of *Cryptosporidium* and *Giardia*. Similarly, *Rhodococcus coprophilus* and sorbitol-fermenting *Bifidobacteria* were not useful across events for identifying potential occurrence of *Cryptosporidium* and *Giardia*. *Rhodococcus coprophilus* and Sorbitol-fermenting *Bifidobacteria* were, however, useful for helping identify potential sources of contamination when *Cryptosporidium* and *Giardia* are detected during individual events. The lack of temporal trends in co-occurrence between alternative indicator organisms, *Cryptosporidium*, and *Giardia* were also noted by LeChevallier *et al.* (1991).

Results presented suggest an elevated risk of *Cryptosporidium* and *Giardia* downstream of beaver impoundments. It is suggested that this risk, however, is more related to dispersal patterns and activity of beaver away from their impoundments. In general, the impoundments were a probable sink of upstream sources of pathogens; it is suggested this potential is due to enhanced opportunities for adsorption and settling. The potential failure of beaver impoundments or the occurrence of events causing scour of sediment stores within impoundments are perhaps the most significant concern in terms of beaver effects on downstream water quality.

Based on the small size of the watersheds and the inherent difficulties in identifying a metric for events other than precipitation, it is felt that the captured events adequately represent the study site watershed response to a wide range of hydrologic conditions. The resulting data are reasonable both in terms of the positive detection ratios and their range of values. The data are useful for identifying the magnitude of isolated contributions (of contaminants) from distinct land uses. The following key insights are drawn from the results:

- *Cryptosporidium* and *Giardia* tend to exhibit strong flushing characteristics in non-urbanized areas, with highest levels occurring during the rising limb of the hydrograph or near the peak.
- In agricultural and wildlife (forested) areas, the highest probability for transport of *Cryptosporidium* and *Giardia* to the stream appears to be during periods of high intensity rain, accompanied by sharp rises in the stream hydrograph.
- In heavily residential settings, baseflow appears to be an important transport mechanism of *Giardia*. No data are available to determine if this also holds for *Cryptosporidium*. This is likely limited to areas where septic systems or combined septic/sewer systems are in use.
- None of the sampled traditional water quality parameters, standard indicator organisms, or alternative indicator organisms are useful in terms of routine monitoring for use in identification of events to be monitored for *Cryptosporidium* and *Giardia*.
- Data suggest that when *Cryptosporidium* and *Giardia* have been detected during an event, more frequent sampling of suspended sediment (and possibly total coliform) has the potential to serve as a surrogate for determining the relative magnitude of *Cryptosporidium* and *Giardia* over the course of the event.
- There appears to be a higher potential for transport of *Cryptosporidium* and *Giardia* during late winter and spring. It is hypothesized that this may be related either to young beaver dispersal timing or to the occurrence of rain-on-snow events.
- Long-term monthly precipitation records (rainfall and snowfall) may be useful in terms of identifying wet-weather events that have a high potential for *Cryptosporidium* and *Giardia* transport. In particular, it appears that monthly accumulations approaching the 75th quartile may have a higher potential for transport. Such a criterion, however, overlooks the need to also monitor storms occurring early in the month when soils are just saturated from precipitation during the prior month. Thus a monthly running average of precipitation may be a better gauge of transport potential. In addition, the 72-hour antecedent precipitation accumulation may provide additional guidance for when monitoring is warranted. Alternatively, remotely sensed soil moisture data indicating saturated or near saturated conditions may be a better criterion for flagging events where sampling for *Cryptosporidium* and *Giardia* is warranted. Regardless of whether such sampling is conducted, utilities should note that high efficiency water treatment is especially important during these periods.
- Ponds, lakes, and beaver impoundments may serve as an along-stream sink of *Cryptosporidium* and *Giardia* from upstream sources, depending on the relative velocity and flow through patterns for a particular event.
- The stream bed near and downstream from agricultural areas may be an important source of *Cryptosporidium* and *Giardia*. Events which result in high stream velocities and erosion in these areas may be particularly important in terms of monitoring.

The points noted above are to a large extent a repeat of the observations made with respect to a detailed review of the events during which *Cryptosporidium* and *Giardia* **were** detected. It should thus be noted that these points can be justified by analysis of both the positive and negative *Cryptosporidium* and *Giardia* analysis results.

Because the results of this study are based on a relatively small sample size, additional data would be beneficial to confirm the conclusions drawn with respect to the occurrence and transport of *Cryptosporidium* and *Giardia*, in addition to the usefulness of common hydrologic, water quality and standard indicator organism data for inferring pathogenic contamination. It should also be noted that the study focused on very small watersheds, and thus results may not hold over larger scales. Lastly, sequential filtering to reach a target volume could verify that the volume filtered had little effect on detection. Such a study would similarly confirm that high suspended sediment concentrations did not result in false negatives.

CHAPTER 7
CONCLUSIONS AND RECOMMENDATIONS

This project examined the occurrence and transport of *Giardia*, *Cryptosporidium*, nutrients, standard microbial indicators, alternative microbial indicators, and physical water quality parameters in three watersheds. An upstream – downstream and paired watershed monitoring strategy was used to isolate the impact of heavily urbanized (residential), agriculture (dairy and beef farm, no manure spreading), and pristine wildlife (with and without beaver) land uses on downstream water quality. The study was designed to attain several research objectives, including:

- Collection of data describing the occurrence and transport of *Giardia* cysts, *Cryptosporidium* oocysts, and other microorganisms of fecal origin from select land use areas during wet weather events (Occurrence and Transport).
- Collection of data characterizing the seasonal and year-to-year variability of pathogen occurrence during wet weather (Temporal and Spatial Variability).
- Evaluation of both the relative magnitude and timing of specific water-quality parameter export from various land-use areas (Temporal and Spatial Variability).
- Evaluation of factors which affect the spatial, seasonal, and annual variability of pathogen occurrence for a range of watershed conditions, including meteorological (precipitation depth, duration, intensity, time since previous precipitation), hydrologic (stream order), basin (size, shape), and/or land use characteristics (Temporal and Spatial Variability).
- Evaluation of relationships between protozoa and standard water quality parameters, standard indicator organisms, and alternative indicator organisms (Relationships Between Protozoa and Water Quality Parameters).
- Evaluation of data to identify the best sampling techniques and number of samples for adequate characterization of pollutant concentrations and loads, particularly human pathogens, during storm events within reasonable economic and technical constraints (Monitoring Scheme Recommendations).

The final database consists of occurrence, concentration and loading data for the standard water quality parameters (nitrate, TKN, TP, TSS, UV254), standard indicator organisms (TC and FC), alternative indicator organisms (*Rhodococcus coprophilus,* Sorbitol-fermenting *Bifidobacteria*, *Clostridium perfringens*, and coliphages), physical water quality (water temperature, pH, and conductivity), and pathogens (*Giardia* cysts and *Cryptosporidium* oocysts) during twenty-one distinct storm events and three dry weather events, in addition to monthly standard water quality data and continuous physical data. The focus of this study is on small (less than 10 km^2) watersheds. A majority of the studies reported in the literature have focused on much larger (order of magnitude) watersheds with mixed land uses. Some studies have utilized upstream – downstream sampling techniques, as herein, to isolate suspect sources. Based on drainage basin area alone, the probability of ***presence*** or ***occurrence***, particularly of *Cryptosporidium* and *Giardia*, is likely lower in watersheds included in this study while the probability of *transport* to the stream and outlet, when present, is likely higher. The following conclusions and recommendations are organized based on the study objectives, as detailed above.

OCCURRENCE AND TRANSPORT OF *GIARDIA* AND *CRYPTOSPORIDIUM*

Historical data on the occurrence of *Giardia* and *Cryptosporidium* was not available for the sampling locations. Positive recovery and observed levels were thus compared with literature values. Detection rates per study site ranged from non-detect (residential study site) to 10% when individual event samples (40 to 42 in total) were considered for *Cryptosporidium*. Detection rates per study site for *Giardia* ranged from 10 to 15% when individual event samples (40 to 42 in total) were considered. In all, *Cryptosporidium* was detected in 14% and *Giardia* in 19% of the events studied. The positive detection rate for this study is on the low end of reported literature values but within the range reported.

Giardia levels ranged from no detect to 256 cysts/100 L in the wildlife watershed (geometric mean of positive values, 46.85 cysts/100L), from no detect to 698 cysts/100 L in the agricultural watershed (geometric mean of positive values, 78.76 cysts/100 L), and from no detect to 208 cysts/100 L in the residential watersheds (geometric mean of positive values, 47.34 cysts/100 L). Although a lower detection rate was observed, when detected, *Giardia* levels tended to be on the high side of those reported for forested areas and exceeded the range of literature values reported for agricultural watersheds and cattle areas, approaching the values reported for much larger mixed land use watersheds. Literature values specific to residential areas are scarce and no direct comparison could be made. *Cryptosporidium* levels ranged from no detect to 2 oocysts/100 L in the wildlife watersheds (geometric mean of positive values, 2 cysts/100 L), from no detect to 100 oocysts/100 L in the agricultural watershed (geometric mean of positive values, 27.14 oocysts/100 L) and were all no detect at the residential site. Values for both the agricultural and pristine watershed were on the lower end of reported literature values.

The resulting data are reasonable both in terms of the positive detection ratios and their range of values. The data are unique in providing loading information from distinct land uses and represent a wide range of hydrologic conditions in the study watersheds. Review of ancillary data and spatial sampling during dry weather suggests that the low positive detections were due to a lack of presence rather than a lack of detection. Lack of presence may be attributable to the small size of the study watersheds and the resulting smaller pool of potential sources of contamination. Cycles of infectivity rather than continuous infectivity are expected in a given population. Since a single population of cattle (agricultural site BBFM) and beaver (wildlife site GRB) were isolated, it is thus expected that only a limited number of infections and thus positive detections would occur.

Future monitoring schemes would benefit by including a number of contingency samples for specifically evaluating the effect of filter clogging on detection. Contingency samples could be used to filter a predetermined, large volume of water by running the filters in sequence (changing when one becomes clogged) and then reporting the results determined across the sample set. Alternatively, spiking individual samples with inactivated cysts to provide a percent recovery value per sample holds promise for identifying the rate of false negative environmental samples.

The data from this study suggest that when present in small order headwater streams, *Cryptosporidium* and *Giardia* have a high potential for transport through the watershed. Even small rainfall events (less than 0.25 inches accumulations) were capable of washing cysts and oocysts into the stream. Saturated or near-saturated ground conditions appear to be an important factor controlling transport to the stream. Long-term monthly precipitation records (rainfall and snowfall) may be useful in terms of identifying wet-weather events that have a high potential for *Cryptosporidium* and *Giardia* transport. In particular, it appears that monthly accumulations approaching the 75th quantile may have a higher potential for transport. Such a criterion, however, overlooks the need to also monitor storms occurring early in the month when soils are just saturated from precipitation during the prior month. Thus a monthly running average of precipitation may be a better gauge of transport potential. In addition, the 72-hour antecedent precipitation accumulation may provide additional guidance for when monitoring is warranted. Alternatively, remotely sensed soil moisture data indicating saturated or near saturated conditions may be a better criterion for flagging events where sampling for *Cryptosporidium* and *Giardia* is warranted. Regardless of whether such sampling is conducted, utilities should note that high efficiency water treatment is especially important during these periods.

While pathogen occurrence patterns appeared dependent on animal infectivity and watershed rainfall-runoff dynamics, the following generalizations with respect to occurrence and transport may be drawn:

- *Cryptosporidium* and *Giardia* tend to exhibit strong flushing characteristics in non-urbanized areas, with highest levels occurring during the rising limb of the hydrograph or near the peak.
- In heavily populated residential settings, baseflow appears to be an important transport mechanism of *Giardia*. No data are available to determine if this also holds for *Cryptosporidium*. This phenomenon may be limited to areas where septic systems or combined septic/sewer systems are in use and thus warrants further study.
- In agricultural and wildlife (forested) areas, the highest probability for transport of *Cryptosporidium* and *Giardia* to the stream appears to be during periods of high intensity rain and saturated or near-saturated antecedent soil conditions, resulting in sharp rises in the stream hydrograph.
- Ponds, lakes, and beaver impoundments may serve as an along-stream sink of *Cryptosporidium* and *Giardia* from upstream sources, depending on the relative velocity and flow through patterns for a particular event.
- The stream bed near and downstream from agricultural areas may be an important source of *Cryptosporidium* and *Giardia*. Events which result in high stream velocities and erosion in these areas may be particularly important in terms of monitoring.
- The first event occurring following attainment of near saturated conditions is likely to exhibit high detection levels. However, pathogen sources are likely to be flushed following such events, resulting in low occurrence levels for subsequent events.

TEMPORAL AND SPATIAL VARIABILITY

Giardia and *Cryptosporidium* appear to have a seasonal pattern, with higher levels of occurrence in the winter and early spring. It is hypothesized that this may be related to young beaver and other wildlife dispersal timing. Cycles of animal birth and infectivity are also likely an important factor. In addition, the potential for rain on snow events during this period may be of particular importance. Such events have a high potential to result in flushing of stored fecal material from the land surface. Many of the documented waterborne *Cryptospordium* outbreaks have also occurred during this period.

Based on the limited data available from this study, inter-annual hydrologic variability appears to be an important consideration in terms of transport to headwater streams. A higher number of positive detections occurred during wetter than average (in terms of precipitation) water years. While *Cryptosporidium* and *Giardia* were detected in one event during a drought year, this was a rain-on-snow event occurring in the wettest part of that year. Transport potential appears elevated when saturated soil conditions exist, and remotely based soil moisture indexes may assist in identifying events at greater risk for pathogen transport. The first event occurring following attainment of near saturated conditions is likely to exhibit high detection levels. However, pathogen sources are likely to be flushed following such events, resulting in low occurrence levels for subsequent events.

Relative risk of contamination to downstream surface waters by land use type was inferred based on study results. The results suggested that the highest risk of contamination by *Cryptosporidium* was associated with agriculture land use (cattle), with moderate risk associated with pristine wildlife areas and low risk associated with residential areas. The highest risk of contamination by *Giardia* was associated with beaver influenced watersheds. Residential areas were classified with a moderate to high risk level, while agricultural and ranging wildlife locations were classified with a moderate to low risk level. Additional focused studies are necessary to better assess such risk.

RELATIONSHIPS BETWEEN PROTOZOA AND WATER QUALITY PARAMETERS

Nutrient, physical water quality, hydrologic, standard indicator organism, and alternative indicator organism data were not found to be promising candidates as stand alone surrogate indicators of pathogenic risk. Therefore to definitively determine presence or absence, sampling must be conducted directly for *Cryptosporidium* and *Giardia.* Based on analysis of individual events, and to a lesser extent correlation analyses, data suggest that when *Cryptosporidium* and *Giardia* have been detected during an event, more frequent sampling of suspended sediment (and possibly total coliform) has the potential to serve as a surrogate for determining the relative magnitude of *Cryptosporidium* and *Giardia* over the course of the event. When only events where *Cryptosporidium* or *Giardia* were detected were analyzed, additional correlations were found. Weak (0.5 to 0.8) negative correlations were found between *Cryptosporidium* and TKN, FC, and temperature. Greater then 0.8 correlations were found between nitrate, TP, TSS, conductivity and *Giardia*. Extrapolation of these correlations to other studies, or use for developing monitoring schemes, is questionable due to the small sample number and lack of a clear understanding of physical processes which would lead to such correlations.

Other studies have suggested turbidity as an indicator of transport risk. In this study, evaluation of the potential for turbidity to act as a surrogate was problematic due to underlying problems with the data. In general, in-situ turbidity probes do not provide the level of sensitivity and confidence necessary to drive monitoring for pathogens. Turbidity as measured by enclosed turbidity meters may have more value, but requires field visits if watershed monitoring, rather than water intake monitoring, is of interest. Future generation in-situ turbidity probes may have the potential to provide remote data suitable for triggering watershed sampling. In-situ depth, temperature, pH, and conductivity probes strategically placed near beaver impoundments have the potential to provide information on when they are spilling, thus perhaps warranting monitoring for pathogen presence downstream.

MONITORING SCHEME RECOMMENDATIONS

Data suggest that supplementation of routine monitoring with event based monitoring is essential for accurate characterization of watershed based pathogenic risk. Hydrologic variability and spatial/temporal patterns of pathogen occurrence in the watershed and near-stream zone make quantification of the long-term risk associated with wet weather difficult. Due to the inherent logistic, personnel and financial costs associated with event based sampling, the goals and methodology for any such required sampling must be carefully addressed. It appears appropriate that wet weather sampling schemes be classified under one of two broadly defined goals: (1) characterization for watershed based controls or (2) characterization for specification of treatment processes, either short- or long-term. Such classification is necessary due to underlying differences in the monitoring strategies most appropriate for quantifying the associated pathogenic risk. For both types of monitoring schemes, the time scales of hydrologic variability alone suggest that emphasis should be placed on developing watershed specific long-term records of event-based risk, rather than monitoring of wet-weather events over a short period of time.

Long-term hydrologic records, both rainfall and streamflow, should be utilized to set goals for capturing events which fall into different precipitation accumulation, rainfall rate, and peak discharge recurrence intervals. Both short (e.g., 1 to 2-year interval) and long (e.g., 25 to 50-year) recurrence interval events should be considered for incorporation. To accomplish capturing of events with long recurrence intervals, time frames for completion of wet weather event sampling must be flexible. While confounding effects caused by land use change and other anthropogenic influences may affect pathogen sources over a long-term sampling period, such a sampling scheme would provide the best overview of long-term risk. Monitoring during winter and spring months appears particularly important. In colder climates, monitoring of a rain-on-snow event should be considered.

Rainfall rate based wet weather sampling may better capture transport in small headwater streams while hydrograph and hydraulic transport based sampling schemes appear more appropriate for larger order streams. A combination of the two holds promise for accurately characterizing wet weather loads. At both scales, soil moisture data and monitoring of animal infectivity may be useful for identifying events with a higher associated pathogenic risk. Remotely sensed soil moisture data may prove to be an appropriate metric for triggering watershed monitoring. In addition, results presented suggest an elevated risk of *Cryptosporidium* (potential) and *Giardia* (confirmed) downstream of beaver impoundments. It is suggested that this risk, however, is more related to dispersal patterns and activity of beaver away from their impoundments. In general, the impoundments were a probable sink of upstream sources of pathogens; it is suggested this potential is due to enhanced opportunities for adsorption and settling. The potential failure of beaver impoundments or the occurrence of events causing scour of sediment stores within impoundments are perhaps the most significant concern in terms of beaver effects on downstream water quality. It is thus suggested that known beaver impoundments in drinking water supply watersheds be monitored. In-situ depth, temperature, pH, and conductivity probes strategically placed near beaver impoundments have the potential to provide information on when they are spilling, thus perhaps warranting monitoring for pathogen presence downstream.

Characterization for Treatment Processes

Results from this study provide only limited insight into the appropriate techniques for event sampling at a utility's raw water intake. Data both from this study and others, however, suggest that once entrained in the stream, *Cryptosporidium* and *Giardia* are mainly transported as wash load. Characterization of dry and wet weather hydraulic transport times from suspected sources of contamination within the water supply watershed thus appears important. It is suggested that event sampling at the intake be designed to capture wet weather samples at time intervals which roughly correspond with arrival of water from these potential source areas. Ancillary data with a short turn around, such as turbidity, may be useful for refining sample timing but could not be directly evaluated in this study due to equipment errors. Monitoring should focus on the rising limb and hydrograph peak, but some falling limb sampling may be necessary to capture transport from remote areas if they have a high pathogenic risk and little opportunity for loss exists.

Characterization for Watershed Based Controls

Because of its focus on small watersheds, the results of this study are most applicable for developing monitoring schemes for characterization of loads from headwater areas or areas of suspected contamination. Negative water quality impacts should be expected as land use changes from pristine forests to agriculture to urban uses. However, collection of at least a limited amount of watershed specific wet weather data appears necessary for developing watershed management plans with a high potential for improving water quality. Such monitoring schemes should take into account variability in precipitation, runoff and watershed characteristics. Co-collection of concentration and discharge data are critical for understanding cumulative impacts on water quality and placing wet weather transport in context with annual loads. These data are often not collected during routine watershed monitoring.

Collection of data on the rising limb of the hydrograph appears particularly important in terms of capturing variability of water quality parameters during wet weather. The use of passive or automated samplers is promising for reducing personnel requirements for wet weather sampling and for reducing missed sampling of the first flush and peak concentration of the contaminant of interest, particularly for nutrient data. Further work is necessary to see if composite samples may reduce analysis cost without reducing the ability of the data to describe the actual transport characteristics.

Experience suggests that preliminary collection of spatially distributed water quality data during dry and wet weather should be conducted prior to selecting locations for temporally varied wet weather data collection. Such data is necessary to identify stream reaches characterized by losses and gains. Sampling locations for wet weather data collection should be located as far downstream as possible from the suspected source of negative impact, yet upstream from tributary confluences and the sensitive stream reach, lake, or reservoir to be protected. This will allow natural stream attenuation processes to be accounted for when setting management goals and will avoid specific targeting of land owners. Bed sediments have been identified as a potential source of many contaminants. While events which result in scour of the beds may result in severe downstream water quality data, monitoring of such events is difficult due to their comparatively low recurrence interval in most watersheds. Passive bed samplers (currently available for other parameters) may be of use.

It is extremely important to systematically document wet weather events in terms of their hydrologic and environmental conditions. Co-location of sampling sites with long-term USGS stream gauges would facilitate this. However, relatively few USGS stream gauges are located within small water supply watersheds. Utility commitment to discharge and precipitation measurement is strongly suggested. Even with the approximately two years of continuous discharge data collected as part of this study, defining the events in terms of their long-term hydrologic and annual load characteristics (such as probability of occurrence) is difficult. This is largely due to inter-annual variability characteristic of the underlying hydrology. The existence of in-house precipitation and discharge data would facilitate development of watershed models based on collected data.

In terms of gaining a better watershed-wide understanding of wet weather effects on water quality, collection of a reduced number of samples over the rise and fall of the hydrograph but at a greater number of spatial locations may be beneficial. Locations of importance in terms of monitoring appear to differ between water quality parameters of interest. In terms of pathogens, beaver impoundments immediately upstream of raw water intakes should be removed; impoundments further upstream should be monitored when feasible. Additional locations to consider for monitoring include the most downstream feedlot or residential areas. In addition, the time of year of most importance for sampling is not consistent across variables. Wet weather monitoring schemes on a per parameter basis may thus be more efficient in terms of personnel, analysis costs, and usefulness. It may also be more effective to routinely monitor for animal infectivity in watersheds with extensive feedlot or beaver activity. In-stream or increased intake wet weather monitoring could then be initiated when significant watershed sources are present and prime hydrologic conditions for transport occur simultaneously.

In headwaters, capture of stream contributions resulting from extreme rainfall may be problematic due the associated rapid response in small order streams. As a result, flushed *Cryptosporidium* and *Giardia* may be missed. Automated or passive sampling devices would assist in better capturing transport associated with such events.

CONCLUSIONS

In summary, the project demonstrated that elevated pathogenic risk may be associated with specific land uses. In small headwater streams, detection appears more directly related to the existence of infected populations within the watershed rather than opportunities for transport in terms of the magnitude of a hydrologic event. Antecedent soil moisture conditions, however, appear to be an important factor dictating transport. In addition, transport mechanisms appear to become more important for in-stream levels as watershed size increases. While no clear surrogate for pathogenic risk was found, data suggest that once *Cryptosporidium* or *Giardia* have been detected during a wet weather event, their relative magnitude may be inferred from total suspended solids or total coliform temporal variability. Routine sampling for *Cryptosporidium* and *Giardia* were not conducted as part of this study, but apparent correlation between the number of cysts or oocysts detected (when present) with discharge over the course of the hydrograph rise and fall suggests that wet weather events pose a greater pathogenic risk than baseflow periods. In particular, when present in the watershed, there appears to be both an increased likelihood of *Cryptosporidium* and *Giardia* occurrence and magnitude during the first flush of storms.

APPENDIX A
ANTECEDENT RAINFALL VERSUS MONTHLY DATA

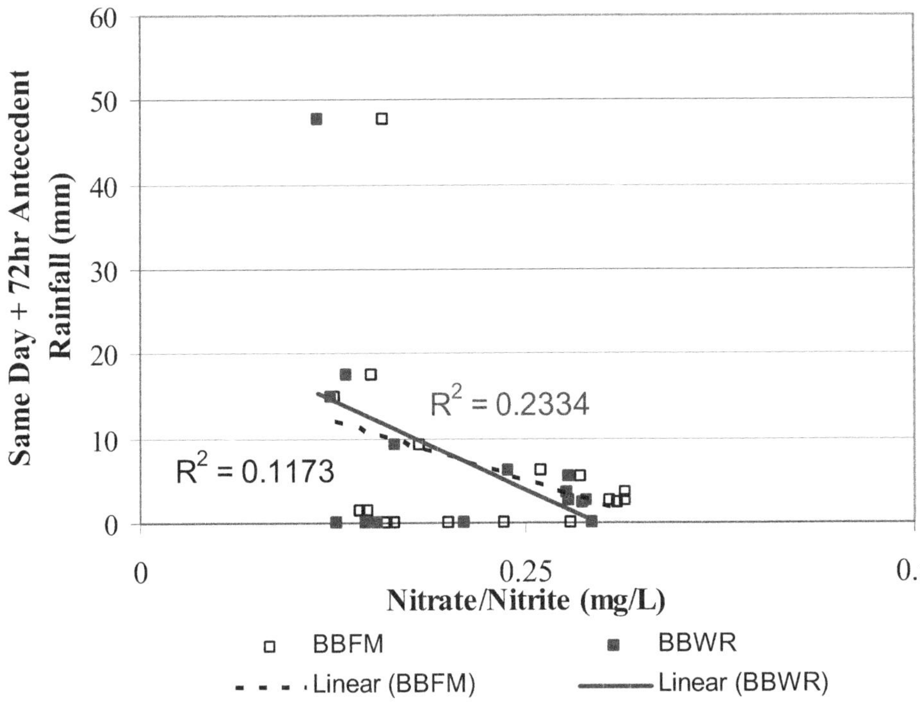

Figure A.1 Same day + 72hr antecedent rainfall versus nitrate/nitrite for BB

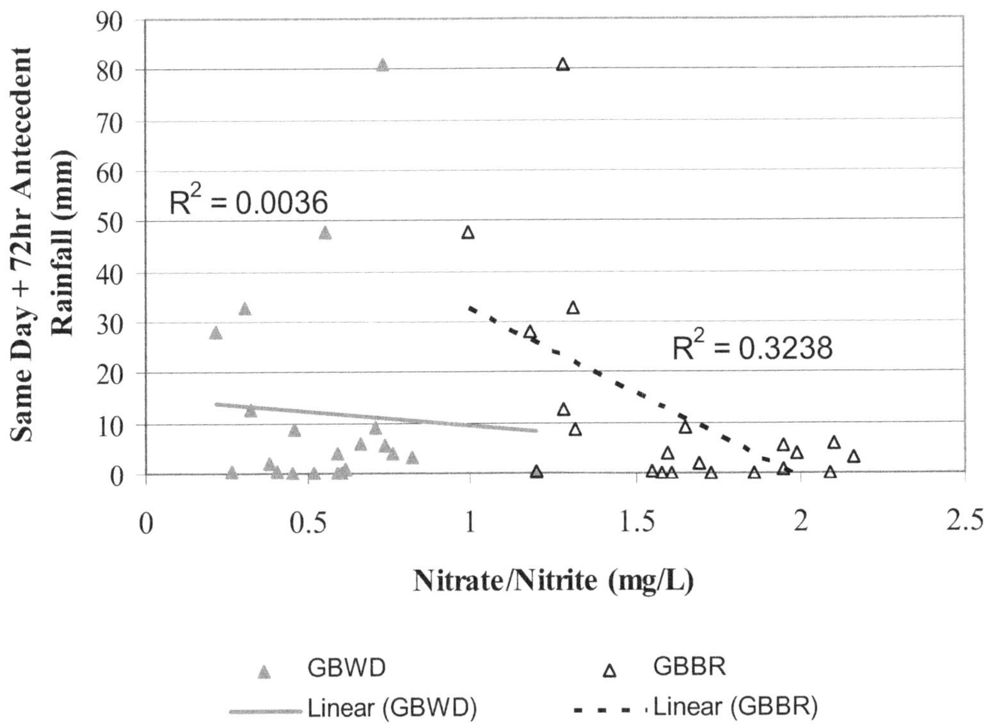

Figure A.2 Same day + 72hr antecedent rainfall versus nitrate/nitrite for GB

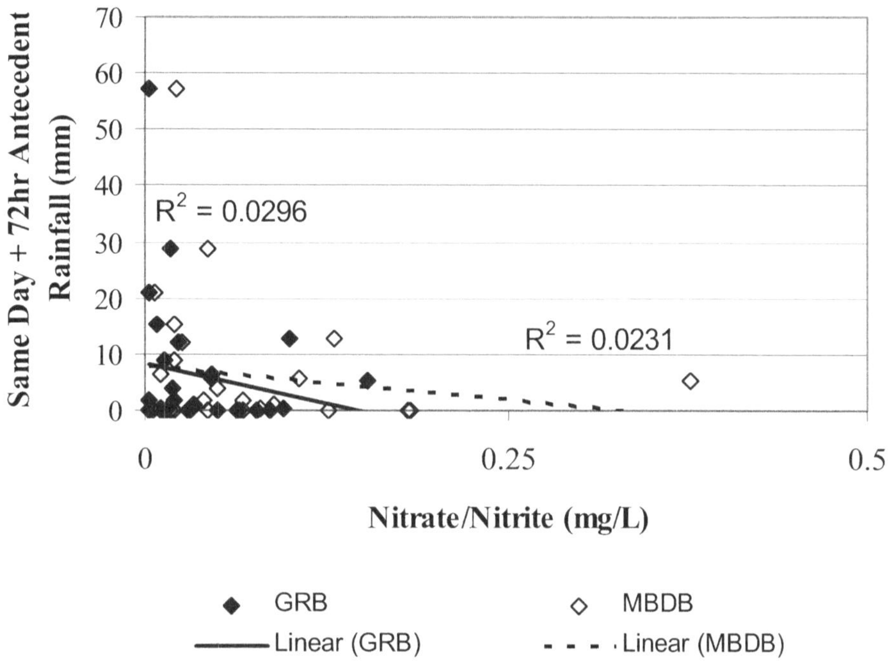

Figure A.3 Same day + 72hr antecedent rainfall versus nitrate/nitrite for PR

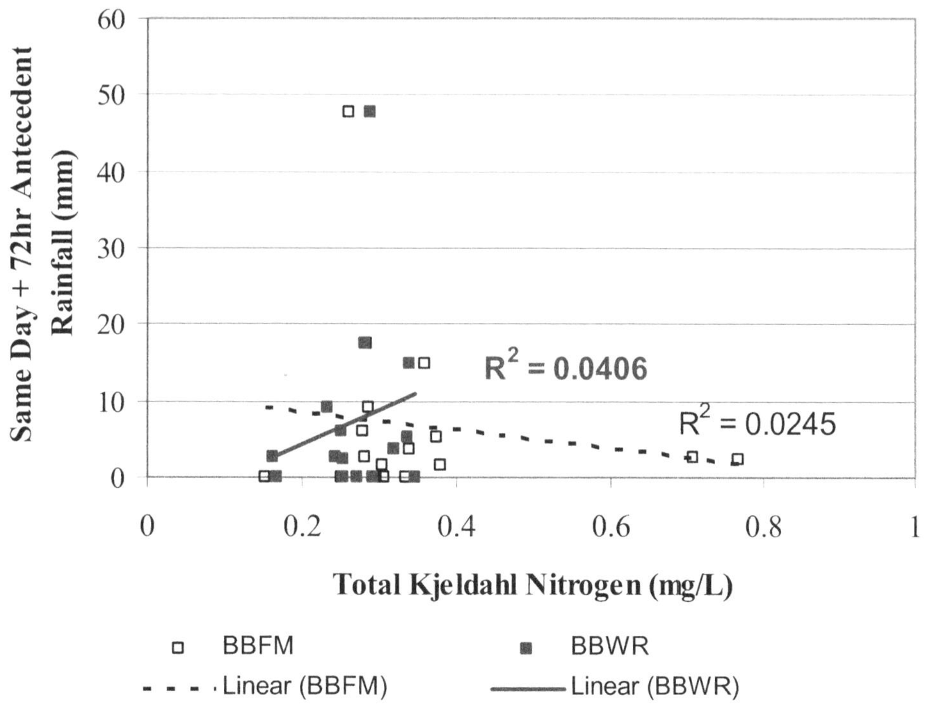

Figure A.4 Same day + 72hr antecedent rainfall versus total Kjeldahl nitorgen for BB

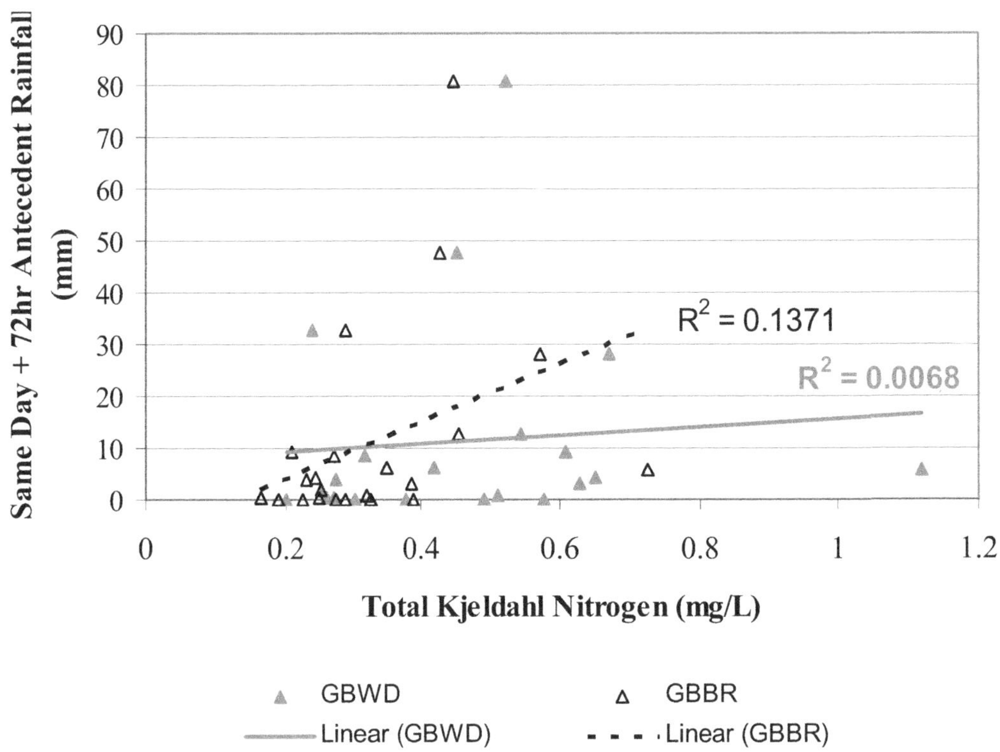

Figure A.5 Same day + 72hr antecedent rainfall versus total Kjeldahl nitorgen for GB

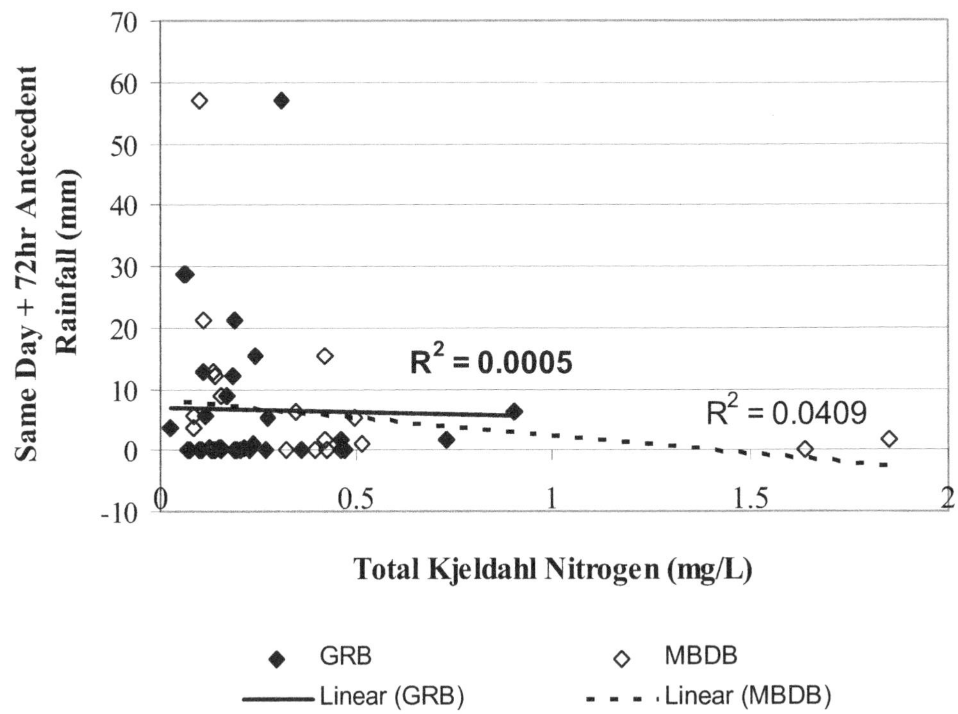

Figure A.6 Same day + 72hr antecedent rainfall versus total Kjeldahl nitorgen for PR

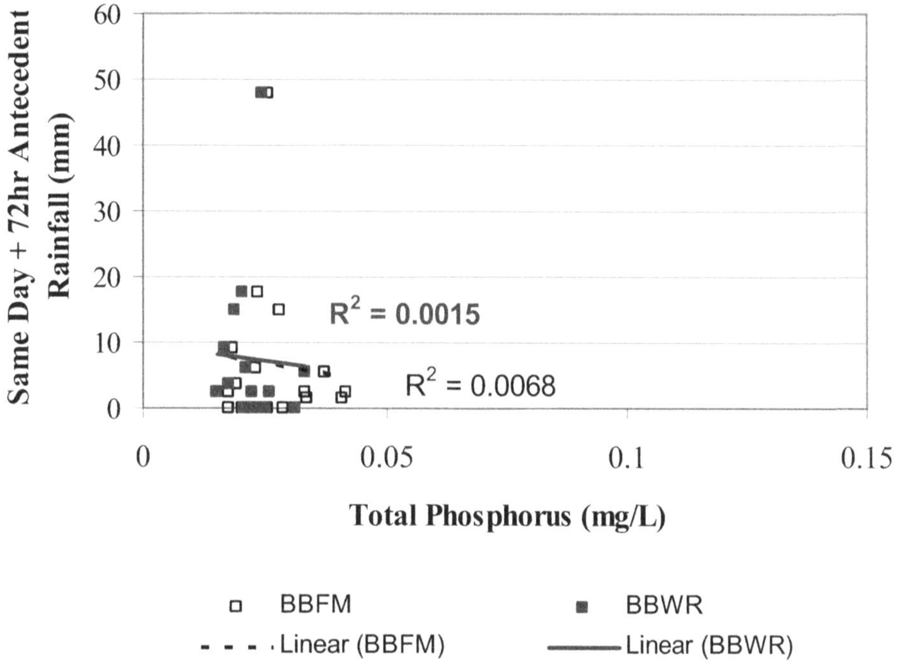

Figure A.7 Same day + 72hr antecedent rainfall versus total phosphorus for BB

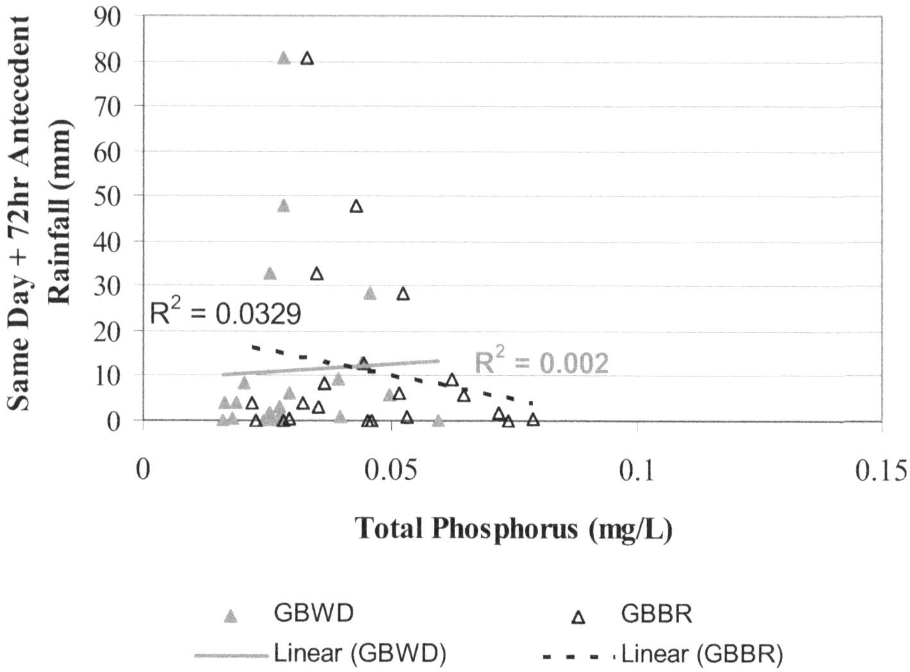

Figure A.8 Same day + 72hr antecedent rainfall versus total phosphorus for GB

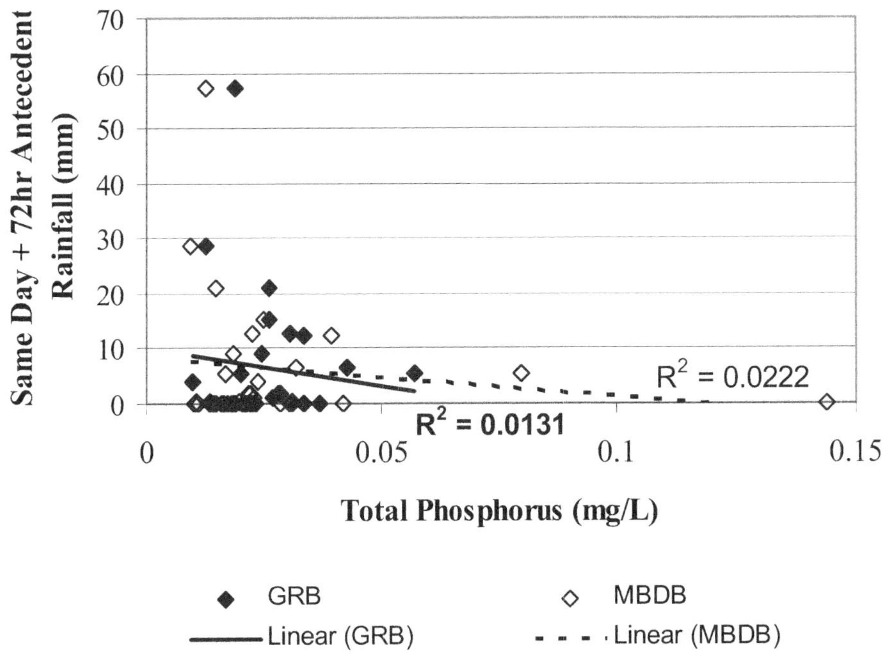

Figure A.9 Same day + 72hr antecedent rainfall versus total phosphorus for PR

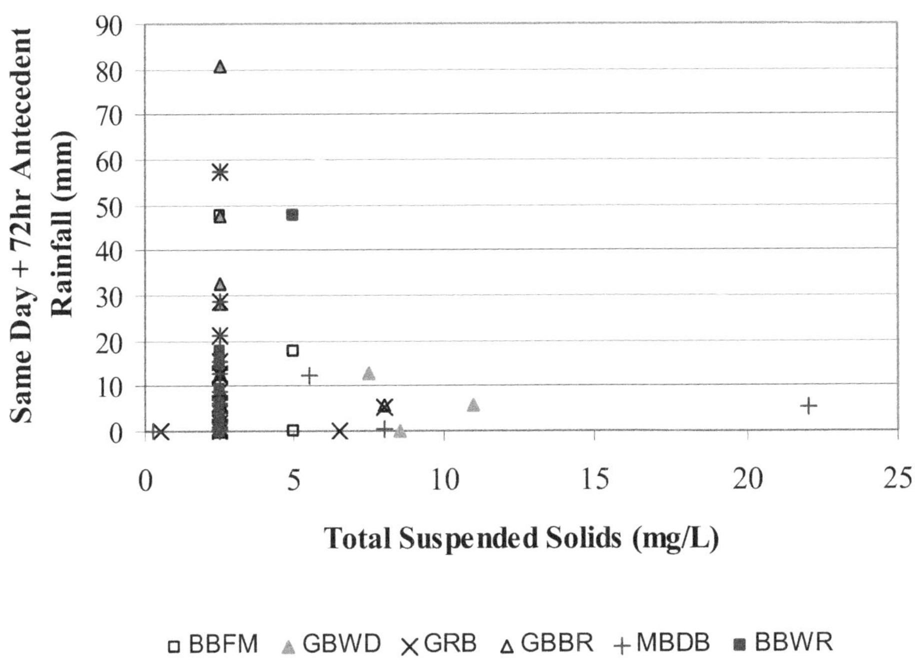

Figure A.10 Same day + 72hr antecedent rainfall versus total suspended solids for all sites

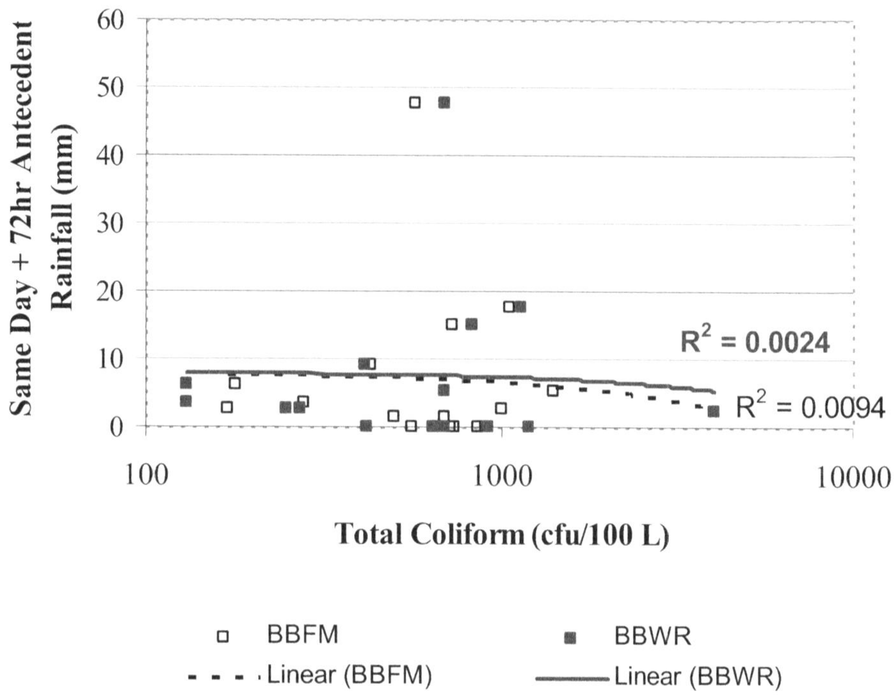

Figure A.11 Same day + 72hr antecedent rainfall versus total coliform for BB

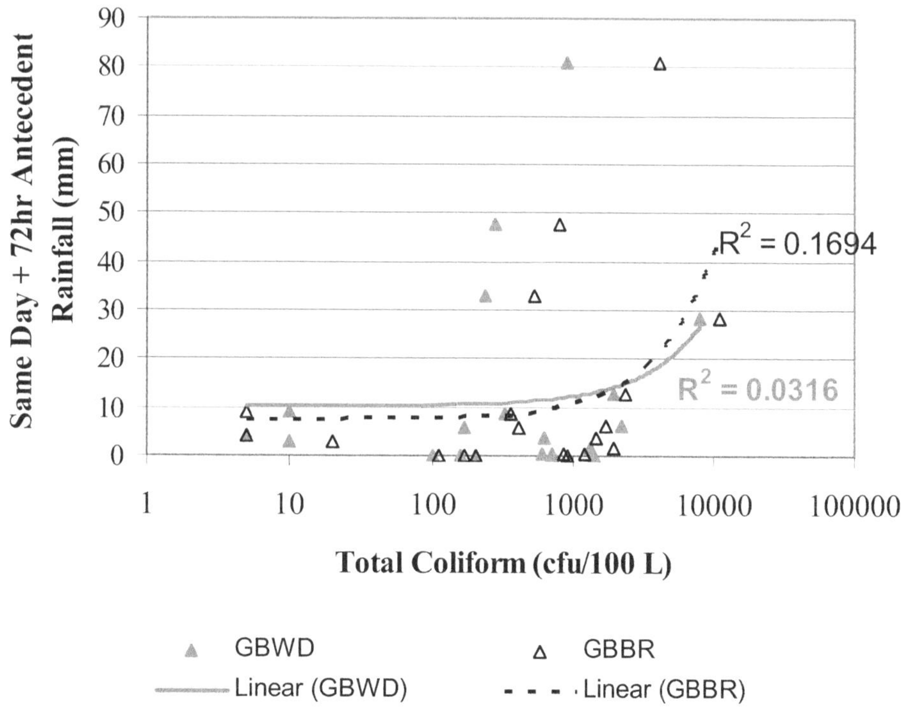

Figure A.12 Same day + 72hr antecedent rainfall versus total coliform for GB

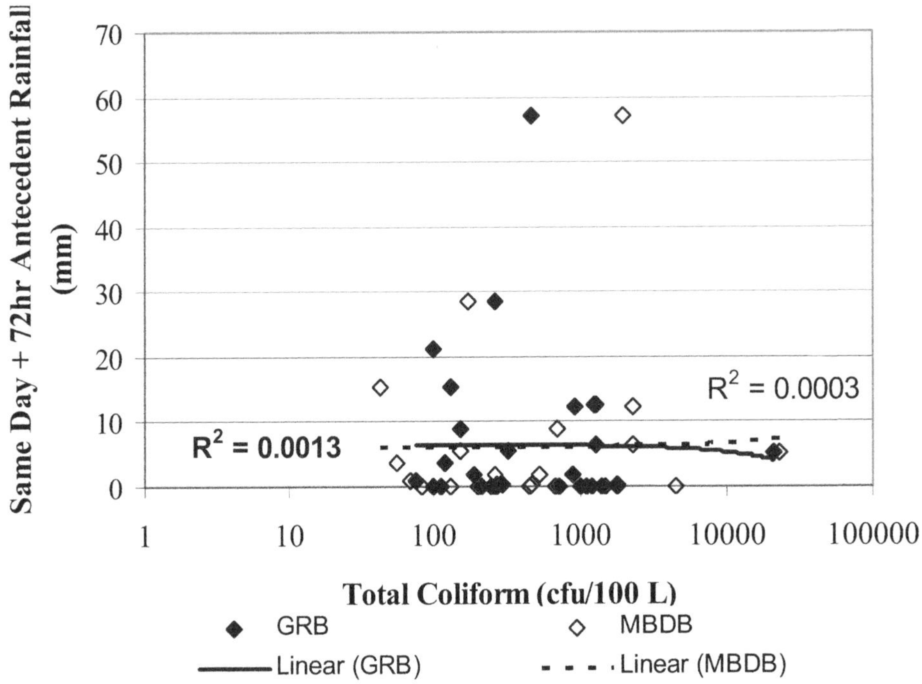

Figure A.13 Same day + 72hr antecedent rainfall versus total coliform for PR

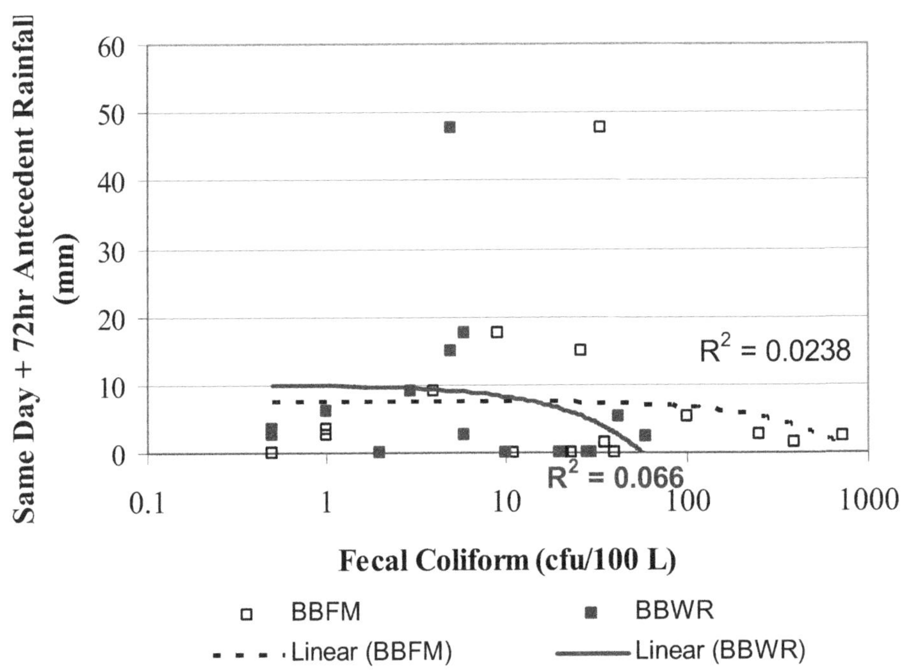

Figure A.14 Same day + 72hr antecedent rainfall versus fecal coliform for BB

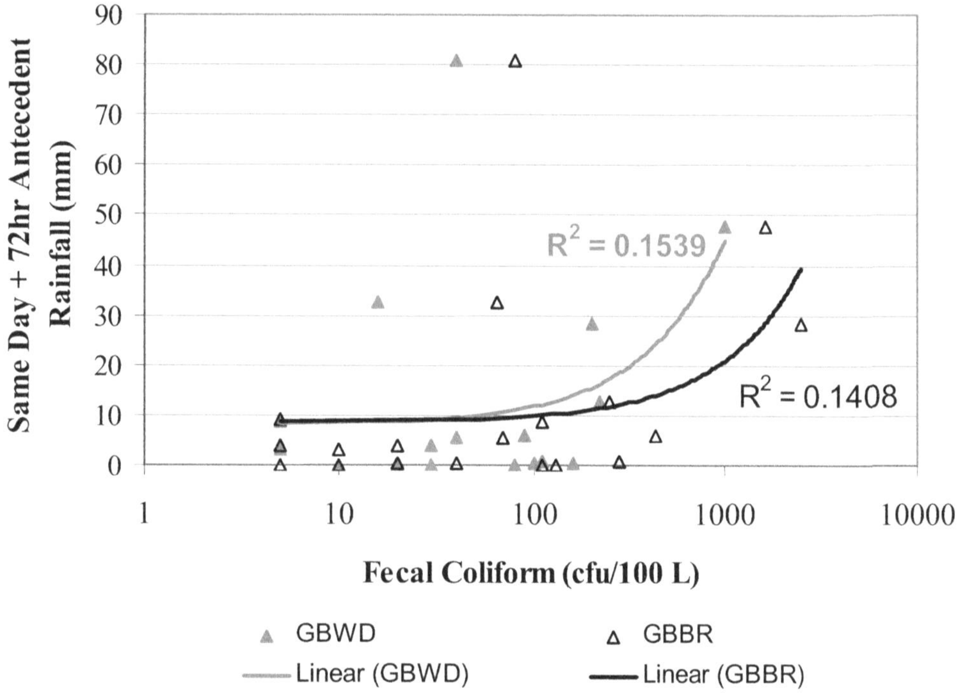

Figure A.15 Same day + 72hr antecedent rainfall versus fecal coliform for GB

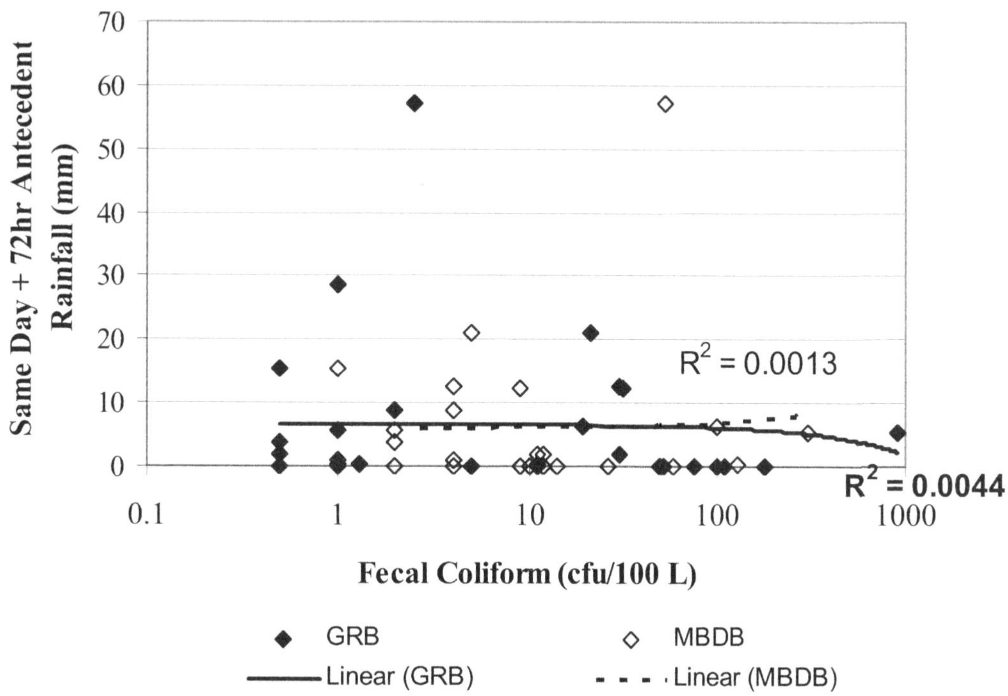

Figure A.16 Same day + 72hr antecedent rainfall versus fecal coliform for PR

APPENDIX B
ANTECEDENT RAINFALL VERSUS EVENT MEAN CONCENTRATION DATA

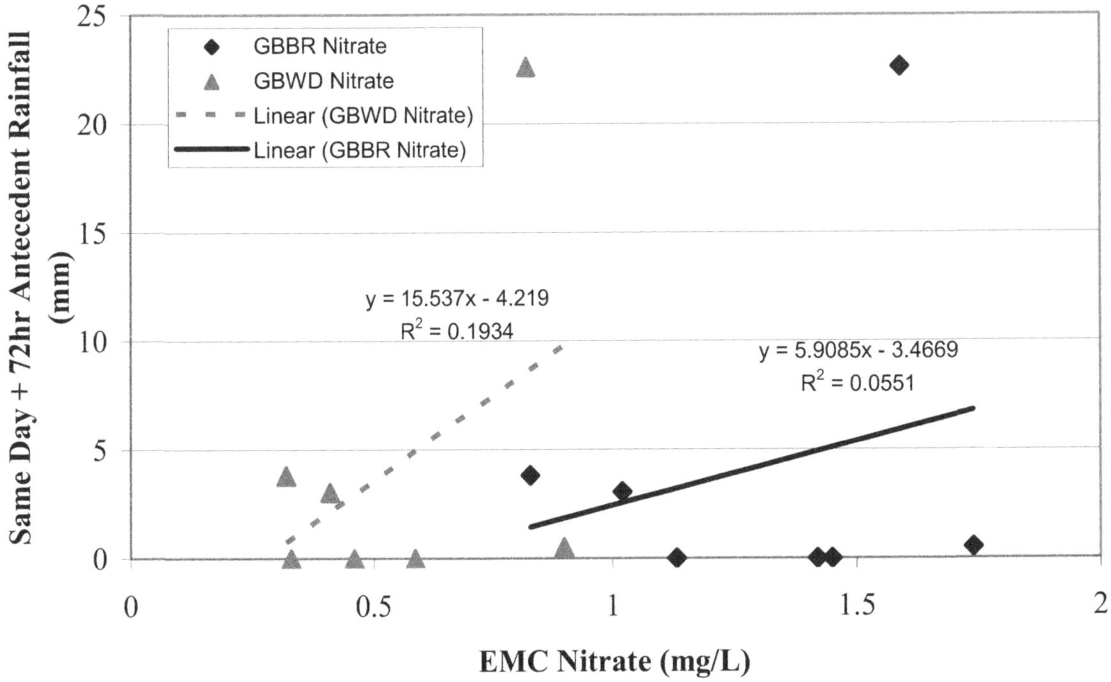

Figure B.1 Same Day + 72hr Antecedent Rainfall versus EMC Nitrate for GB

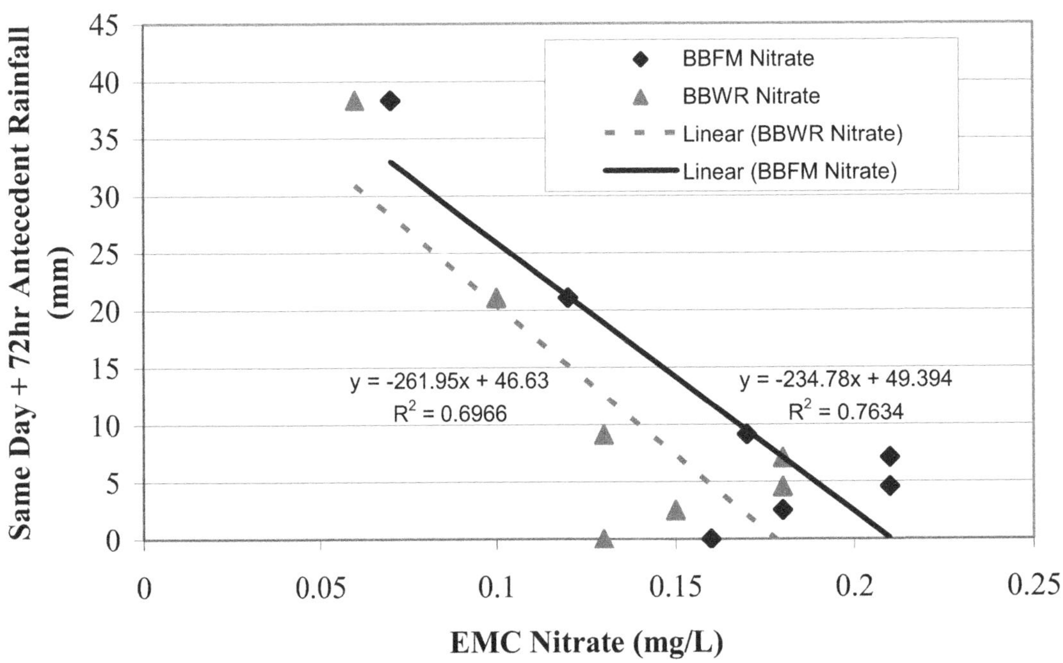

Figure B.2 Same Day + 72hr Antecedent Rainfall versus EMC Nitrate for BB

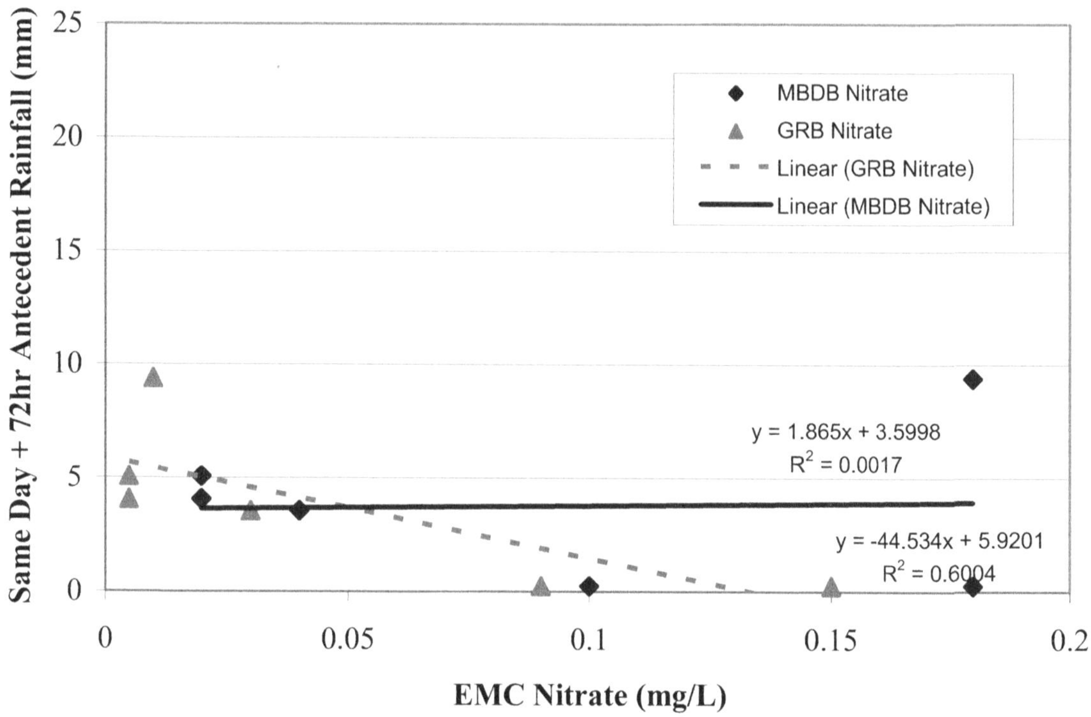

Figure B.3 Same Day + 72hr Antecedent Rainfall versus EMC Nitrate for PR

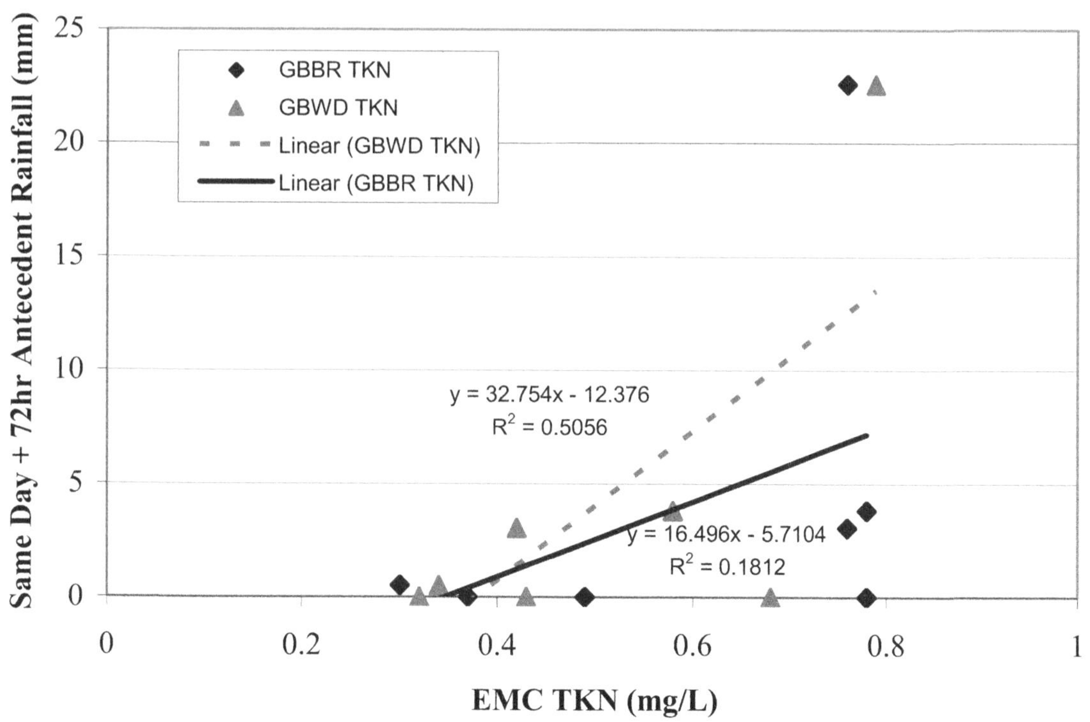

Figure B.4 Same Day + 72hr Antecedent Rainfall versus EMC TKN for GB

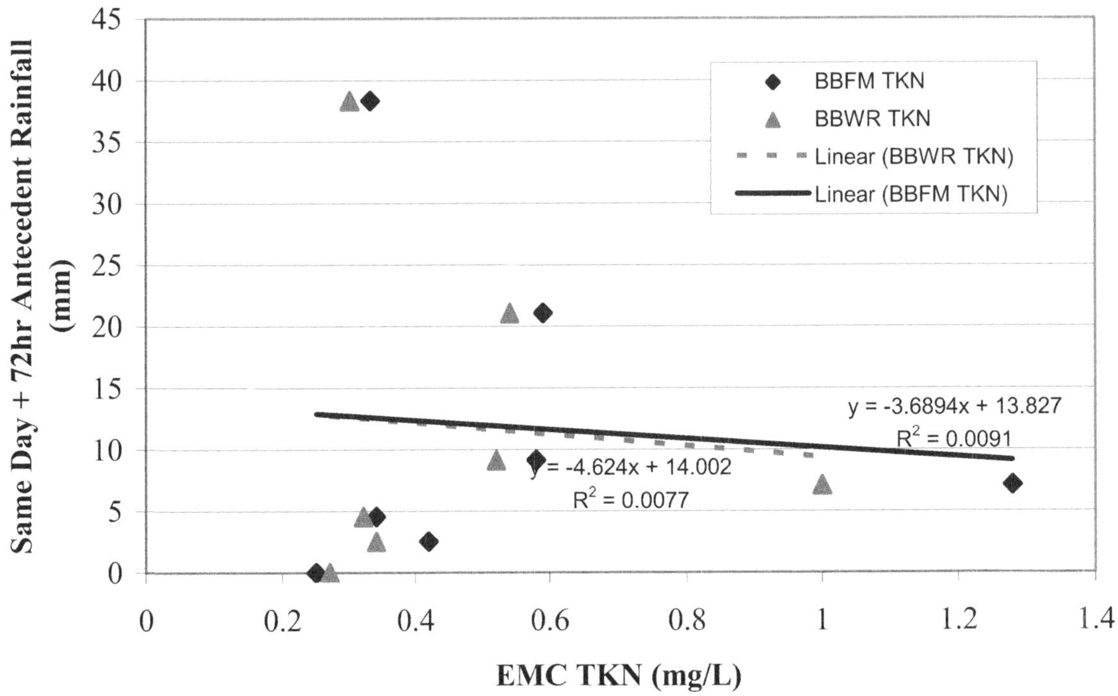

Figure B.5 Same Day + 72hr Antecedent Rainfall versus EMC TKN for BB

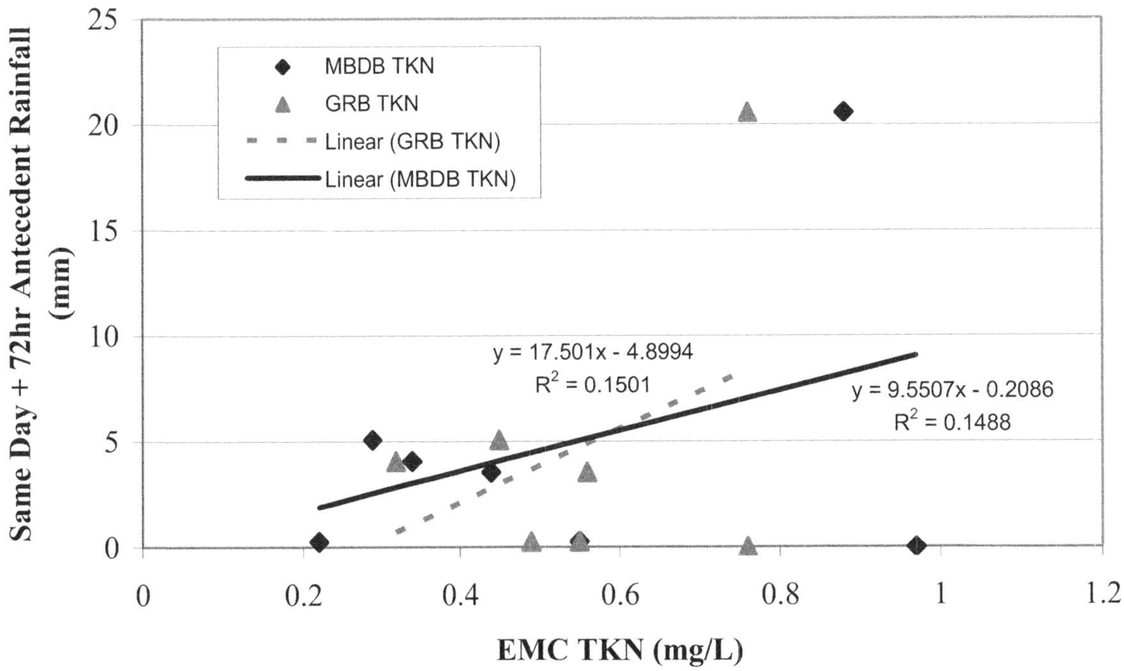

Figure B.6 Same Day + 72hr Antecedent Rainfall versus EMC TKN for PR

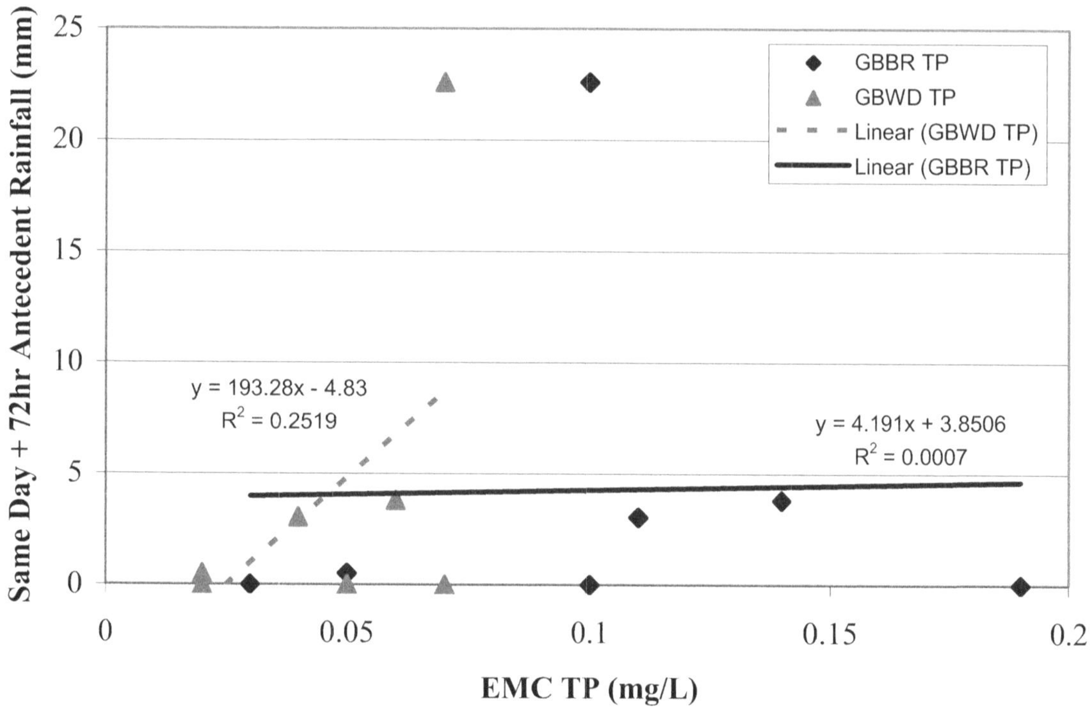

Figure B.7 Same Day + 72hr Antecedent Rainfall versus EMC TP for GB

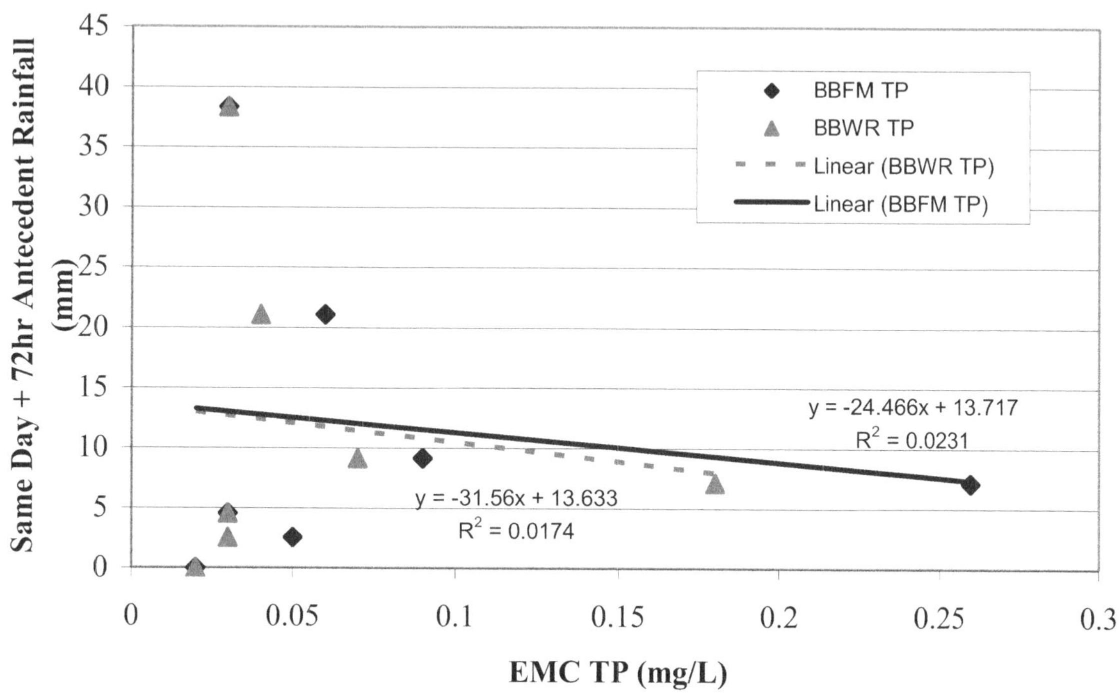

Figure B.8 Same Day + 72hr Antecedent Rainfall versus EMC TP for BB

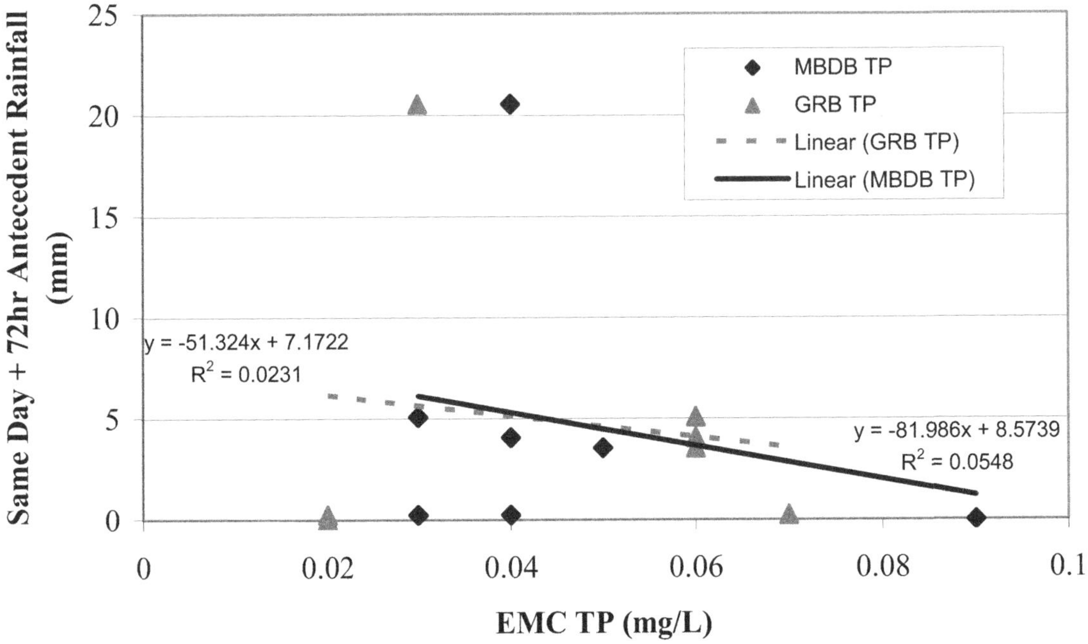

Figure B.9 Same Day + 72hr Antecedent Rainfall versus EMC TP for PR

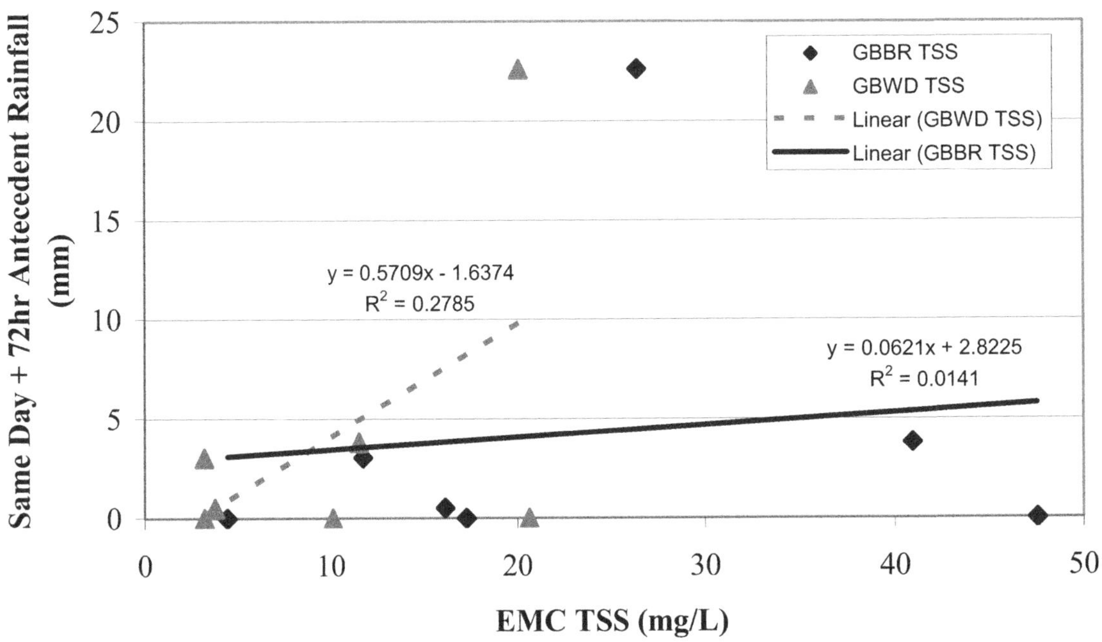

Figure B.10 Same Day + 72hr Antecedent Rainfall versus EMC TSS for GB

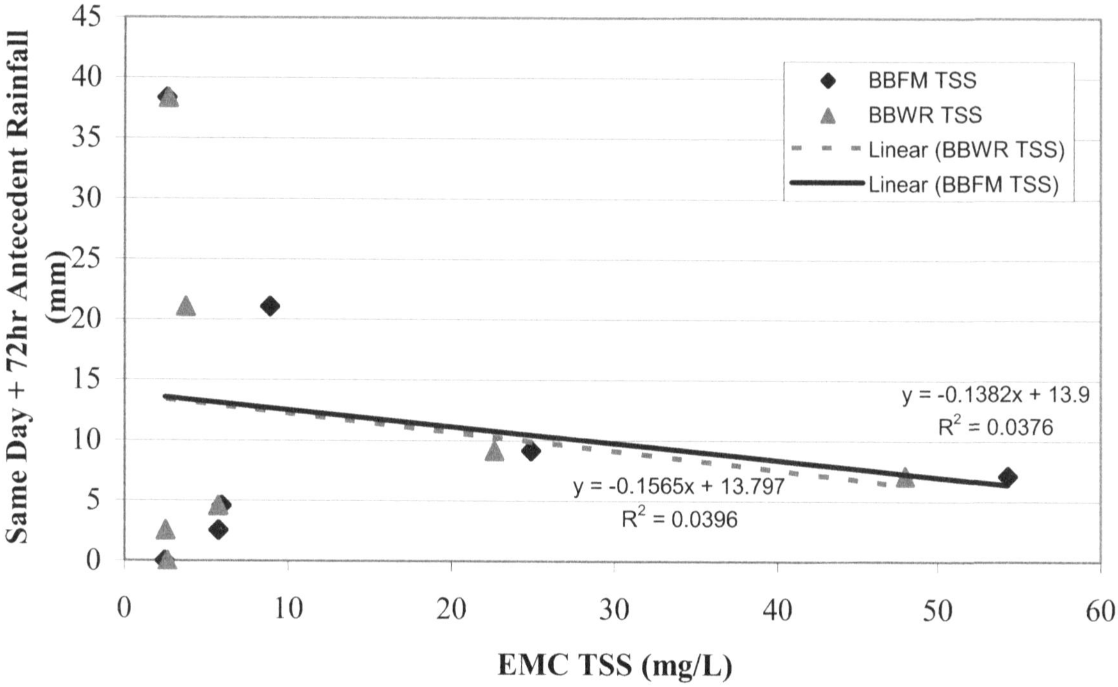

Figure B.11 Same Day + 72hr Antecedent Rainfall versus EMC TSS for BB

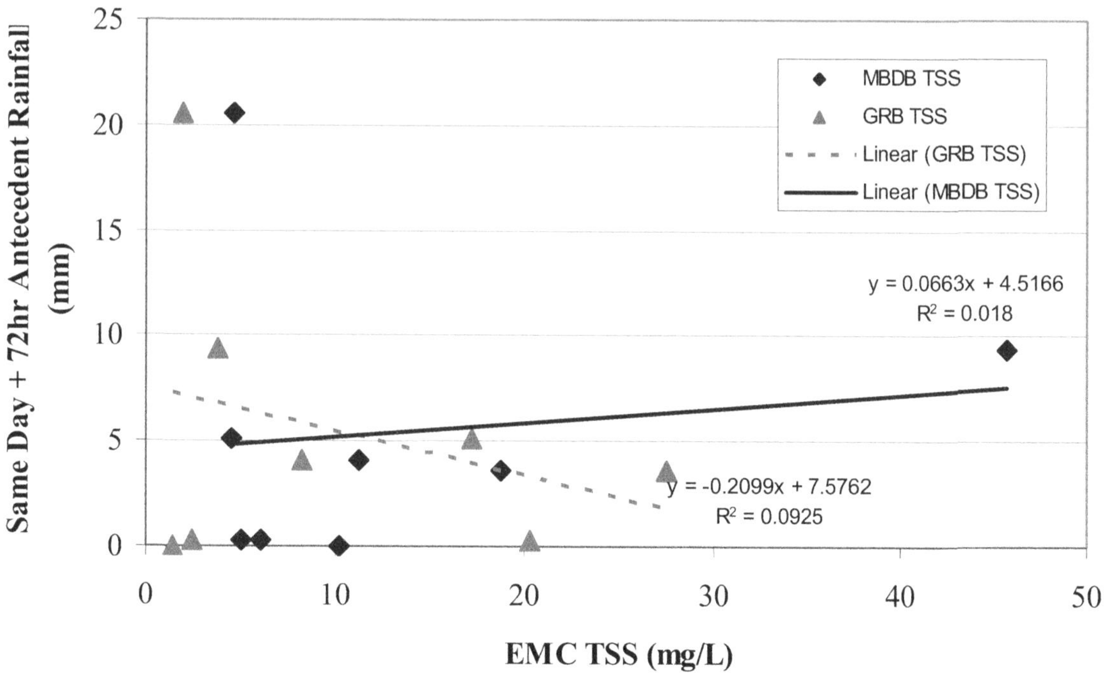

Figure B.12 Same Day + 72hr Antecedent Rainfall versus EMC TSS for PR

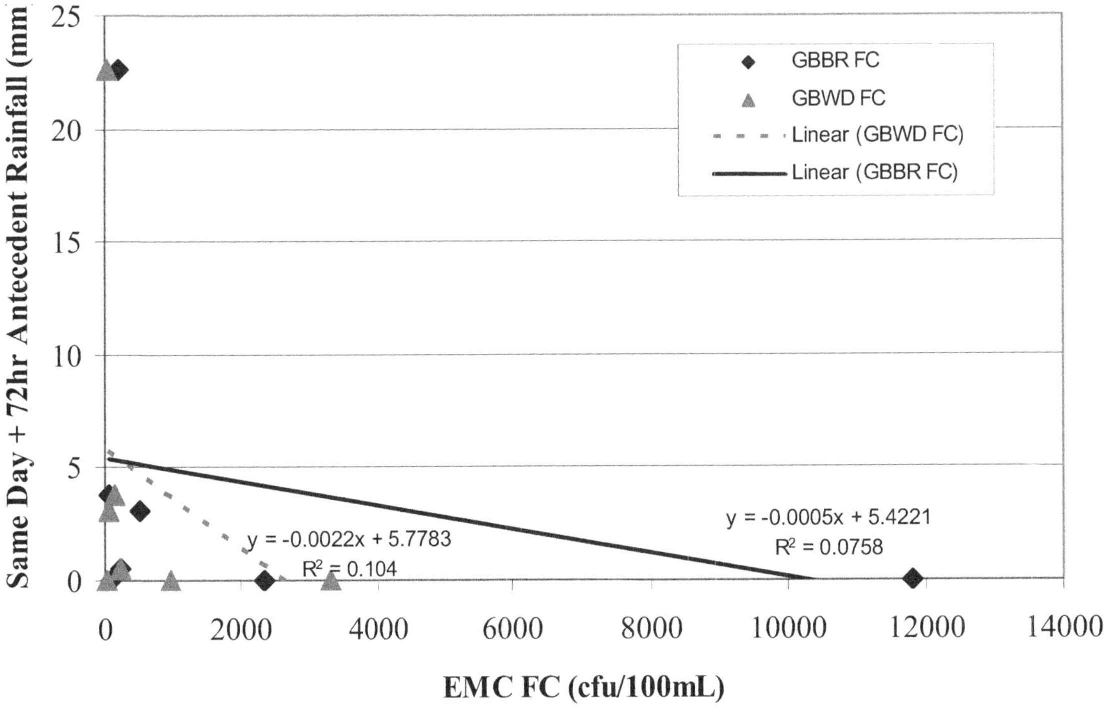

Figure B.13 Same Day + 72hr Antecedent Rainfall versus EMC FC for GB

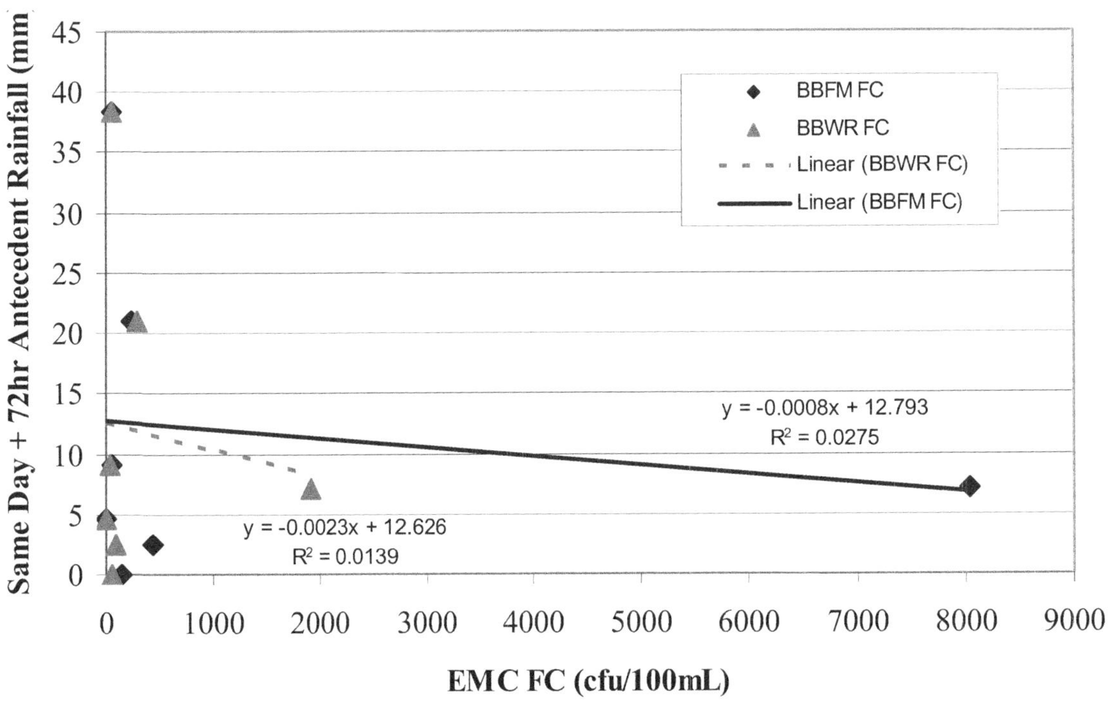

Figure B.14 Same Day + 72hr Antecedent Rainfall versus EMC FC for BB

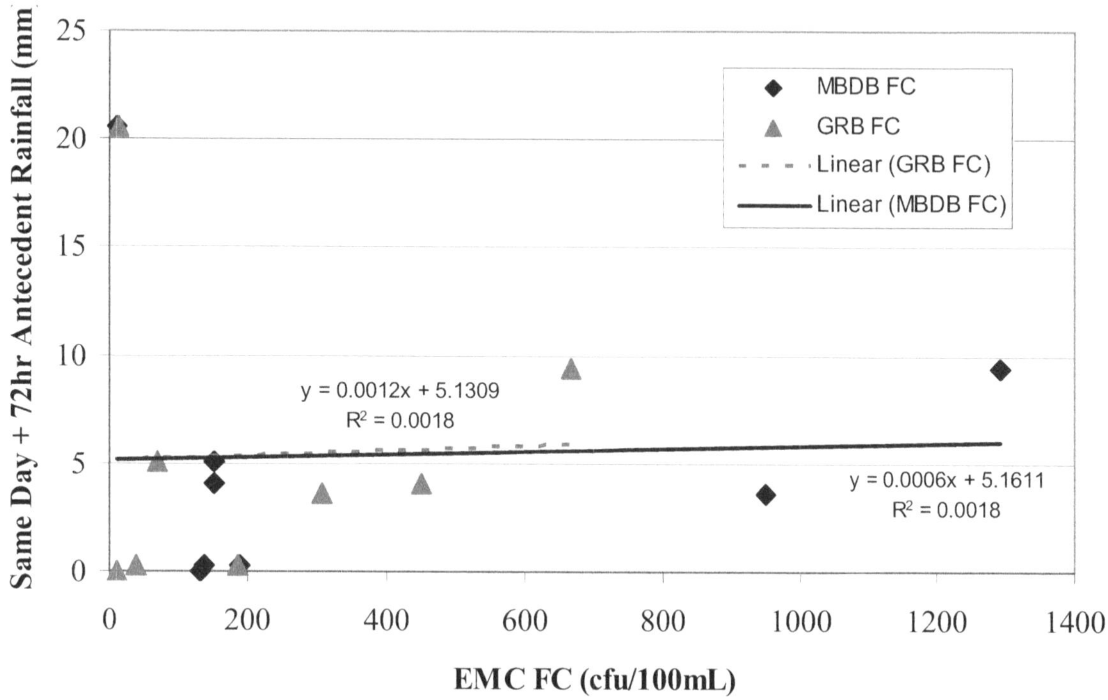

Figure B.15 Same Day + 72hr Antecedent Rainfall versus EMC FC for PR

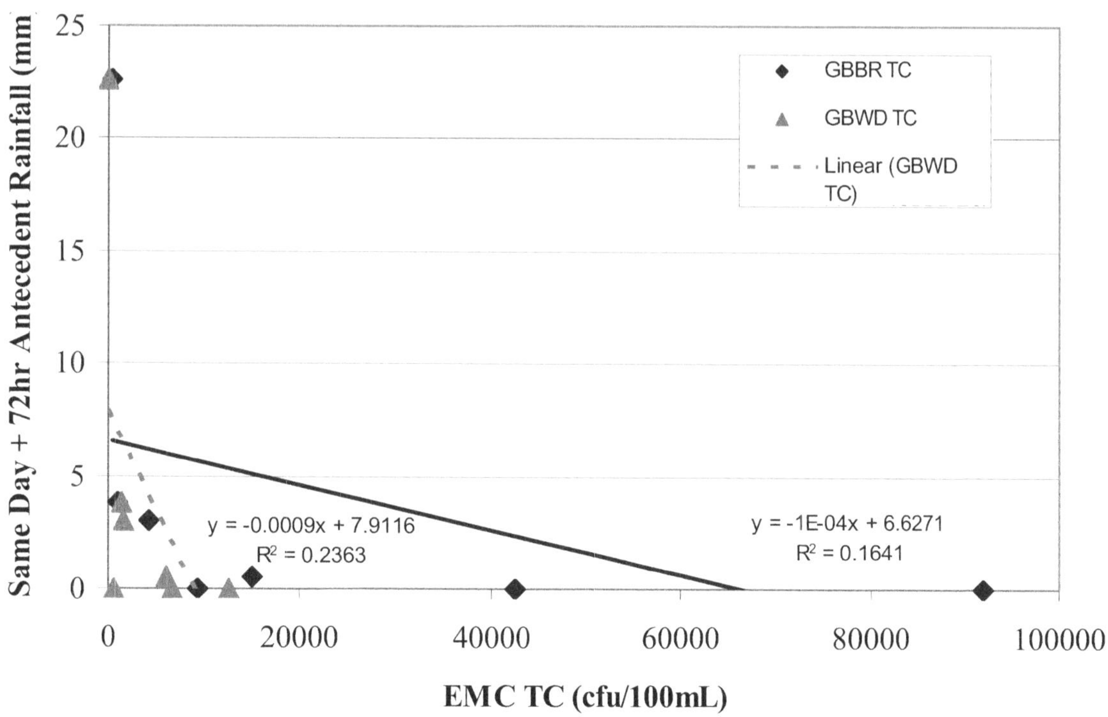

Figure B.16 Same Day + 72hr Antecedent Rainfall versus EMC TC for GB

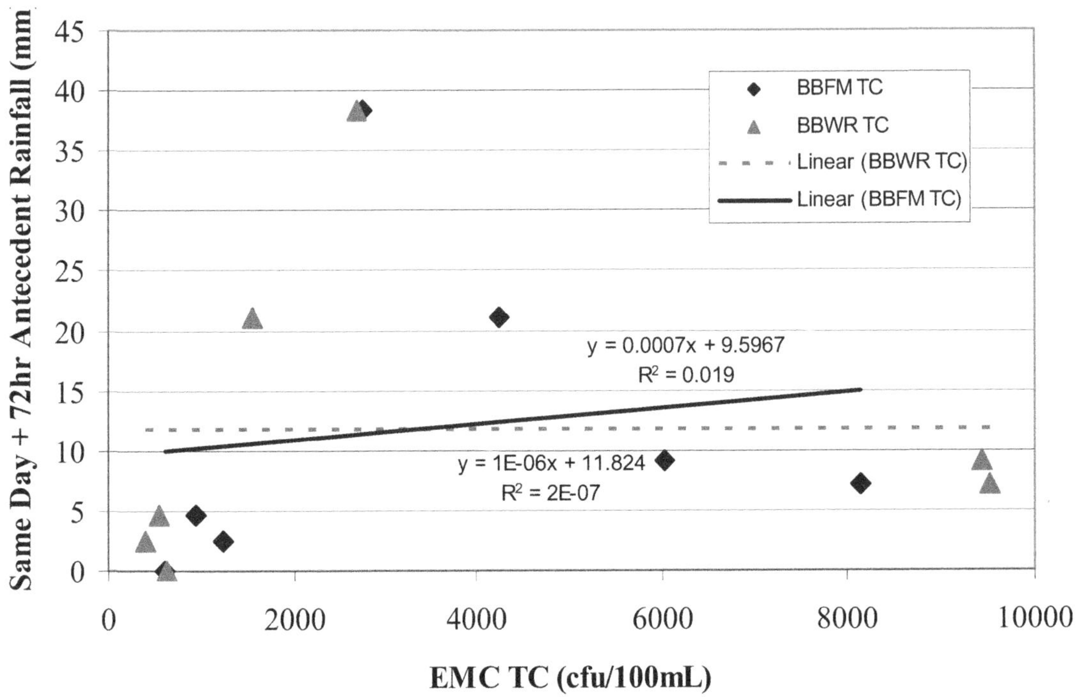

Figure B.17 Same Day + 72hr Antecedent Rainfall versus EMC TC for BB

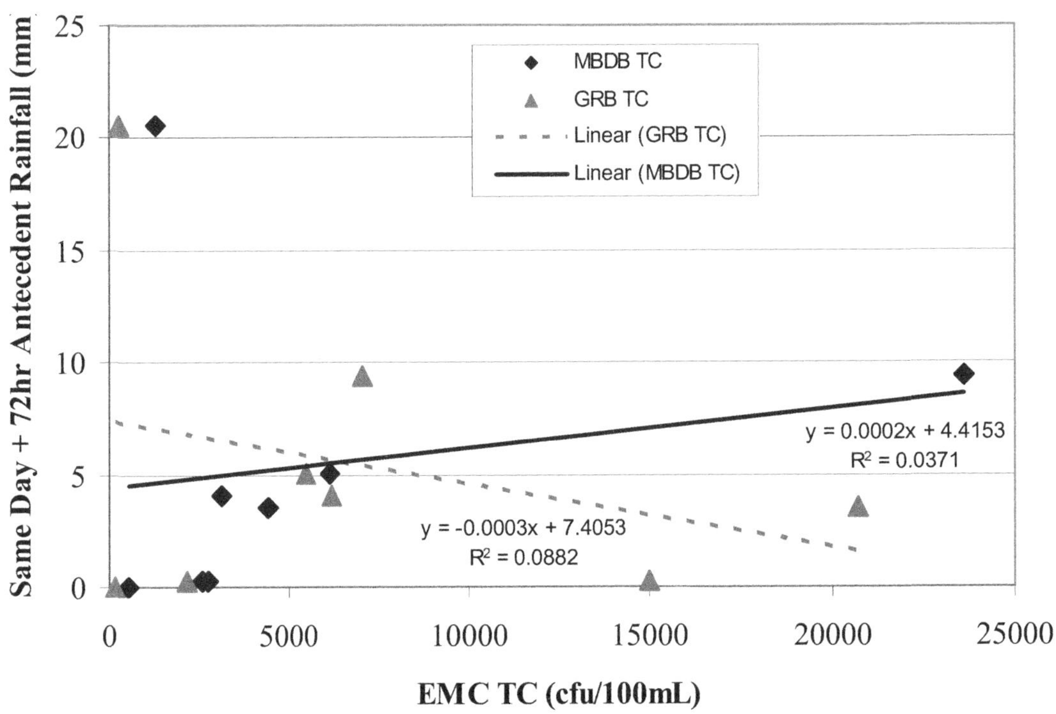

Figure B.18 Same Day + 72hr Antecedent Rainfall versus EMC TC for PR

REFERENCES

Anderson, S.A, S.J. Turner and G.D. Lewis. 1997. Enterococci in the New Zealand Environment: *Implications for Water Quality Monitoring. Water Science and Technology,* 35: (11-12): 325-331.

APHA, AWWA, WEF. 1998. *Standard Methods for the Examination of Water and Wastewater*, 20th edition. Washington, DC: American Public Health Association.

Arango, C. 2000. Evaluation and Optimization of Detection Methods for *R. coprophilus* and Sorbitol-Fermenting *Bifidobacteria* as Source-Specific Indicator Organism for Drinking Water Sources. Doctoral dissertation, Department of Civil and Environmental Engineering, University of Massachusetts, Amherst, MA.

Arango P.C., and S.C. Long. 1998. Alternative Bacterial Indicators of Sources of Microbial Inputs to Watersheds of Drinking Water Sources. In *Proceedings of the American Water Works Association Annual Conference and Exposition*, *Volume C*, June 21-15 (1998), Dallas, TX, pp. 679-687.

Ashbolt, N.J., G.S. Grohmann, and C.S.W. Kueh. 1993. Significance of Specific Bacterial Pathogens in the Assessment of Polluted Receiving Waters of Sydney. *Water Science and Technology,* 27 (3-4): 449-452.

Ashendorff, A, Principe, M.A., Seeley, A., LaDuca, J., Beckhardt, L., Faber Jr., Walter, Mantus, J. 1997. Watershed protection for New York City's supply. *Journal of AWWA*, 89 (3):75-88.

Atherholt, T.B., M.W. Lechevallier, and J.S. Rosen. 1998. Effect of Rainfall on *Giardia* and *Crypto*. *Journal of AWWA*, 90 (9): 66-80.

Atherton F., C.P. Newman, and D.P. Casemore DP. 1995. An Outbreak of Waterborne Cryptosporidiosis Associated with a Public Water Supply in the UK. *Epidemiology and Infection*, 115 (1):123-31.

Auer, M.T., and S.L. Niehaus. 1993. Modeling Fecal Coliform Bacteria I. Field and Laboratory Determination of Loss Kinetics. *Water Research,* 27 (4): 693-701.

AWWA. 1999a. *Waterborne Pathogens, AWWA Manual M48*. Denver, CO: American Water Works Association.

AWWA. 1999b. *Water Quality and Treatment 5th Edition*. Denver, CO: American Water Works Association, Denver.

Badenoch. 1990. *Cryptosporidium* – A Water-borne hazard. *Letters in Applied Microbiology,* 11(6):269-270.

Bales, R.C, S.M. Li, K.M. Maguire, M.T. Yahya, C.P. Gerba, and R.W. Harvey. 1995. Virus and Bacteria Transport in a Sandy Aquifer, Cape Cod, MA. *Groundwater,* 33: 653-661.

Bannerman, R.T., D.W. Owens, R.B. Dodds, and N.J. Hornewer. 1993. Sources of Pollutants in Wisconsin Stormwater. *Water Science and Technology,* 28 (3-5): 241-259.

Barbe, D.E., J.C. Francis, and M. Gunta. 1998. Seasonal Microbial Levels in Adjacent Watersheds. In *Proceedings of Water Resources Engineering* August 3-7: 1625-1630.

Barten, P.K., T. Kyker-Snowman, P.J. Lyons, T. Mahlstedt, R. O'Connor, and B.A. Spencer. 1998. Managing a Watershed Protection Forest. *Journal of Forestry,* 96(8):10-15.

Baudart, B., J. Grabulos, J.P. Barusseau, and P. Lebaron. 2000. *Salmonella spp.* and Fecal Coliform Loads in Coastal Waters from a Point vs. Nonpoint Source of Pollution. *Journal of Environmental Quality,* 29 (January-February): 241-250.

Baudisova, D. 1997. Evaluation of *Escherichia coli* as the Main Indicator or Fecal Pollution. *Water Science and Technology,* 35 (11-12): 333-336.

Beerens, H. 1991. Detection of Bifidobacteria by Using Propionic Acid as a Selective Agent. *Applied and Environmental Mictobiology,* 57: 2418-2419.

Berka, C., H. Schreier, and K. Hall. 2001. Linking Water Quality With Agricultural Intensification in a Rural Watershed. *Water, Air, and Soil Pollution,* 127 (May): 389-401.

Bisson, J.W., and V. J. Cabelli. 1979. Membrane Filter Enumeration Method For Clostridium Perfringens. *Applied and Envrionmental Mcrobiology*, 37(1), p. 55-66.

Bisson, J.W., and V.J. Cabelli. 1980. *Clostridium perfrigens* as a Water Pollution Indicator. *Journal of the Water Pollution Control Federation,* 52 (2): 241-248.

Bolstad, P.V., and W.T. Swank. 1997. Cumulative Impacts of Land Use on Water Quality in a Southern Appalachian Watershed. *Journal of the American Water Resources Association,* 33 (3): 519-533.

Bourdeau, D.H. 2004. Effect of Beaver (*Castor canadensis*) on Stream Water Quality: a Paired Watershed Study. Masters Thesis, Department of Civil and Environmental Engineering, University of Massachusetts, Amherst, MA.

Brock, T.D., M.T. Madigan, J.M. Martinko and J. Parker. 1994. *Biology of Microorganisms.* 7th Edition. Englewood Cliffs, NJ: Prentice Hall.

Burby, R.J., E.J. Kaiser, T.L. Miller, D.H. Moreau. 1983. *Drinking Water Supplies, Protection through Watershed Management.* Ann Arbor, MI: Ann Arbor Science.

Carter, R.W. and J. Davidian. 1968. *General Procedures for Gaging Streams: U.S. Geological Survey Techniques Water Resources Inv., Book 3.* Washington, D.C.: Unites States Geological Survey.

Camp Dresser McKee. 2002. *Merrimack River Watershed Assessment Study.* Prepared for the New England District U.S. Army Corps of Engineers.

Chapra, S.C. 1997. *Surface Water Quality Monitoring, 1st ed.* New York, New York: McGraw Hill.

Collins, C.H., P.M. Lyne, and J.M. Grange. 1995. *Collins and Lyne's Microbiological Methods*, 7th Edition. London: Butterworth and Heinemann.

Conroy, R.M., M. Elmore-Meegan, T. Joyce, K.G. McGuigan, and J. Barnes. 1996. Solar Disinfection of Drinking Water and Diarrhoea in Maasai Children: A controlled field trial. *The Lancet,* 348 (21/28): 1695-1697.

Crabill, C., Donald, R., Snelling, J., Faust, and Soltham, G. 1999. The Impact of Sediment Fecal Coliform Reservoirs on Seasonal Water Quality in Oak Creek, Arizona. *Water Resources,* 33 (9):2163-2171.

Craun, G.F., Hubbs, S.A., Frost, F., Calderon, R.L., and Via, S.H. 1998. Waterborne Outbreaks Of Cryptosporidiosis. *Journal of the American Water Works Association,* 90 (9):81-91.

Curtis, T.P., D.D. Mara, and S.A. Silva. 1992. Influence of pH, Oxygen and Humic Substance on Ability of Sunlight to Damage Fecal coliforms in Waste Stabilization Pond. *Water. Applied and Environmental Microbiology,* 58 (4): 1335-1343.

Davies, C.M., J.A.H. Long, D. M. Donald, and N.J. Ashbolt. 1995. Survival of Fecal Microorganisms in Marine and Freshwater Sediments. *Applied and Environmental Microbiology*, 61 (5): 1888-1896.

DeNileaon, G.P. 1998. Water Supply Indicated in *E. coli* Outbreak. *Main Stream,* 42 (8): 1.

DiGiovanni, G.D. 2003. Update on LT2ESWTR *Cryptosporidium* Monitoring and Genetic Characterization. Presented at the Early Warning and Source Monitoring Workshop, 2003 Source Water Protection Symposium, Albuquerque, NM, January 19, 2003.

Donnelly J.K., and E. I. Stentiford. 1997. The *Cryptosporidium* Problem in Water and Food Supplies. *Food Science and Technology - Lebensmittel-Wissenschaft & Technologie,* 30 (2): 111-120.

Doran, J.W., and D.M. Linn. 1979. Bacteriological Quality of Runoff Water from Pasture Land. *Applied and Environmental Microbiology,* 37 (5): 985-991.

Doran, J.W., J.S. Schepers, and N.P. Swanson. 1981. Chemical and Bacteriological Quality of Pasture Runoff. *Journal of Soil and Water Conservation,* 36 (3): 166-171.

Dorner, S.M., P.M. Huck, R.M. Slawson. 2004. Estimating potential environmental loadings of Cryptosporidium spp. and Campylobacter spp. from livestock in the Grand River Watershed, Ontario, Canada. *Environmental Science and Technology*, 38(12): 3370-3380.

Dunlap, B.G. and Thies, M.L. 2002. Giardia in Beaver (*Castor canadensis*) and Nutria (Myocastor Coypus) from East Texas. *Journal of Parasitology,* 88: 1254-1258.

Edberg, S.C., and D.B. Smith. 1989. Absence of Association between Total Heterotrophic and Total Coliform Bacteria from a Public Water Supply. *Applied and Environmental. Microbiology,* 55 (2): 380-384.

Erkenbrecher, C.W. 1981. Sediment Bacterial Indicators in an Urban Shellfishing Subestuary of the Lower Chesapeake Bay. *Applied and Environmental Microbiology,* 42 (3): 484-492.

Faust, M.A., A.E. Aotaky, and M.T. Hargadon. 1975. Effect of Physical Parameters on the in Situ Survival of *Escherichia coli* MC-6 in an Estuarine Environment. *Applied Microbiology,* 30 (5): 800-806.

Faust, M.A. 1982. Relationship between Land Use Practices and Fecal Bacterial in Soils. *Journal of Environmental Quality,* 11 (1): 141-147.

Fayer , R., L. Gasbarre, P. Pasuali, A. Canals, S. Almeria, and D. Zarlenga. 1998. Cryptosporidium parvum infection in bovine neonates-dynamics clinical, parasitic, and immunologic patterns. *International Journal of Parasitology,* 28 (1): 49-56.

Fayer, R. and R.G. Leek. 1984. The Effects of Reducing Conditions, Medium, pH, Temperature and time on In-vitro Excystation of *Cryptosporidium*. *Journal of Protozoology*, 31(4): 567-569.

Ferguson, C.M., B.G. Coote, N.J. Ashbolt, and I.M. Stevenson. 1996. Relationships Between Indicators, Pathogens and Water Quality in an Estuarine System. *Water Research,* 30 (9): 2045-2054.

Fisher, K.R., 1995. *Manual of Sediment Transport in Rivers*. Wallingford, UK: HR Wallingford.

Fraser, R.H., P.K. Barten and D.A.K. Pinney. 1998. Predicting Stream Pathogen Loading from Livestock using a Geographical Information System-Based Delivery Model. *Journal of Environmental Quality,* 27: 935-945.

French, R.H. 1985. *Open Channel Hydraulics*. New York, New York: McGraw-Hill Book Company.

Fujioka, R.S., H.H. Hashimoto, E.B. Siwak, and R.H. Young. 1981. Effect of Sunlight on Survival of Indicator Bacteria in Seawater. *Applied and Environmental Technology,* 41 (3): 690-696.

Fujioka, R.S. and L.K. Shizumura. 1985. *Clostridium perfringens*, a Reliable Indicator of Stream Water Quality. *Journal of Water Pollution Control Federation,* 57 (10): 986-992.

Furuse, K., T. Sakurai, Y. Inokuchi, H. Inoko, A. Ando, and I. Watanabe. 1978. Distribution of Ribonucleic Coliphages in South and East Asia. *Applied and Environmental Microbiology,* 35 (6): 995-1002.

Gerba, C.P., and J.S. McLeod. 1976. Effects of Sediments on the Survival of *Escherichia coli* in Marine Waters. *Applied and Environmental Microbiology,* 32: 114-120.

Gordon, N.D., T.A. McMahon, and B.L. Finlayson. 1992. *Stream Hydrology: An Introduction for Ecologists*. New York: John Wiley & Sons.

Graczyk, T.K., R. Fayer, J.M. Trout, E.J. Lewis, C.A. Farley, I. Sulaiman, and A.A. Lal. 1998. *Giardia sp*. Cysts and Infectious *Cryptosporidium parvum* Oocysts in the Feces of Migratory Canada Geese (*Branta Canadensis*). *Applied and Envrionmental Microbiology*, 64 (7): 2736-2738.

Griffin, D.W., R. Stokes, J.B. Rose, and J.H. Paul III. 2000. Bacterial Indicator Occurrence and the use of F+specific RNA Coliphage Assay to Identify Fecal Sources in Homosassa Springs, Florida. *Microbial Ecology,* 39, 56-64.

Grimes, D.J. 1975. Release of Sedimental Bound Fecal Coliforms by Dredging. *Applied Microbiology,* **29**: 109-111.

Gyllenberg, H., S. Niemela, T. Sormunen. 1960. Survival of Bifid Bacteria in Water as Compared with that of Coliform Bacteria and Enterococci. *Applied Microbiology,* 8 (1): 20-23.

Haavelar A.H., W.M. Pot-Hogeboom, K. Furuse, R. Pot, M.P. Hormann. 1990. F-specific RNA Bacteriophages and Sensitive Host Trains in Faeces and Wastewater of Human and Animal Origin. *Journal of Applied Bacteriology,* 69: 30-37.

Hardina C.M. and R.S. Fujioka. 1991. Soil: The Environmental Source of *Escherichia coli* and Enterococci in Hawaii's streams. *Environmental Toxicology and Water Quality: An International Journal,* 6: 185-195.

Harms, L.L., P. Middaugh, J.N. Dornbush, and J.R. Anderson. 1975a. Agricultural Runoff Pollutes Surface Waters, part I. *Water Sewage Works,* 122 (October): 85-86.

Harms, L.L., P. Middaugh, J.N. Dornbush, and J.R. Anderson. 1975b. Bacteriological Quality of Surface Runoff From Agricultural Land, part II. *Water Sewage Works,* 122 (November): 71-73.

Harvey, R.W. and S.P. Garabedian, 1991. Use of Colloid Filtration Theofy iIn Modeling Movement of Bacteria Through a Contaminated Sandy Aquifer. *Environmental Science and Technology,* 25(1):178-185.

Hazen, T.C., and G.A. Toranzos. 1990. Tropical Source Water. In *Drinking Water Microbology*, Edited by G.A. McFeters, New York, NY: Springer-Verlag.

Hem, J.D. 1985. *Study and Interpretation of the Chemical Characteristics of Natural Waters, 3rd edition.* Washington, D.C.: U.S. Geological Survey .

Henderson, F.M. 1966. *Open Channel Flow*. New York, New York: Macmillan Publishing Company.

Hodgdon, H.E. 1978. Social Dynamics and Behavior within an Unexploited Beaver (Castor Canadensis) Population. Ph.D. thesis, Univeristy of Massachusetts, Amherst..

Holt, J.G., N.R. Krieg, P.H.A. Sneath, J.T. Staley, and S.T. Williams. 1994. *Bergey's Manual of Determinative Bacteriology, 9th Edition*. Philadelphia, PA: Williams and Wilkins.

Howell, J.M., M.S. Coyne, and P. Cornelius. 1995. Fecal Bacteria in Agricultural Waters of the Bluegrass Region of Kentucky. *Journal of Environmental Quality,* 24 (May-June): 411-419.

Howell, J.M., M.S. Coyne, and P.L. Cornelius. 1996. Effect of Sediment Particle Size and Temperature on Fecal Bacteria Mortality Rates and the Fecal Coliform/Fecal Streptococci Ratio. *Journal of Environmental Quality* **25**: 1216-1220.

Hunter, C., and A. McDonald. 1991. Seasonal Changes in the Sanitary Bacterial Quality of Water Draining a Small Upland Catchments in the Yorkshire Dales. *Water Research,* 25 (4): 447-453.

Hunter, C., J. Perkins, J. Tranter, and J. Gunn. 1999. Agricultural land use effects on the indicator bacterial Quality of an Upland Stream in the Derbyshire Peak District in the U.K. *Water Research,* 33 (17): 3577-3586.

Ingraham, J. 1987. Effect of Temperature, pH, Water Activity, and Pressure on Growth. In *Escherichia coli and Salmonella typhimurium - Cellular and Molecular Biology*, Edited by R. C. Neidhardt, Washington, D.C.: ASM Press.

Irvine, K.N., and G.W. Pettibone. 1996. Planning Level Evaluation of Densities and Sources of Indicator Bacteria in a Mixed Land Use Watershed. *Environmental Technology,* 17 (1): 1-12.

Jagals, P., W.O.K. Grabow, J.C. De Villiers. 1995. Evaluation of Indicators for Assessment of Human and Animal Fecal Pollution of Surface Runoff. *Water Science and Technology,* 31 (5-6): 235-341.

Jagals, P., and W.O.K. Grabow. 1996. An Evaluation of Sorbitol-fermenting *Bifidobacteria* as Specific Indicators of Human Faecal Pollution of Environmental Water. *Water SA,* 22 (3): 235-238.

Jawson, M.D., L.F. Elliott, K.E. Saxton, and D.H. Fortier. 1982. The Effects of Cattle Grazing on Indicator Bacterial in Runoff from a Pacific Northwest Watershed. *Journal of Environmental Quality,* 11 (4): 621-627.

Jenkins, S.H., and P.E. Busker. 1979. Castor Canadensis. *Mammalian Species*, 120:1-9.

Joyce, T.M., K.G. Mcguigan, M. Elmore-Meegan, and R.M. Conroy. 1996. Inactivation of Fecal Bacteria in Drinking Water by Solar Heating. *Applied and Environmental Microbiology,* 62 (2): 399-402.

Kay, D., and A. McDonald. 1980. Reduction of Coliform Bacteria in Two Upland Reservoirs: The Significance of Distance Decay Relationships. *Water Research,* 14 (4): 305-318.

Kebabjian, R., 1994. Monitoring the Effects of Urban Runoff on Recreational Waters. *Journal of Environmental Health*, 56: 15-18.

Kelch, W.J., J.S. Lee. 1978. Modeling Techniques for Estimating Fecal Coliforms in Estuaries. *Journal of Water Pollution Control Federation,* 50 (May): 862-868.

Kellogg, R.L., M.S. Maizel, and D.W. Goss. 1994. The Potential for Leaching of Agrichemicals use in Crop Production: A National Perspective. *J. Soil and Water Conservation,* 49 (3): 294-298.

King, H.W., and E.F. Brater. 1963. *Handbook of Hydraulics for the Solution of Hydrostatic and Fluid-flow Problems*. New York: McGraw-Hill Book Company.

Kistemann, T., T. ClaBen, C. Kock, F. Dangendorf, R. Fischeder, J. Gebel, V. Vacata, and M. Exner. 2002. Microbial Load of Drinking Water Reservoir Tributaries during Extreme Rainfall and Runoff. *Applied and Environmental Microbiology*, 68(5): 2188-2197.

Klein, R.D. 1979. Urbanization and Stream Quality Impairment. *Water Resources Bulletin,* 15 (4): 948-963.

Klock, J.W. 1971. Survival of Coliform Bacteria in Wastewater Treatment Lagoons. *Journal Water Pollution Control Federation,* 43 (10): 2071-2083.

Kocasoy, G. 1995. Effects of Tourist Population Pressure on Pollution of Coastal Seas. *Environmental Management,* 19 (1): 75-79.

Krumperman, P.H. 1983. Multiple Antibiotic Resistance Index of *Escherichia coli* to Identify High-risk Sources of Fecal Contamination of Foods. *Applied and Environmental Microbiology,* 46 (1): 165-170.

Kuhn, I., G. Allestam, T.A. Stenstrom, and R. Mollby. 1991. Biochemical Fingerprinting of Water Coliform Bacteria, a New Method for Measuring Phenotypic Diversity and for Comparing Different Bacterial-Populations. *Applied and Environmental Microbiology,* 57 (11): 3171-3177.

Lamka, K.G., M.W. LeChevallier, and R.J. Seidler. 1980. Bacterial Contamination of Drinking Water Supplies in a Modern Rural Neighborhood. *Applied and Environmental Microbiology,* 39: 734-738.

LeChevallier, M.W., W.D. Norton and R.G. Lee. 1991. Occurrence of Giardia and Cryptosporidium spp. in Surface Water Supplies. *Applied and Environmental Microbiology* 57 (9): 2610.

LeChevallier, M.W. and W.D. Norton. 1995. Giardia and Cryptosporidium in Raw and Finished Water. *Journal of the American Water Works Association,* 87 (9): 54.

Levine, ND, 1984. Taxonomy and Review of the Coccidian Genus *Cryptosporidium* (Protozoa, Apicomplexa). *Journal of Protozoology,* 31(1): 94-98, 1984.

Lim, C.-H., and K.P. Flint. 1989. The Effects of Nutrients on Survival of *Escherichia coli* in Lake Water. *Journal of Applied Bacteriology,* 66: 559-569.

Lisle JT and JB Rose. 1995. Cryptosporidium Contamination of Water in the USA and UK – A Minireview. *Journal of Water Supply Research andTechnology – AQUA*, 44(3): 103-117.

Long, S.C. 1998. Project Report: Development of Methods to Differentiate Microorganisms in MDC Reservoir Watersheds. Civil and Environmental Engineering, University of Massachusetts, Amherst, MA.

Long, S.C. and C. Arango P. 1998. Evaluation of Enumeration Media for Sorbitol-Fermenting Bifidobacteria as Indicators of Human Fecal Contamination in Drinking Water. Presented as the 98th General Meeting of the American Society for Microbiology, Atlanta, GA, May 17-21, 1998.

Long, S.C. 1999. Project Report for Project Continuation: Development of Methods to Differentiate Microorganisms in MDC Reservoir Watersheds. Civil and Environmental Engineering, University of Massachusetts, Amherst, MA.

Long, S.C., E.J. Mahar, R. Pei, C. Arango, E. Shafer, and T.H. Schoenberg. 2002. *Development of Source-Specific Indicator Organisms for Drinking Water.* AwwaRF, Denver, CO.

Long, S.C., E.J. Shafer, C. Arango P., and D. Siraco. 2003. Evaluation of Three Source-Specific Indicator Organisms for Watershed Management. *Journal of Water Supply: Research & Technology – AQUA,* 52 (8): 565-576.

Madore, M.S., J.B. Rose, M.J. Arrowood, and C.R. Sterling. 1987. Occurrence of Cryptosporidium oocysts in Sewage Effluencts and Selected Surface Waters. *Journal of Parasitology,* 72 (4): 702.

Mahar, E.J. 2002. Sorbitol-fermenting *Bifidobacteria* as An Indicator of Human Fecal Contamination in Source Waters for Watershed Management Applications. Masters Thesis, Department of Civil and Environmental Engineering, University of Massachusetts, Amherst.

Maiolo, J.R., and P. Tschetter. 1981. Relating Population Growth to Shellfish bed closures: A Case Study from North Carolina. *Coastal Zone Management Journal,* 9: 1-18.

Mallin, M.A., K.E. Williams, E.C. Esham, and R.P. Lowe. 2000. Effect of Human Development on Bacteriological Water Quality in Coastal Watersheds. *Ecological Applications,* 10 (4): 1047-1056.

Mancini, J. L. and A. H. Plummer. 1995. Effects of Temperature on Infectivity of Cryptosporidium Oocysts.

Mara, D.D., and J.I. Oragui. 1981. Occurrence of *Rhodococcus coprophilus* and Associated Actinomycetes in Feces, Sewage, and Freshwater. *Applied and Environmental Microbiology,* 42 (6): 1037-1042.

Mara, D.D., and J.I. Oragui. 1983. Sorbitol-fermenting *Bifidobacteria* as Specific Indicator of Human Fecal Pollution. *Journal of Applied Bacteriology,* 55 (March): 349-357.

Mara, D.D., and J.I. Oragui. 1985. Bacteriological Methods for Distinguishing Between Human and Animal Faecal Pollution of Water: Results of Filed Work in Nigeria and Zimbabwe, *Bulletin of the World Health Organization,* 63: 773-783.

MWRA. 2005. Massachusetts Water Resources Authority Water System [online]. Available: http://www.mwra.com. [August, 2005]

MassGIS 2001. Massachusetts Geographic Information System: Download Free Data [online]. Available: http://www.mass.gov/mgis. [June, 2005]

McCuen, R.H.. 1998. *Hydrologic Analysis and Design*. Upper Saddle River, NJ: Prentice Hall.

McDonald, A., and D. Kay. 1981. Enteric Bacterial Concentration in Reservoir Feeder Streams: Baseflow Characteristics and Response to Hydrograph Events. *Water Research,* 15 (8): 961-968.

McDonald, A., D. Kay, and A. Jenkins. 1982. Generation of Fecal and Total Coliform Surges by Stream Flow Manipulation in the Absence of Normal Hydrometeorological Stimuli. *Applied & Environmental Microbiology*, 44: 292+.

McFeters, G.A., G.K. Bissonnette, J.J. Jezeski, C.A. Thomson, and D.G. Stuart. 1974. Comparative Survival of Indicator Bacteria and Enteric Pathogens in Well Water. *Applied. Microbiology,* 27 (5): 823-829.

McKay, M and D.S. Wilks. 1995. *Atlas of Short-Duration Precipitation Extremes for the Northeastern United States and Southeastern Canada*. Ithaca, NY: Northeast Regional Climate Center Research Series Publication No. RR 95-1.

Medema G.J., Schets FM , Teunis PFM, Havelaar AH. 1998. Sedimentation of free and attached *Cryptosporidium* oocysts and *Giardia* cysts in Water. *Applied and Environmental Microbiology*, 64(11):4460-4466.

Miller, D.A. and R.A. White. 1998. A Conterminous United States Multi-Layer Soil Characteristics Data Set for Regional Climate and Hydrology Modeling. *Earth Interactions*, 2. [Available on-line at http://EarthInteractions.org]

Montogmery J.M.M., Consulting Engineers. 1985. *Water Treatment Principles and Design,* John Wiley and Sons, NY.

Moorhead, D.L., Davis, C.F. Wolf. 1998. Coliform Densities in Urban Waters of West Texas. *Environmental Health*, 60 (7): 14-18.

Morinigo, M.A., D. Berr, C. Jone, M.A. Munoz, R. Cornax, J.J. Borrego. 1992. Evaluation of Different Bacteriophage Groups as Faecal Indicators in Contaminated Natural Waters in Southern England. *Water Resources,* 26 (3):267-261.

Moulton, L.A. 2001. Downstream Influences of Beaver Impoundments on the Biogeochemical Characteristics of a New England Stream. Senior thesis. Mt. Holyoke College, Holyoke, Massachusetts.

Murphy, S. 2002. *General Information on Dissolved Oxygen.* Boulder, CO: US Geological Survey.

Nelson, M.A., T.S. Soerens, K.F. Steele, and P.F. Vendrell. 1998. Investigation of optimum sample number and timing for determining pollution loads [Online]. Available: <http://water.usgs.gov/wrri/98grants/Arkansas.htm>. [cited June 16, 2005]

Okun, D.A., G.F. Craun, J.K. Edzwald, J.B. Gilbert, and J.B. Rose. 1997. New York City: To Filter or Not to Filter? *Journal of the American Water Works Association,* 89 (3):62.

Olson, B.H., E.C. Chern, L.A. Khatib, and R.K. Oshiro. 2000. Factors Affecting Validity and Representation of Biomarkers to Differentiate Fecal Source in Environmental Waters. In *Proceedings of the AWWA Water Quality Technology Conference*, Paper No. TU9-2 CD Rom, Salt Lake City, UT, Nov. 5-9, 2000.

Ong, C, W. Moorehead, A. Ross, and J. Isaac-Renton. 1996. Studies of *Giardia* spp. and *Cryptosporidium* spp. in Two Adjacent Watersheds. *Applied and Environmental Microbiology*, 62(8): 2798-2805.

Parker, G., 2000. Stormwater and Mainstem Loads of Bacteria, Nutrients, and Selected Metals, Lower Charles River Watershed, Massachusetts. Massachusetts and Rhode Island U.S. Geological Survey Publication MA154.

Payne, J.F., A.E.J. Morris, P.Beers. 1999. Note: Evaluation of Selective Media for the Enumeration of *Bifidobacterium* sp. in Milk. *Journal of Applied Microbiology,* 86 (2): 353-358.

Perez, P.F., Y.I. Minnaard, E.A. Disalvo, and G. L. de Antoni. 1998. Surface Properties of *Bifidobacteria* Strains of Human Origin. *Applied and Environmental Microbiology*, 64:. 21-26.

Potter, W.R., and M.W. Gilliland. 1988. Bacterial Pollution in Runoff from Agricultural Lands. *Journal of Environmental Quality,* 17 (1): 27-34.

Rees, P.L.S., R. Baker, S.C. Long, P.K. Barten. 2005. Transport of *Cryptosporidium* and *Giardia* in Massachusetts Watersheds During Storm Events. In *Managing Watersheds for Human and Natural Impacts: Engineering, Ecological, and Economic Challenges*. Williamsburg, Virginia: ASCE.

Resnick, I.G. and M.A. Levin. 1981. Assessment of *Bifidobacteria* as Indicators of Human Fecal Pollution. *Applied and Environmental Microbiology,* 42 (3): 433-438.

Rijal, G.K., and R.S. Fujioka. 2001. Synergistic Effect of Solar Radiation and Solar Heating to Disinfect Drinking Water Sources. *Water Science and Technology,* 43 (12): 155-162.

Robbins, J.W.D., D.H. Howells, and G.U. Kriz. 1972. Stream Pollution from Animal Production units. *Journal of Water Pollution Control Federation,* 44 (8): 1536-1544

Roberson, J.A., J.J. Cassidy, and M.H. Chaudhry. 1998. *Hydraulic Engineering*. Second Edition. New York: John Wiley & Sons, Inc.

Roberston, L.J., A.T. Campbell, and H.V. Smith. 1992. Survival of *Cryptosporidium* parvum Oocysts under Various Environmental Pressures. *Applied and Environmental Microbiology*, 58(11): 3494-3500.

Robertson J.B. and S.C. Edberg. 1997. Natural Protection of Spring and Well Drinking Water Against Microbial Contamination. *Critical Reviews in Microbiology,* 23(2): 143-178..

Roll, B.M. and R.S. Fujioka. 1997. Sources of Fecal Iindicator Bacteria in a Brackish, Tropical Stream and their Impact on Recreational Water Quality. *Water Science Technology*, 35(11-12): 179-186.

Rose, J.B., C.P. Gerba, S.N. Singh, G.A. Toranzos, and B. Keswick. 1986. Isolation of Entero and Rotaviruses from a Drinking Water Facility. *Journal of the American Water Works Association,* 78: 56-61.

Rose, J.B., D. Kayed, M.S. Madore, C.P. Gerba, M. J. Arrowood, C.R. Sterling, and J. L. Riggs. 1988. Methods of Recovedry of *Giradia* and *Cryptosporidium* from Environmental Waters and their Comparative Occurrence. In *Advances in Giardia research*, Edited by P.M Wallis and B.R. Hammond, Calgary, Alberta, Canada: U of Calgary Press

Rose, J.B. 1988. Occurrence and Significance of *Cryptosporidium* in Water. *Journal of the American Water Works Association,* 80 (2): 53.

Rose, J.B., C.P. Gerba and W. Jakubowski. 1991. Survey of Potable Water Suupplies for Cryptosporidium and Giardia. *Environmental Science & Technology,* 25 (8):1393.

Rose, J.B. 1993. Tampa Water Resource Recovery Pilot Project Microbiological Evaluation in Tampa Water Resource Recovery Project Pilot Studies. Final Report to the City of Tampa. Tampa: CH2M Hill.

Rose, J.B. and E.K. Lipp. 1997. A study on the Presence of Human Viruses in Surface Waters of Sarasota County. St. Petersburg, FL: University of South Florida.

Rowbotham, T.J., T. Cross. 1977a. *Rhodococcus coprophilus* sp. nov.: An Anaerobic Nocardioform Actinomycete Belonging to the 'rhodochrous' Complex. *Journal of General Microbiology,* 100 (May): 123-138.

Rowbotham, T.J., T. Cross. 1977b. Ecology of *Rhodococcus coprophilus* and Associated Actinomycetes in Fresh Water and Agricultural Habitats. *Journal of General Microbiology,* 100 (Jun): 231-240.

Sattar SA, C. Chauret, VS Springthorpe, D Battigelli, Morteza Abbaszadegan and M LeChevallier. 1999. *Giardia Cyst and Cryptosporidium oocysts Ssurvival in Watersheds and Factors Aaffecting Inactivation*. Denver, Colo.: AwwwRF.

Sawyer, C.N., P.L., McCarty, and G.F. Parkin. 1994. Chemistry for Environmental Engineering, 4th Edition. New York, New York: McGraw-Hill, Inc.

Sayler, G.S., J.D. Nelson. Jr., A. Justice, and R.R. Colwell. 1975. Distribution and Significance of Fecal Indicator Organisms in the Upper Chesapeake Bay. *Applied Microbiology,* 30 (4): 625-638.

Scardovi, V. 1981. The Genus *Bifidobacterium*, In *The Prokaryotes: a Handbook on Habitats, Isolation, and Identification of Bacteria*, *Vol II*, Edited by M.P. Starr, M.P., Berlin, West Germany: Springer-Verlag.

Scherer, B.M., J.R. Miner, J.A. Moore, and J.C. Buckhouse. 1992. Indicator Bacterial Survival in Stream Sediments. *Journal of Environmental Quality,* 21 (4): 591-595.

Senus, D.A. 2002. Master's project, Data Analysis in Support of Event Based Contamination Transport, Department of Civil and Environmental Engineering, University of Massachusetts, Amherst, Massachusetts.

Sexton, L., T. Pass and R. Wright. 1995. *The Prevelance of Giardia and Cryptosporidium in Eastern Kentucky Water Supplies*. University of Kentucky – Kentucky Water Resources Research Intitute.

Shelton, L.R., 1994. *Field Guide for collecting and processing stream-water samples for the national water-quality assessment program.* U.S. Geological Survey Open-File Report 94-455, Washington, D.C.: U.S. Geological Survey.

Sherer, B.M. 1988. Resuspending Organisms from a Rangeland Stream Bottom. *Transactions of the American Society of Agricultural Engineers*, 31: 1217.

Sinton, L.W., R.K. Finlay, and D.J. Hannah. 1998. Distinguishing Human from Animal Fecal Contamination in Water: A Review. *New Zealand Journal of Marine and Freshwater Research,* 32: 323-348.

Skinner, Q.D., J.C. Adams, P.A. Rechard, and A.A. Beetle. 1974. Effect of Summer Use of a Mountain Watershed on Bacterial Water Quality. *Journal of Environmental Quality,* 3 (4): 329-335.

Smith, H.V., Robertson, L.J., Ongerth, J.E. 1995. *Cryptosporidiosis* and *Giardiasis*: the Impact of Waterborne Transmission, *AQUA*, 44 (6): 258-274.

Smith, H.V. and Rose, J.B. 1998. Waterborne Cryptosporidiosis: Current Status, *Parasitology Today,* 14 (1): 14-22.

Sobsey, M.D., K.J. Schwab, and T.R. Handzel. 1990. A Simple Membrane Filter Method to Concentrate and Enumerate Male-Specific RNA Coliphages. *Journal of AWWA,* Sept.: 52-59.

Sobsey, M.D., D.A. Battigelli, T.R. Handzel and K.J. Schwab. 1995. *Male-specific Coliphages as Indicators of Viral Contamination of Drinking Water.* Denver, Colo.: AwwaRF.

Stukel, T.A., E.R. Greenberg, B.J. Dain, F.C. Reed, and N.J. Jacobs. 1990. A Longitudinal Study of Rainfall and Coliform Contamination in Small Community Drinking Water Supplies. *Environmental Science and Technology.* 24 (4): 571-575.

Stuntebeck, T.D., 1995. *Evaluating Barnyard Best Management Practices in Wisconsin Using Upstream-downstream Monitoring.* U.S. Geological Survey Fact Sheet FS-221-95, Washington, D.C.: U.S. Geological Survey.

Stuntebeck, T.D., and Bannerman, R.T., 1998. *Effectiveness of Barnyard Best Management Practices in Wisconsin.* U. S. Geological Survey Fact Sheet FS-051-98, Washington, D.C.: U.S. Geological Survey.

Sun, L., D. Muller-Schwarze and B.A. Schulte. 2000. Dispersal Pattern and Effective Population Size of the Beaver, *Canadian Journal of Zoology,* 78: 619-626.

Toranzos, G.A., and G.A. McFeters. 1997. Detection of Indicator Microorganisms in Environmental Freshwaters and Drinking Waters. In *Manual of Environmental Microbiology*, Edited by Hurst, Washington, D.C.: ASM Press.

Tunnicliff, B. and Brickler, S.K. 1984. Recreational Water Quality Analyses of the Colorado River Corridor in the Grand Canyon. *Applied Environmental Microbiology*, 48: 909-917.

USEPA. 1996. *Environmental Indicators of Water Quality in the United States.* EPA 841-R-96-002. Washington, D.C.: U.S. Environmental Protection Agency.

USEPA. 1999. *Method 1623: Cryptosporidium and Giardia in Water by Filtration/IMS/FA.* EPA-821-R-99-006, Washington, D.C.: U.S. Environmental Protection Agency.

USEPA. 2002. U.S. Environmental Protection Agency, Safe Water Regulations and Guidance, Longterm 1 Enhanced Drinking Water Treatment Rule [online]. Available: http://www.epa.gov/safewater. [June, 2005].

USEPA. 2004. U.S. Environmental Protection Agencey, Safe Water Regulations and Guidance, Proposed Rules, Long Term 2 Enhanced Surface Water Treatment Rule – Proposed Rule [online]. Available: http://www.epa.gov/safewater. [April, 2005]

Vennard, J.K., and R. L. Street. 1982. *Elementary Fluid Mechanics. Sixth Edition*, New York, New York: John Wiley & Sons.

Walker M.J., C.D. Montemagno, and M.B. Jenkins. 1998. Source Water Assessment and Nonpoint Sources of Acutely Toxic Contaminants: A Review of Research Related to Survival and Transport of *Cryptosporidium* parvum. *Water Resources Research,* 34(12): 3383-3392.

Walker Jr., F. Russell and J.R. Stedinger. 1999. Fate and Transport Model of Cryptosporidium. *Journal of Environmental Engineering,* 125 (4): 325-333.

Weidner, R.B., A.G. Christianson, S.R. Weibel, and G.G. Robeck. 1969. Rural Runoff as a Factor in Stream Pollution. *Journal Water Pollution Control Federation*, 41 (3): 377-385.

Weiskel, P.K., B.L. Howes, and G.R. Heufelder. 1996. Coliform Contamination of a Coastal Embayment: Sources and Transport Pathways. *Environmental Science and Technology.* 30 (6): 1872-1881.

Whitman, R.L., A.V. Gochee, W.A. Dustman, K.J. Kennedy. 1995. Use of Coliform Bacteria in Assessing Human Sewage Contamination. *Natural Areas Journal*, 15 (3): 227-233.

Wilkinson, J.A., J.M. Wyer, and D. Key. 1995. Modeling Fecal Coliform Dynamics in Streams and Rivers. *Water Research*, 29 (3): 847-855.

Wilks, D.S., and R.P. Cember. 1993. *Atlas of Precipitation Extremes for the Northeastern United States and Southeastern Canada*. Northeast Regional Climate Center Research Series, Publication No. RR 93-5, Ithaca, New York: Northeast Regional Climate Center.

Woodruff, D.L. 1992. *Clostridium perfringens* as a Sewage Indicator in Coastal Sediments: Development and Comparison of an Alternative Method of Analysis. Master's thesis. Department of Environmental Science and Engineering, University of North Carolina, Chapel Hill, NC.

Wright, R.M., P.M. Nolan, D. Pincumbe, E. Hartman, O.J. Viator. 2001. *Blackstone River Initiative: Water Quality Analysis of the Blackstone River Under Wet and Dry Weather Conditions*. Boston, Massachusetts: EPA New England.

ABBREVIATIONS

AwwaRF	Awwa Research Foundation
BB	Burrow Brook
Bk	Brook
BMID	Black Mountain Irrigation District
BMP	best management practice
BOD	biological oxygen demand
BBFM	Burrow Brook Farm
BBWR	Burrow Brook Woods Road
°C	degrees Celsius
CFU	colony forming units
CFS	cubic feet per second
CG	*Cryptosporidium-Giardia*
cm	centimeter
COD	chemical oxygen demand
CV	coefficient of variation
CVA	Chicopee Valley Aqueduct
DCR	Department of Conservation and Recreation
DEM	Digital Elevation Model
DEP	Department of Environmental Protection
EA labs	Environmental Associates Laboratory
EFWMC	event flow weighted mean concentrations
EMC	event mean concentrations
°F	degrees Fahrenheit
FC	fecal coliform
FS	Fecal Streptococci
ft	foot
FWMC	flow weighted mean concentration
g	gram
GB	Gates Brook
GBBR	Gates Brook Brayton Road
GBWD	Gates Brook Woodland Street
GIS	geographical information system
GPM	gallons per minute
GRB	Griswold Brook
HBSA	human bific sorbitol agar
HPC	heterotrophic plate count

hr	hour
ICR	Information Collection Rule
ID_{50}	infective dose
IESWTR	Interim Enhanced Surface Water Treatment Rule
in	inches
km	kilometer
L	Liter
LT1ESWTR	Long-Term 1 Enhanced Surface Water Treatment Rule
LT2ESWTR	Long-Term 2 Enhanced Surface Water Treatment Rule
m	meter
Max	Maximum
MBDB	Middle Branch Dickey Brook
MCLG	maximum contaminant level goal
MD61	Middle Branch Dickey Brook
MD62	Griswold Brook
MDC	Metropolitan District Commisssion
mL	milliliter
mm	millimeter
mpn	???
MST	microbial source tracking
MWRA	Massachusetts Water Resources Authority
μm	micrometer
μS	micro-siemens
NA	not applicable or not available
NADP	National Acid Deposition Program
NCDC	???
NED	National Elevation Data
ND	non detect
NPS	non-point source
NTU	nephelometric turbidity units
NURP	Nationwide Urban Runoff Program
NWS	National Weather Service
nm	nano-meters
PBS	???
PCR	polymerase chain reaction
pfc	plack forming colonies
pfu	???
PI	principle investigator
PR	Prescott Peninsula
PVC	polyvinyl chloride

PWS	public water system
Q	flow
QA/QC	Quality Assurance/Quality Control
QPF	quantitative precipitation forecasts
RCA-S	reinforced clostidial agar
s	second
SD	standard deviation
SDWA	Safe Drinking Water Act
sq	square
SM	Standard Method
SS	suspended sediment
SWTR	Surface Water Treatment Rule
TC	total coliform
THPC	total heterotrophic plate counts
TKN	total Kjeldahl nitrogen
TMDL	total maximum daily load
TOC	total organic carbon
TP	total phosphorus
TSS	total suspended solids
TWMC	time weighted mean concentration
UMass	University of Massachusetts
US	United States
USEPA	United States Environmental Protection Agency
USGS	Unites States Geological Survey
UV254	ultraviolet absortption at 254 nm
VID	Vernon Irrigation District

IWA Publishing's authorised EU representative for General Product Safety Regulations is Diane D'Arras, 15 rue Duret, 75116 Paris, France, e-mail: safety@iwap.co.uk.

Printed and bound by CPI Group (UK) Ltd, Croydon, CR0 4YY
17/09/2026
02218810-0005

Cancer Risks Associated With Elevated Levels of Drinking Water Arsenic Exposure

Awwa Research Foundation

About the Awwa Research Foundation

The Awwa Research Foundation (AwwaRF) is a member-supported, international, nonprofit organization that sponsors research to enable water utilities, public health agencies, and other professionals to provide safe and affordable drinking water to consumers.

The Foundation's mission is to advance the science of water to improve the quality of life. To achieve this mission, the Foundation sponsors studies on all aspects of drinking water, including supply and resources, treatment, monitoring and analysis, distribution, management, and health effects. Funding for research is provided primarily by subscription payments from approximately 1,000 utilities, consulting firms, and manufacturers in North America and abroad. Additional funding comes from collaborative partnerships with other national and international organizations, allowing for resources to be leveraged, expertise to be shared, and broad-based knowledge to be developed and disseminated. Government funding serves as a third source of research dollars.

From its headquarters in Denver, Colorado, the Foundation's staff directs and supports the efforts of more than 800 volunteers who serve on the board of trustees and various committees. These volunteers represent many facets of the water industry, and contribute their expertise to select and monitor research studies that benefit the entire drinking water community.

The results of research are disseminated through a number of channels, including reports, the Web site, conferences, and periodicals.

For subscribers, the Foundation serves as a cooperative program in which water suppliers unite to pool their resources. By applying Foundation research findings, these water suppliers can save substantial costs and stay on the leading edge of drinking water science and technology. Since its inception, AwwaRF has supplied the water community with more than $300 million in applied research.

More information about the Foundation and how to become a subscriber is available on the Web at **www.awwarf.org**.